T0305566

# Thermal Safety Margins in Nuclear Reactors

This book presents an overview of state-of-the-art approaches to determine thermal safety margins in nuclear reactors. It presents both the deterministic aspects of thermal safety margins of nuclear reactors and a comprehensive treatment of aleatory and epistemic uncertainties to facilitate the understanding of these two difficult topics at various academic levels, from undergraduates to researchers in nuclear engineering.

It first sets out the theoretical background before exploring how to determine thermal safety margins in nuclear reactors, through examples, problems and advanced state-of-the-art approaches. This will help undergraduate students better understand the most fundamental aspects of nuclear reactor safety. For researchers and practitioners, this book provides a comprehensive overview of the most recent achievements in the field, offering an excellent starting point to develop new methods for the assessment of the thermal safety margins.

This book is written to bridge the gap between deterministic methods and appropriate treatment of uncertainties to assess safety margins in nuclear reactors, presenting these approaches as complementary to each other. Even though these two approaches are frequently used in parallel in real-world applications, there has been a lack of a consistent teaching approach in this area.

This book is suitable for readers with a background in calculus, thermodynamics, fluid mechanics, and heat transfer. It is assumed that readers have previous exposure to such concepts as laws of thermodynamics, enthalpy, entropy, and conservation equations used in fluid mechanics and heat transfer.

**Key Features:**
- Covers the theory, principles, and assessment methods of thermal safety margins in nuclear reactors whilst presenting state-of-the-art technology in the field
- Combines the deterministic thermal safety considerations with a comprehensive treatment of uncertainties, offering a framework that is applicable to all current and future commercial nuclear reactor types
- Provides numerous examples and problems to be solved

**Henryk Anglart** is Professor Emeritus of Nuclear Engineering at the KTH Royal Institute of Technology, Stockholm, Sweden, and at the Warsaw University of Technology (WUT), Warsaw, Poland. He received his MSc from WUT and his PhD from the Rensselaer Polytechnic Institute, Troy, NY. After his eighteen-year career as a research and development engineer at Westinghouse in Sweden, he accepted a tenure position at KTH, where he has supervised many PhD students and post-doctoral fellows, and has taught several courses in nuclear engineering. In addition to research and teaching, Prof. Henryk Anglart was serving for a long time as head of Reactor Technology Division, Deputy Director of the Physics Department, and Director of Nuclear Technology Center at KTH. Prof. Henryk Anglart authored and coauthored over 200 journals, conference and other scientific publications. He is also an author of three textbooks used in teaching of nuclear engineering courses at WUT and KTH, and two CRC Press books: Multilingual Dictionary of Nuclear Reactor Physics and Engineering, and Introduction to Sustainable Energy Transformation.

# Thermal Safety Margins in Nuclear Reactors

## Henryk Anglart

### CRC Press
Taylor & Francis Group
Boca Raton London New York

CRC Press is an imprint of the
Taylor & Francis Group, an **informa** business

Designed cover image: © Shutterstock_2192659579

First edition published 2024
by CRC Press
2385 NW Executive Center Drive, Suite 320, Boca Raton FL 33431

and by CRC Press
4 Park Square, Milton Park, Abingdon, Oxon, OX14 4RN

*CRC Press is an imprint of Taylor & Francis Group, LLC*

---

**Library of Congress Cataloging-in-Publication Data**

---

Names: Anglart, Henryk, author.
Title: Thermal safety margins in nuclear reactors / Henryk Anglart.
Description: First edition. | Boca Raton : CRC Press, 2024. | Includes
bibliographical references and index. | Summary: "This book presents an
overview of state-of-the art approaches to determine thermal safety margins
in nuclear reactors. It presents both the deterministic and
probabilistic aspects of thermal safety margins of nuclear reactors to
facilitate the understanding of these two difficult topics at various
academic levels, from undergraduates to researchers in nuclear
engineering"--
Provided by publisher.
Identifiers: LCCN 2023054406 | ISBN 9781032171050 (hardback) | ISBN
9781032185316 (paperback) | ISBN 9781003255000 (ebook)
Subjects: LCSH: Nuclear power plants--Thermodynamics. | Nuclear
reactors--Cooling. | Nuclear power plants--Safety measures.
Classification: LCC TK9212 .A54 2024 | DDC 621.48/3--dc23/eng/20240208
LC record available at https://lccn.loc.gov/2023054406

---

ISBN: 978-1-032-17105-0 (hbk)
ISBN: 978-1-032-18531-6 (pbk)
ISBN: 978-1-003-25500-0 (ebk)

DOI: 10.1201/9781003255000

Typeset in Nimbus Roman font
by KnowledgeWorks Global Ltd.

*Publisher's note:* This book has been prepared from camera-ready copy provided by the authors.

# Dedication

*In Memory of My Parents*

# Contents

# SECTION II  Reactor Core Thermal-Hydraulics

# Preface

Nuclear power engineering has undergone remarkable transformations and expansion over the past eight decades. While the CP-1 experiment, carried out at the University of Chicago in December 1942, already integrated principles such as safety margins, defense in depth, and safety culture, their application within the nuclear power sector has evolved significantly. It is evident that our comprehension of these principles has profoundly shaped the progression of nuclear engineering in recent times. Undeniably, the future trajectory of nuclear engineering will be heavily reliant on the efficacious incorporation of safety margins in novel reactor designs.

Maintaining thermal safety margins in nuclear reactors is crucial to prevent potential damage to the reactor core structure from excessive heat. Accurately predicting these margins is vital for the safe and cost-effective operation of a nuclear power plant, both during normal operations and anticipated operational occurrences. However, predicting thermal safety margins is a complex task. It typically requires a substantial database to provide a statistical foundation for margin estimates. When the database is limited, other methods such as bounding, engineering judgment, or analytical adjustment must be employed. It goes without saying that these factors contribute to the complexity and difficulty of predicting thermal safety margins.

This book has a twofold objective. The first two sections aim to provide a comprehensive review of the current state of nuclear engineering in general and in nuclear reactor thermal hydraulics in particular. The third section offers insights into topics such as the assessment of thermal performance of fuel elements, the handling of uncertainties, and the analysis and prediction of thermal safety margins.

The book is versatile in its usage and does not presume any prior knowledge of the subject from the reader. Portions of the book, especially the first two sections, can serve as foundational material for teaching undergraduate courses at both B.S. and M.S. levels. This content can also be beneficial for most practicing nuclear engineers. Chapters 4 and 8 and parts of the third section are tailored for graduate students, code developers, and researchers.

Given the central role that thermal-hydraulic design and analysis of the core of a nuclear reactor play in predicting thermal margins, the book places significant emphasis on both the theoretical and practical foundations for advanced computational methods for reactor core thermal-hydraulics. While the book does not include descriptions of reactor thermal-hydraulics codes, it provides sufficient material to facilitate a basic understanding of code manuals and code development.

The topics explored in this book were conceived and developed by various individuals. Numerous exceptional texts have guided me in shaping the content of this book. Throughout my career, I have had the privilege of engaging with many brilliant minds and discussing the subject matter. Much of the material stems from lectures on reactor thermal-hydraulics, nuclear reactor technology, and nuclear reactor dynamics, delivered as part of the nuclear engineering program at the KTH Royal Institute of Technology in Stockholm, Sweden.

Completing this book would not be possible without the support and inspiration of many people. First and foremost I thank my wife Ewa for her continued support in writing this book. It is challenging to recall all the individuals who have influenced me, and any list of names would likely omit someone. However, I would like to express my gratitude to my teachers, mentors, colleagues, fellow educators and researchers, and undergraduate and graduate students for their profound and inspiring discussions. Special acknowledgments go to C. Adamsson, I.G. Anghel, M. Bergagio, D. Caraghiaur, J. Dufek, J.M. Le Corre, W. Fan, K. Fu, S. Hedberg, G.F. Hewitt, K. Karkoszka, N. Kurul, H. Li, O. Nylund, R. Pegonen, M.Z. Podowski, A. Riber-Marklund, B.R. Sehgal, R. Thiele, and G. Yadigaroglu. Finally, I would like to acknowledge encouragement and support from Danny Kielty at CRC Press, Taylor & Francis Group.

Stockholm, Sweden
December 2023
Henryk Anglart

# Section I

## Background Overview

# 1 Overview of Nuclear Reactor Safety and Design

Nuclear power is employing nuclear fission to produce carbon-free electricity and heat. Since about 200 MeV= $3.2 \times 10^{-11}$ J of energy is released per fission of a single nucleus of $^{235}$U, one mole (235 g) of $^{235}$U, containing $N_A = 6.02 \times 10^{23}$ nuclei, provides during fission enormous amount of energy equal to $1.93 \times 10^{13}$ J, while releasing about 235 g of radioactive fission products. The same amount of energy could be obtained from the combustion of more than 580 tons of coal, emitting to the atmosphere about 1160 tons of $CO_2$. Thus, compared to conventional power plants, nuclear power plants require much less fuel and generate much less waste per unit of final energy produced, but require more attention as far as safety precautions are concerned due to the high power density and the high toxicity of the radioactive wastes.

Radioactivity is a natural phenomenon in which particles are emitted from nuclei as a result of nuclear instability. Natural radiation comprises cosmic radiation and the radiation arising from the decay of naturally occurring radionuclides. In fact, radiation and radioactive substances have many beneficial applications, ranging from power generation to uses in medicine, industry, and agriculture. However, any exposure to radiation may be potentially harmful. Already in 1928, in response to the growing recognition of the hazards of radiation, the Second International Congress of Radiology established the International Commission on Radiological Protection (ICRP) to set standards of permissible exposure to radiation.

From the very beginning of the peaceful use of nuclear power, stringent safety standards have been established to protect humans and the environment from the harmful effects of ionizing radiation. These standards established fundamental safety principles, requirements, and measures to control the radiation exposure of humans and the release of radioactive material to the environment.

A very comprehensive set of safety standards and guides has been established and adopted by the International Atomic Energy Agency (IAEA). The *IAEA safety standards* include

1. *Safety Fundamentals*, describing the fundamental safety objective and principles.
2. *General and Specific Safety Requirements*, establishing the requirements that must be met to ensure the protection of humans and the environment.
3. *General and Specific Safety Guides*, providing recommendations and guidance on how to comply with the safety requirements.

In this chapter we present an overview of nuclear power safety and design. In §1.1 we recall some most important facts from the history of nuclear power development,

DOI: 10.1201/9781003255000-1

such as defining the early safety research goals and establishing design principles for major nuclear power plant components and systems. Nuclear safety standards are shortly introduced in §1.2, whereas the concepts of safety margins and safety analyses are explained in §1.4 and §1.5, respectively. The causes and consequences of major nuclear power plant accidents are described in §1.7.

## 1.1   HISTORICAL REVIEW

Early in 1942, a group of scientists led by Enrico Fermi gathered at the University of Chicago to transform scientific theories into technological reality. They designed and constructed the first nuclear reactor known as *Chicago Pile-1* (CP-1), containing 40 tons of natural uranium, distributed throughout 385 tons of graphite. The natural uranium served as fuel and the graphite as moderator and reflector. The reactor also contained control rods made of cadmium. On December 2, 1942, at 3:25 p.m. Chicago time, after a very slow withdrawal of control rods that took several hours, the nuclear reaction became self-sustaining. The world had entered the age of nuclear power.

To address the possibility of a failure, multiple safety precautions were designed into the CP-1 experiment. Three sets of control rods were employed. Along with the primary set used to fine-control the nuclear chain reaction, two other sets were designed with safety in mind. One of the safety sets was automatically operated by an electric motor and responded to an instrument reading from a radiation counter. The other system consisted of an emergency safety rod, which was withdrawn from the pile and tied down by a rope. It was the job of the "Safety Control Rod Axe Man" (*scram*) to stand by ready to cut the rope with an axe should something unexpected happen. In addition to the two mechanical systems, a "liquid-control squad" was organized to pour a cadmium-salt solution over the pile in case of a failure of these systems.

The CP-1 experiment demonstrated that nuclear reactor safety has been an important consideration from the very beginning of the development of nuclear reactors. Even though not expressed in these words, the experiment gave birth to such safety concepts as the *defense* in-depth approach and the safety system redundancy and diversification. Another important safety aspect, termed today as the *safety culture*, was initiated by Fermi himself. Just before the reactor was expected to reach criticality for the first time, Fermi directed that the experiment be shut down and that all adjourned for lunch. This attitude helped to calm down and release the tension of the crew and to carry out the experiment without any incident.

Before 1957, nuclear power safety had not reached full recognition, independent from nuclear development. In the US, the Atomic Energy Act was signed by President Harry S. Truman on August 1, 1946, and the principal functions of the Atomic Energy Commission (AEC) were defined: to produce fissionable material for weapons and to develop and manufacture weapons as military requirements dictated. The 1946 act also encouraged AEC to develop peaceful uses of atomic energy, though this function remained secondary to weapons production. The nuclear technology was opened to commercial enterprise nine years later, by the 1954 Atomic

Energy Act, and since then AEC started to carry out its role more publicly. In closing his February 1954 message to Congress proposing new atomic-energy legislation, President Eisenhover recommended the AEC "to establish minimum safety and security regulations to govern the use and possession of fissionable material". Throughout the 1954 Act sections on licensing and regulation, the objectives "to protect the health and safety of the public" were frequently and clearly stated. In that the dangerous nature of the technology was recognized and the basic goal of the AEC's regulatory function was underscored [154].

Already from the beginning of the commercial nuclear power development, it was clear that the main danger of a nuclear reactor core is not the possibility that it might explode but the fact that it contains radioactive fission products. In a nuclear accident, the fission products might be liberated into the atmosphere or the ground water. The first document that described the method to calculate off-site doses due to postulated release of fission products from the core of an LWR into the containment atmosphere (so called *source term*) was published by AEC in 1962 [55]. The source term was postulated from an accident that resulted in a substantial meltdown of the reactor core. Since that time substantial information was developed updating the knowledge about severe LWR accidents and the resulting behavior of the released fission products [194].

International cooperation in the development of atomic energy for peaceful purposes was proposed by President Eisenhower on December 8, 1953, in his famous speech "Atom for Peace", delivered before the United Nations. Four years later, in 1957, the United Nations set up the International Atomic Energy Agency (IAEA) with the primary function to act as an auditor of world nuclear safety. Currently (2020s) the IAEA's role is significantly strengthened, including such activities as prescription of safety procedures, reporting of nuclear reactor incidents and accidents, and setting international safety benchmarks to which participating States would subscribe. The IAEA Convention on Nuclear Safety (CNS) was drawn up during a series of expert meetings from 1992 to 1994 and finally entered into force in October 1996. The CNS is based on the Parties' common interest to achieve higher levels of safety that will be developed and promoted through regular meetings. It obliges Parties to submit reports on the implementation of their obligations for "peer review" at meetings that are normally held at IAEA headquarters. The reports are publicly available at the IAEA website.[1]

## 1.1.1 EARLY SAFETY RESEARCH GOALS

Intensive nuclear research program started in the second half of the 1950s. Soon the focus converged on such topics as the plant siting and the reactor core design. Reactor siting affects safety since, in the event of a large uncontrolled release of radioactivity resulting from a serious reactor accident, a significant area surrounding the reactor building would be contaminated.

The key safety issue in the core design was to maintain the integrity of the structure under the stress resulting from high temperatures and irradiation due to fission.

---

[1] https://www.iaea.org/topics/nuclear-safety-conventions/convention-nuclear-safety/documents

Since the fuel cladding represents the first safety barrier containing the hazardous radioactive fission products, finding a suitable material became the first challenge. Various materials including stainless steel, aluminum, niobium, and beryllium were examined as potential cladding materials. It was found, however, that zirconium alloys are best suited as cladding material due to a small capture cross section for thermal neutrons and due to high resistance to corrosion by water at temperatures typical of operating nuclear reactors.

The coolability of nuclear reactor cores under all conceivable operating conditions is one of the main research topics in nuclear engineering. In the 1950s very little was known about single-phase convective and boiling heat transfer at conditions typical of nuclear reactors. New research was needed to predict heat transfer intensity and limiting hot spots in the core. Soon various boiling heat transfer regimes were identified, including subcooled boiling, bulk boiling, nucleate boiling, and film boiling.

Nucleate boiling was found to be an effective means of removing heat from fuel elements. However, at certain conditions the boiling heat transfer rate could dramatically decrease, resulting in an increase in clad temperature. This condition, which could threaten the structural integrity of a fuel element, was called *burnout* or *boiling crisis*, and the maximum heat flux just before reaching the condition was called the *critical heat flux* (CHF). Further research revealed that CHF could be reached within a wide range of two-phase flow patterns, starting from a very low content of the vapor phase and ending with a relatively high content of the vapor phase in the boiling channel. Since the CHF mechanisms in both situations were found to be different, the former type of CHF was called *departure from nucleate boiling* (DNB), and the latter one was called *dryout*.

Early reactor core designers realized that to ensure safety, it must be required that none of the thermal limitations on the core behavior were exceeded. One of the limitations arises from requiring that the clad surface heat flux always remains below its CHF limit. Clearly, due to nonuniform power distribution in a core, the "distance" to the CHF limit varies at different locations. The channel with the shortest distance to the CHF limit was termed *hot channel*, and the location where it occurred was called *hot spot*.

## 1.1.2 CONTAINMENT DEVELOPMENT

The role of containment is to protect the public from uncontrolled release of fission products following a severe accident. The earliest reactors did not have any containment, and it was first when designing the Shippingport Plant in 1954 that the decision was made that commercial plants would be required to have this type of reinforced enclosure surrounding the primary system. That was the beginning of the *defense in depth* philosophy according to which there should be several independent barriers to prevent a free spread of radioactivity to the atmosphere. The containment would be the fourth and last such barrier, after the uranium dioxide pellets, the Zircaloy cladding of the fuel pellets, and the thick-walled primary system.

During the 1960s, the containment research and design focused on the enhancing heat removal by the containment wall and on the removal of fission products from the steam-air atmosphere. The basic experiments that were performed included the iodine transport tests, condensate-driven fission product removal studies, and large scale tests with simulated fission products that allowed increased concentrations without increasing the radioactivity.

Pressure suppression containment system for BWRs was developed by General Electric to reduce the containment size through exploiting the high energy absorption capacity of a water pool. According to this design, the containment contains a drywell that surrounds the reactor and provides a primary barrier to coolant release during an accident. Discharge of drywell air and the water–steam mixture occurs through a vent system into a large water pool placed in a wetwell. The water pool serves as a large passive heat sink and absorbs and retains radioactive material released during a severe core accident.

For PWRs the containment volume is determined by the large size of a primary system that includes a reactor pressure vessel, main circulation pumps, steam generators, and the primary system pipings. During the 1960s Westinghouse developed the ice condenser concept so that steam released from a primary system rapture must flow through ice condensers before escaping into the free volume of the containment. This arrangement helps to reduce the pressure and temperature of a post-accident containment atmosphere, which are the main driving forces for containment leakage and release of fission products to the environment. In continuation, active containment systems were designed, such as containment spray cooling systems, recirculation air cooling systems, and other heat removal systems. Further improvement of PWR containment was achieved in the 2000s by the introduction of passive safety systems in the AP1000 reactor developed by Westinghouse. The AP1000 plant passive containment cooling system requires no operator actions and uses only natural forces such as gravity, natural circulation, and compressed gas to achieve its safety function [48].

### 1.1.3 REACTOR CORE DEVELOPMENT

Reactor types are mainly determined by the material composition of their cores. In particular, a combination of fuel and coolant material determines neutronic properties of the core and its ability to maintain criticality. For thermal reactors, the addition of a moderator is necessary to slow down fast fission neutron to the thermal energy spectrum. For fast reactors, in which fissions are achieved with high-energy neutrons, materials that slow down neutrons should be avoided. A general requirement for any material present in the core is that it should not parasitically absorb neutrons. The reactor core coolant should be able to easily remove heat generated in the core and transport it to the secondary system. For this reason, fluid coolants can be used in a liquid form (mainly water and liquid metals) or as a gas (e.g., carbon dioxide or helium). For fuel and moderator materials, fluid and solid forms can be considered, such as metal, oxide or molten salt for the fuel and water, heavy water or graphite for the moderator.

Already in the 1950s many different reactor core types were examined. These reactors were small research reactors without the necessary balance of plant to generate electricity.

First commercial-scale reactors used graphite as moderator to reduce the need for uranium enrichment. In 1954, the graphite-moderated light water reactor (LWGR) with 6 MWe electric output commenced operation in Obninsk, USSR. Two years later a 60 MWe graphite-moderated carbon dioxide gas-cooled reactor (GCR) commenced operation in Calder Hall in the United Kingdom.

In 1956 in the United States, a legislation initiative was undertaken to direct AEC to construct six pilot nuclear plants using different design principles. It was expected that in this way a faster development toward generation of electricity from nuclear fission would be achieved. The proposed legislation was never enforced, however, and only two reactor types had been developed: the light-water-cooled and moderated reactor and the liquid-metal-cooled breeder reactor. First 5 MWe boiling water reactor (BWR), supplied by General Electric, commence operation in Vallecitos in 1957. A year later Westinghouse supplied a pressurized water reactor (PWR) in Shippingport. The reactor was designed to accommodate three different cores with natural uranium "blanket" and high-enriched (93% $^{235}$U) uranium "seed". The third core was a light water breeder and used pellets made of thorium dioxide and uranium-233 oxide.

A natural-uranium fuel cycle with a heavy-water moderator reactor for civilian electricity production was under development in Canada already during the 1950s. First 20 MWe nuclear power demonstration reactor commenced operation in 1962, followed by a ten-fold larger prototype of CANDU (short for Canada Deuterium Uranium) in Douglas Point, which commenced operation in 1967.

Since the 1970s, light water reactors (LWRs) have been dominating in most countries. Notable exceptions were the United Kingdom, which commercialized GCRs and later developed advanced gas-cooled reactors (AGRs), and Canada, which has continued development of the CANDU reactor. The reason for the LWR dominance was two-fold. Firstly, low-enriched uranium that is needed by LWRs became much cheaper when innovative centrifuge separation technology was introduced. Secondly, the reactor design has been simplified and several vendors in the United States and in Europe were able to supply new nuclear power plants.

## 1.1.4 FUEL ROD DEVELOPMENT

The details of the fuel rod design for commercial power plants were established during the 1950s and 1960s. A fuel element was constructed from a stack of $UO_2$ fuel pellets enclosed in a hollow Zircaloy tube, sealed from both sides. The Zircaloy cladding was designed to contain the radioactive fission products and to protect the fuel from the severe steam-water environment. A gas gap between the clad and the fuel pellets was built in to prevent cracking or fragmenting the ceramic fuel pellets. To improve heat transfer in fuel elements, the gas gap was filled with pressurized helium, which is an inert gas with high thermal conductivity. Much of the research work was performed to investigate thermal properties of the fuel and gap, and an oxidation and heat generation of zirconium alloy cladding.

## 1.1.5 EMERGENCY CORE COOLING SYSTEMS

Cooling of reactor cores received the greatest attention in discussions regarding light water reactor safety. Early calculations showed that even following a scram, a large break in the reactor coolant system could leave the core vulnerable to failure and melting. Results from the semiscale research facility indicated that much of the water used for core cooling could leave the reactor pressure vessel rather than reflood the core immediately [151]. This rather pessimistic scenario was later rejected based on the *Loss-of-Fluid Test* (LOFT) experiments, which showed that the emergency core cooling injected into the coolant loop cold legs would flow down the reactor vessel downcomer, while the steam in the reactor vessel would be vented up the downcomer section near the broken cold leg. LOFT experiments not only demonstrated this behavior but were also able to quantify the amount of bypass which actually would occur. This information, supported by detailed computer simulations, is currently used in designing efficient *emergency core cooling systems* (ECCSs) [161].

The ECCS is designed to deliver water to the core in case of a hypothesized *large-break loss of coolant accident* (LB-LOCA). For a break of a cold leg, steam generated in the core would reverse direction and flow up in the downcomer. Simultaneously, the ECCS would deliver water toward the same region with intended flow into the lower plenum, with the task to refill the plenum region and then reflood the core. When high upward steam velocity would be reached in the downcomer, part or all of the emergency cooling water would be prevented from getting to the lower plenum. This limitation on the rate of liquid flowing down the downcomer is known as *counter-current flow limitation* (CCFL). This phenomenon has been intensively investigated experimentally and analytically and various analytical models have been proposed [228].

Current LWR nuclear power plants use *redundant* (i.e., having parallel divisions with the same safety functions) and *multi-level* (i.e., operating at different conditions, such as pressure level) systems to mitigate loss of coolant accidents of all sizes, and specific non-LOCA events, such as main steam line breaks. The *safety injection system* (SIS) of EPR consists of the *medium-head safety injection* (MHSI) system, the *low-head safety injection* (LHSI) system, and accumulators which contain large amounts of water under gas pressure [70]. The AP1000 reactor is equipped with core make-up tanks that will be drained by gravity in case of large-break LOCA, accumulators that will be activated at intermediate pressure, *in-containment refueling water tank* (IRWST) that will provide long-term water injection at low pressure, and an automatic depressurization system, which will depressurize the primary system to near containment pressure during small-break LOCA [48].

## 1.2 NUCLEAR SAFETY STANDARDS

**Nuclear safety**, often abbreviated to *safety* in various publications on nuclear power plants, is the "achievement of proper operating conditions, prevention of accidents or mitigation of accident consequences, resulting in protection of workers, the public and the environment from undue radiation hazards" [107]. Safety standards are needed for efficient protection of health and minimization of danger to life and property. Usually safety standards are covering a variety of application areas, including nuclear safety, radiation safety, transport safety, waste safety, and general safety.

### 1.2.1  NUCLEAR SAFETY OBJECTIVES

The fundamental *safety objective* is to protect people and the environment from harmful effects of ionizing radiation. To this end, the following measures are needed: (1) To control the radiation exposure of people and the release of radioactive material to the environment; (2) to restrict the likelihood of events that might lead to a loss of control over a nuclear reactor core, nuclear chain reaction, or any source of radiation; (3) to mitigate the consequences of such events if they were to occur [99].

### 1.2.2  NUCLEAR SAFETY PRINCIPLES

One of the basic *safety principles* states that the prime responsibility for safety of a nuclear installation rests with the person or organization responsible for the installation and for activities that give rise to radiation risks. To make this responsibility effective, a legal and governmental framework for safety, including an independent regulatory body, must be established and sustained. People and the environment must be protected against radiation risks and this protection must be optimized to provide the highest level of safety that can reasonably be achieved. Nuclear or radiation accidents must be prevented and mitigated employing all practically available means. In particular, arrangements must be made for emergency preparedness and response for nuclear or radiation incidents.

### 1.2.3  NUCLEAR SAFETY ASSESSMENT

Safety of a nuclear power plant has to be assessed on a regular basis. Safety assessment involves an analysis of normal operation of the plant and its consequences, but it also includes analysis of paths to failures and consequences of such failures. The goal of an assessment is to provide information that is necessary to make a decision on whether or not something is satisfactory. This process typically requires that specific analyses are performed and actual operating parameters are compared to their safety limit values. Examples of safety limit values and analyses that are typically performed are discussed in §1.4 and §1.5.

## 1.3   CATEGORIES OF PLANT STATES

*Nuclear power plant states* are grouped into a limited number of categories primarily on the basis of their frequency of occurrence. Typically the following five basic categories are distinguished [106]:

1. *Normal operation* states, which are expected to occur over the entire lifetime of the nuclear power plant unit.
2. *Anticipated operational occurrences* (AOO), which are plant operational states that are expected to occur one or more times during the life of the nuclear power plant unit.

3. *Design basis accident*, which is a postulated accident leading to accident conditions for which the nuclear power plant is designed in accordance with established design criteria and conservative methodology, and for which releases of radioactive material are kept within acceptable limits.
4. *Design extension conditions* without significant fuel degradation that are not considered for design basis accident, but that are considered in the design process for the facility in accordance with best estimate methodology, and for which releases of radioactive material are kept within acceptable limits.
5. Design extension conditions similar as described above, but with core melting.

The first two categories (normal operation and anticipated operational occurrences) are also termed as *operational states*, whereas the last three categories belong to *accident conditions*.

A plant is considered to be in a *controlled state* when the fundamental safety functions can be ensured and the plant can be brought to a safe state following an anticipated operational occurrence or accident conditions.

Normal operation of a nuclear power plant usually comprises startup, power operation, shutting down, shutdown, maintenance, testing, and refueling. Any deviation from normal operation that is expected to occur at least once during the plant lifetime, but which does not lead to accident conditions, is considered to be an anticipated operational occurrence.

Examples of anticipated operational occurrences in light water reactors (PWRs and BWRs) are as follows:

- Loss of offsite power.
- Inadvertent control rod group withdrawal.
- Inadvertent moderator cooldown.
- Depressurization by spurious operation of an active element, such as a relief valve.
- Blowdown of reactor coolant through a safety valve.
- Loss of normal feedwater.
- Loss of condenser cooling.
- Reactor-turbine load mismatch, including loss of load and turbine trip.

Examples of anticipated operational occurrences that are specific to PWRs are as follows:

- Inadvertent chemical shim dilution.
- Loss or interruption of core coolant flow, excluding reactor coolant pump locked rotor.
- Steam generator tube leaks.
- Control rod drop (inadvertent addition of absorber).
- Minor secondary system break.

The following anticipated operational occurrences are specific to BWRs:

- Trip of any of recirculation pumps.

- Inadvertent blowdown of reactor coolant system.
- Inadvertent pump start in a hot recirculation loop.
- Condenser tube leak.
- Startup of an idle recirculation pump in a cold loop.

Accident conditions are all deviations from normal operation that are less frequent and more severe than anticipated operational occurrences. They are generally divided into design basis accidents and design extension conditions. The former refer to postulated accidents leading to accident conditions for which a plant is designed in accordance with established design criteria, and for which releases of radioactive material are kept within acceptable limits. The latter refer to accident conditions that are not considered as design basis, but which are analyzed during the design process, and which, similarly to design basis accidents, cause radioactive releases within acceptable limits. However, they comprise events with melting of the reactor core and with insignificant fuel degradation.

*Anticipated transients without scram* (ATWSs) are special types of anticipated operational occurrences in which a reactor scram is demanded but fails to occur due to a common-mode failure in the reactor scram system. Such transient can have severe consequences and are addressed separately beyond design basis.

Accidents that are unanticipated occurrences, i.e., they are postulated but not expected to occur during the lifetime of the nuclear power plant, are termed as *postulated accidents*. In light water reactors, the most frequently considered postulated accidents are

- Loss of coolant accident (LOCA) in PWRs and BWRs.
- Ejection of a control rod assembly in PWRs.
- Control rod drop accident in BWRs.
- Major secondary system pipe rupture up to and including double-ended rupture in PWRs and BWRs.
- Single reactor coolant pump locked rotor in PWRs.
- Seizure of one recirculation pump in BWRs.

## 1.4  SAFETY MARGINS

In the 1970s, when the fleet of nuclear power plants in operation increased significantly, rules and criteria for the design and construction of safe plants were established. A concept of the design basis accident (DBA) was introduced, and plants were required to satisfy the safety criteria under such extreme hypothetical conditions. The criteria such as $1204°C$ peak cladding temperature, 1% mean equivalent cladding oxidation, and 17% local maximum cladding oxidation were introduced with varying degrees of conservatism to compensate for limited level of knowledge and weaknesses in the experimental database. For each particular plant a question then could be raised whether it indeed satisfies the criteria and how large is the "distance" between the plant's most severe state and the criteria. This led to a commonly understood concept of an "adequate" *safety margin* when clearly defined safety limits were defined and the plant was shown to always stay under these limits.

However, very soon it became clear that this approach was problematic since it was not known how far the plant was from the safety limit. For that the real state of the plant must be known and the crude approximations used for conservative estimates were not accurate enough. Thus it was necessary to develop new models based on physical laws of conservation and validate the models against proper experimental data. This was achieved through an intensive research program in thermal hydraulics, which was launched in the mid-1970s and lasted up to the 1990s. The main outcome of the program was the validated *best estimate* computational codes and methods for evaluation of the safety margins.

At the beginning of 1990s the usage of best estimate codes became well-established. However, it was still not clear how to estimate and deal with the uncertainties that were present in calculations. To support the new development, the Nuclear Regulatory Commission in the United States and its contractors and consultants developed and demonstrated a method called the *code scaling, applicability, and uncertainty* (CSAU). The main goal of the work was to demonstrate that uncertainties in complex phenomena could be quantified. The attention was turned to the scalability and applicability of computational codes to safety studies of postulated safety scenarios in nuclear power plants. A general two-step methodology to quantify the uncertainty of calculated results was proposed. A "top-down" approach was used to define the dominant phenomena and a "bottom-up" approach was followed to quantify the uncertainties [23, 143, 236, 238, 245, 246].

The CSAU method initiated the development of modern *best-estimate plus uncertainty* (BEPU) methods that are currently used by the nuclear power industry around the world for power upratings, license renewals, and new design certifications. The original CSAU approach suffered from several drawbacks such as the lack of objectivity, high cost, lack of clear separation of numerical errors from other uncertainties, and limits on uncertainty propagation. The development of efficient BEPU methods is still ongoing and focused mainly on codes with an internal assessment of uncertainties. The industrial practice shows that uncertainty analyses with a random sampling of input parameters to computer codes and nonparametric statistical tolerance limits for estimating the uncertainty of output parameters have been widely accepted.

## 1.4.1 TRADITIONAL VIEW ON SAFETY MARGINS

A precise definition of a safety margin is straightforward for simple systems only. In fact this concept was developed and formalized in civil engineering applications through the work on load-strength interference. The general definition of safety margin, called *margin to damage* in civil engineering applications, takes into consideration the mean values of load $L$ and strength $S$, along with their corresponding probability density functions. The margin to damage can be then obtained as:

$$MD = \frac{\bar{S} - \bar{L}}{\sqrt{\sigma_S^2 + \sigma_L^2}}, \tag{1.1}$$

where $\sigma_S^2$ is the strength's $S$ standard deviation and $\sigma_L^2$ is the load's $L$ standard deviation. The margin to damage $MD$ is an indirect measure of the overlap in the probability density functions and can be used to estimate the probability that the load does not exceed the strength:

$$p(S > L) = \int_0^\infty f_L(L) \left[ \int_L^\infty f_S(S)dS \right] dL, \tag{1.2}$$

where $f_S(S)$ and $f_L(L)$ are the probability density functions for strength and load, respectively. The above relationship suggests that, given sufficient information about load, strength, and their standard deviations, the reliability of a system can be precisely computed. However, such information is often beyond the current state of the art in the nuclear industry, since the probability functions for strengths of various components are prohibitively expensive to obtain.

### Example 1.1: Evaluation of the Margin to Damage

The damage temperature of a fuel cladding material is 900 K and the standard deviation of the damage temperature is 3.5%. The maximum temperature of the fuel cladding in the reactor core during normal operation is estimated as 823 K and the corresponding estimated standard deviation is 15 K. Calculate the margin to damage for the cladding material during reactor normal operation.

* * *

*Solution*: The standard deviation of the clad damage temperature is equal to $900 \times 0.035 = 31.5$ K. Thus, the margin to damage becomes $MD = (900 - 823)/(31.5^2 + 15^2)^{0.5} = 2.207$.

A difference between the established acceptable limit of a safety parameter and the calculated operational value of the parameter is termed a *safety margin*. Safety margins are expressed in the same physical units that are used for the relevant safety parameter. Clearly, in a complex nuclear installation such as a nuclear power plant, there will be as many safety margins as barriers or systems whose loss is considered to be a safety issue.

The concept of safety margin is not new and specific to nuclear power plants. It is applicable to virtually all systems where damage is possible and there is some uncertainty about when and why the damage occurs. Therefore there is a need for clear identification of safety parameters and their relation to the possible damage. In nuclear application the damage is mostly related to unacceptable radiological releases. The safety margin concept for conservative and best estimate calculations with uncertainty quantification is illustrated in Fig. 1.1. This concept is further developed and discussed in §12.

### 1.4.2 PHYSICAL BARRIERS TO LIMIT RADIOACTIVE RELEASE

In complex systems like nuclear power plants, there are many paths to failure or damage. Due to this, various redundant safety systems and radiation release barriers are

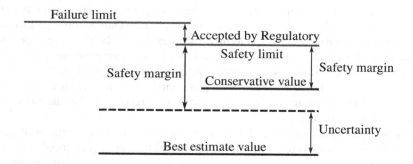

**Figure 1.1**  Safety margin determined by conservative and best estimate calculations.

introduced. Since the ultimate goal of nuclear safety is to prevent unacceptable radiological release to the public or to the environment, safety margins should be considered at least for those systems and barriers whose failure could potentially contribute to unacceptable radiological releases. Furthermore, for each barrier or safety system, a set of safety variables and their relation with the barriers and system function losses, must be clearly identified.

Concerning the public protection in case of an accident, three successive barriers to limit radioactive release are considered: the fuel cladding, the primary system boundary, and the containment. For each of the barriers, the **defense in-depth** principle is used during the system design. The three levels of the defense in depth are:

1. Prevention of departure from normal operation.
2. Detection of departure from normal operation and protective systems to cope with this deviation.
3. Safety, protective systems, and operator actions to mitigate accident consequences.

### 1.4.3  SAFETY VARIABLES AND SAFETY LIMITS

**Safety variables** are defined as such operating parameters, for which safety system settings are specified in the safety analysis report. These parameters vary according to reactor type and design, but in general, they are concerned with such physical quantities as neutron flux, power, pressure, temperature, flow, and radioactivity. They also include various safety-related events or conditions, such as loss of normal electrical power supply or emergency power supply [110].

The neutron flux density is strictly related to the reactor fission power and thus safety system settings are necessary for the flux level and distribution, rate of change, and oscillation. To avoid sudden power increases, the safety system settings should also include reactivity protection devices.

The temperature level and its increase rate belong to the most important safety variables, since too high temperatures may cause damage to the various radioactivity-release barriers, such as fuel pellets and fuel cladding. Thus safety settings are imposed on temperatures of fuel cladding, fuel channel coolant, and reactor core coolant. Inlet water and outer steam temperatures for the steam generator are important parameters to determine the safe operation of the nuclear power plant.

Pressure needs to be maintained within predefined ranges in various components and systems such as the reactor coolant system, the steam lines and turbines, the drywell, and the containment.

Sufficient flow rates and water levels are essential for the safe operation of such components as a reactor core and a steam generator. Thus settings are needed for reactor coolant flow and its rate of change. Tripping of the primary coolant circulation pump is a related safety condition since it directly affects the flow rate of coolant through the core.

The radioactivity levels should be monitored in the primary circuit, in the steam lines, and in the reactor building.

Safety variables are characterizing onset of some type of damage. Class-specific acceptance criteria are set in terms of acceptable extreme values of safety variables. Given a particular plant scenario, the evolution of the safety variable must be calculated to determine whether it remains below the *safety limit* or not.

The selection of safety variables and safety limits is based on the analysis of the barrier failure modes. For each accident frequency class, a set of design basis transients is selected. For each considered barrier degradation mode there should be at least one protection and for each pair of degradation mode/protection, there should be at least one design basis transient considered.

As an example, for a core damage event, three safety variables can be considered: the *peak cladding temperature* (PCT), the *enthalpy deposition rate*, and the total *cladding oxidation*. While the *embrittlement* damage mechanism occurs as a consequence of an increase in the PCT, the too high enthalpy deposition rate governs the initiation of cracking of the cladding. However, both PCT and the total cladding oxidation can be tracked if the subject damage mechanism is embrittlement.

## 1.4.4  SAFETY CRITERIA

The general principles for deriving the *safety criteria* are as follows. Category 1 and 2 events (normal operation and anticipated operational occurrences, respectively) have a high probability that very drastic reference values are defined for the radiological consequences to be acceptable. For category 2 the limit is typically bounded by the integral annual limit of activity release for normal operation (category 1). The phenomena which endanger the fuel rod integrity are thermal and thermomechanical loads to the cladding and the loss of integrity of fuel pellets by melting. The following measures are taken to prevent those damages:

- Prevention of the *critical heat flux* (CHF) to avoid a large temperature rise in the cladding. A typical requirement is that the probability to remain below CHF limit in the hottest point is 95% with a 95% level of confidence.

- Prevention of fuel melting by limiting the fuel maximum linear power.
- Prevention of cladding embrittlement by forcing a maximum value of the allowable cladding temperature not to be exceeded. In addition, limits are given for the cladding oxidation and the hydrogen pick up, characterizing the metallurgical state, which may induce cladding embrittlement.

Other criteria, related to the mechanical loads of the cladding, are defined, including: cladding circumferential deformation, rod internal pressure, cladding stress, cladding fatigue, total strain in category 1 and 2 transients, and fretting wear of cladding. Cladding thermal loads are limited by fixing the maximum metal oxide interface temperature.

Additional fuel safety criteria are introduced for events of categories 3 and 4. Due to the lower occurrence probability of category 3, limited fuel damage in some fuel rods is allowed. However, the prescribed reference values of radiological consequences shall be met and the fuel damage shall not degrade the reactor core cooling function. In particular, it is required that the core geometry remains coolable under such conditions.

Category 4 includes some specific low-probability accident types, such as the large-break loss of coolant accident, the reactivity initiated accident, and sometimes the main steam line break accident. For this category, significant damage for a few fuel elements can be accepted. However, similarly as for category 3, the core geometry should still be preserved to allow long-term coolability.

For both categories 3 and 4 the following is valid:

- Prevention of critical heat flux phenomena.
- Prevention of fuel melting.
- Prevention of cladding embrittlement.

Widely used criteria for category 4 LOCAs are:

- Maximum cladding oxidation including corrosion before and during an accident shall not exceed 17% of the clad wall thickness for a Zircaloy cladding material.
- Maximum cladding temperature during transient shall not exceed 1477 K (1204°C).

Various safety criteria are used to assess the level of safety of nuclear power plants. The term "limit" is used to indicate that a specific value of a certain quantity must not be exceeded since otherwise some legal sanctions would be invoked. Criteria used for other purposes, for example as a threshold value indicating a need for further investigations, are using other terms, such as *reference level.*

*Safety limits* are limits on operational parameters within which a nuclear power plant has been shown to be safe. Thus safety limits are operational limits and conditions beyond those for normal operation. A limit acceptable to the regulatory body is called an *acceptable limit.*

Safety limits are applicable to operational parameters that have an influence on nuclear reactor safety. Complete or sufficient information is not available for all such operational parameters. However, most of the current safety criteria were established during the 1960s and early 1970s, using available experimental data for verification. Further development of the criteria was continued during decades of operational experience. Currently well-established lists of safety criteria exist for nuclear power plants [172].

Limiting criteria imposed in safety analyses can be divided into the following three categories:

1. *Safety criteria*, which are criteria imposed by the regulator. If the criteria are preserved during reactor operation, safety criteria ensure that the impact of a design basis accident on the environment is acceptable.
2. *Operational criteria*, which are criteria specific to the nuclear power plant design and provided by the plant vendor as part of the licensing basis. Operational criteria ensure that safety criteria are not violated.
3. *Design criteria*, which represent limits employed by vendors or utilities for the design. Design criteria are preserved during normal operation and anticipated operational occurrences.

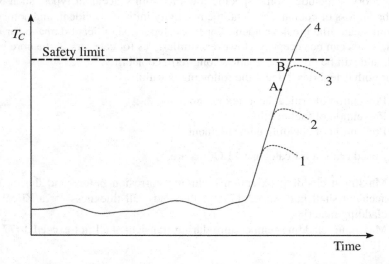

**Figure 1.2** Illustration of a safety limit, a safety system setting, and an operational limit using the fuel cladding temperature as the critical safety parameter.

The concept of safety limit and its relation to safety system settings and to an operational limit are illustrated in Fig. 1.2. The figure shows the case in which the critical safety parameter of concern is the fuel cladding temperature. It is assumed that the fuel cladding temperature is obtained by correlation with the monitored coolant temperature. If the monitored parameter exceeds an alarm setting, the operator will be alerted and will activate an automatic system to reduce the temperature to the previous steady-state value. The delay in the operator's response should be taken into consideration so that the temperature will not reach the operational limit for normal operation. Curve 1 shows the condition when the alarm setting is exceeded. Curve 2 represents the transient when the operational limit is exceeded and the operator is able to take corrective action to prevent the safety system setting from being reached. If a malfunction of the control system or operator error occurs, the monitored parameter might reach the safety setting at point A on curve 3. As a consequence, after some inherent delay in the instrumentation and equipment, the correction action becomes effective at point B. In case of a more severe failure, which goes beyond the most severe one that the plant was designed for, the temperature of the cladding might exceed the value of the safety limit, as demonstrated by curve 4. Since under such circumstances significant amounts of radioactive material could be released, measures for accident management should be activated to mitigate the possible consequences [110].

A set of *operational limits and conditions* for a nuclear power plant shall be developed by the plant operating organization. The operational limits should be accessible as a single document to control room personnel. They should clearly state the plant conditions that must be met to avoid situations that might lead to accident conditions. Typically the document contains information on the safety limits, limiting safety system settings, limits and conditions for normal operation, and surveillance requirements. It should also provide information on actions to be taken when deviations from the stated operation limits and conditions occur.

## 1.5  SAFETY ANALYSES

The primary goal of a *safety analysis* is a proper understanding of the processes that take place during normal operation and accidental conditions of a nuclear power plant. When the analysis also includes determinations or judgments of the plant state acceptability, it is called a *safety assessment*. A safety analysis is typically used to evaluate the potential hazards associated with various operational states and accident conditions of a plant. These operational states are broadly divided into normal operations and anticipated operational occurrences, as discussed below.

### 1.5.1  ACCIDENT SCENARIOS

In the context of safety analysis of a nuclear power plant, an **event** is any occurrence unintended by the operator. It includes such occurrences as operating errors, equipment failures, or any other mishap, the consequences of which are not negligible from the point of view of protection and safety. A set of postulated or assumed events or conditions is called a **scenario**.

One of the methods for performing uncertainty analysis is based on the use of *bounding scenario* (also called *enveloping scenario*) calculations. The bounding scenarios should be chosen so that they include cases presenting the greatest possible challenges to each of the relevant acceptance criteria. Several postulated initiating events may be combined and the safety analysis should confirm that the grouping of initiating events is acceptable [109].

### 1.5.2 EFFECT OF REACTOR INPUT PARAMETERS AND STATE

Reactor state and operating conditions are known only with limited precision. For example, a coolant flow rate through each individual channel can fluctuate with time due to the turbulent character of the flow, or due to a reactor power time variations. The fuel state is also changing with an increasing burn-up, and the tolerances may no longer be as fabricated. Realistic reactor state variations can be determined by examination of the most probable condition and the distribution around this condition. To this end both experimental data and analytic studies can be used.

## 1.6 REACTOR CORE DESIGN

A reactor core is the central system of a nuclear power plant and thus its design must be performed with the highest possible safety standards. Since fuel elements and assemblies contain radioactive material, their structural integrity has to be maintained to avoid radioactivity release. The design of fuel elements has to assure that they withstand the anticipated radiation and heat flux levels without significant deterioration of their mechanical properties. The deterioration could result from, e.g., differential expansion and deformation, internal and external pressure changes, irradiation, and thermal loads acting on the fuel elements.

The design of the reactor core should be done together with the design of reactor cooling systems and the reactor control and reactor protection systems to assure effective core cooling under operational conditions and accidents without significant fuel degradation. The design contains three main components: the neutronic design, the thermal-hydraulic design, and the mechanical design. Each of these design components is addressing specific design considerations, as discussed in the following sections.

### 1.6.1 NEUTRONIC DESIGN

One of the primary goals of the *neutronic design* of a nuclear reactor core is to ensure that the feedback characteristics of the core rapidly compensate for an increase in reactivity. This goal can be achieved by a combination of the inherent neutronic characteristics of the reactor core and its thermohydraulic characteristics. The inherent neutronic characteristics of the reactor core include such features as negative temperature coefficients of reactivity for the fuel and the moderator, the flat power distribution, and reduction of variations in reactivity during fuel burnup. Other frequently defined key safety parameters include the shutdown margin, the maximum linear heat generation rate, the maximum reactivity insertion rate, and the void coefficient of reactivity. The key safety parameters should be re-defined and justified in case of any major modifications to the reactor core design.

### 1.6.2 THERMOHYDRAULIC DESIGN

The purpose of the *thermohydraulic design* of a nuclear reactor core is to ensure that specified thermohydraulic design limits are not exceeded in normal reactor operation and during anticipated operational occurrences. In case of a design basis accident or design extension conditions without significant fuel degradation, the design should ensure that the failure rates of fuel rods remain within acceptance levels.

Specific design limits should be established for the thermohydraulic design safety parameters. These parameters and their adequate safety margins should be predictable by validated computational procedures. The most frequently defined parameters that are tracked in the thermohydraulic design are the *minimum critical power ratio* (MCPR) for boiling water reactors, the *minimum departure from nucleate boiling ratio* (MDNBR) for pressurized water reactors, and the *dryout power ratio* for pressurized heavy water reactors. Other safety parameters of interest include maximum linear heat generation rate, the peak fuel temperature or enthalpy, and the *peak cladding temperature* (PCT).

The design analysis should address the influence of various parameters that determine the values of the predicted safety margins. This includes the uncertainties in the values of process parameters (e.g. reactor power, coolant flow rate and distribution, core bypass flow, inlet temperature and pressure, and power peaking factors), core design parameters, and calculation methods including code uncertainties.

Various approaches should be taken to demonstrate the fulfillment of the safety recommendations. For pressurized water reactors, the limiting MDNBR should be established at a such level that the rod with the lowest margin in the core does not experience DNB during normal operation or anticipated operational occurrences with a 95% probability at the 95% confidence level. For boiling water reactors the limiting MCPR should be established such that the number of fuel rods that experience dryout does not exceed a very small fraction (e.g. less than 0.1%) of the total number of fuel rods in the core. For pressurized heavy water reactors, if the maximum fuel cladding the temperature remains below a certain limit (e.g. 873 K) and the duration of the post-dryout heat transfer regime is limited (e.g. less than 60 s), it is considered that the fuel deformation is small and a failure of the pressure tube is avoided [108]. For gas-cooled reactors specific safety criteria are established by means of fuel performance models for the time at temperature histories of the core and fuel [101].

Critical heat flux limits should be applied in the safety analysis to ensure that the potential for cladding failure is avoided. To this end the limiting MDNBR and MCPR values have to be determined using the critical heat flux correlations derived from experimental data obtained at steady state conditions and other idealized circumstances departing from the reactor operation conditions. As a consequence, adequate margins or provisions should be added to MDNBR and MCPR values to take into account additional factors, not included in the correlation derivation. These include such factors as, e.g., spatial and temporal variation in the power distribution and the impact resulting from the potential presence of *crud* in the core.

### 1.6.3  THERMOMECHANICAL DESIGN

Thermomechanical analysis is used to study the properties of materials as they change with temperature. Many materials during heating or cooling undergo changes in their thermomechanical properties. Thermomechanical analysis can provide valuable insight into the structure, composition, and application possibilities of various materials. This technique is used to measure a variety of material properties, including the coefficient of thermal expansion, melting temperature, and elevated temperature creep or stress relaxation behavior.

The thermomechanical design of fuel rods and fuel assemblies should ensure that their structural integrity is maintained for normal operation and anticipated operational occurrences. Only a limited number of fuel rod failures should be allowed during accident conditions, such as design basis accidents. The allowable number of failed fuel rods depends on the frequency of the accident. For more frequent accidents this number should be the lowest.

## 1.7   MAJOR NUCLEAR POWER PLANT ACCIDENTS

Nuclear power plants are designed with their safe and effective operation in mind, and with making all necessary efforts to minimize the likelihood of accidents, and to ensure that their consequences can be reliably mitigated. The primary means to achieve these goals is the defense in depth approach, consisting of the implementation of consecutive and independent levels of protection. Yet, three major commercial nuclear power plant accidents have occurred since 1979. Even though it would be most desirable to avoid these accident to occur, they provided new knowledge that can be used in designing future and safer nuclear reactors. The accidents revealed several design flaws, lack of safety culture, inadequate supervision by regulatory bodies, and erroneous human actions. In this section we make an overview of the three major accidents with a focus on their root causes and proposed safety improvements.

### 1.7.1   THREE MILE ISLAND UNIT 2 ACCIDENT

The most serious accident in U.S. commercial nuclear power plant operating history occurred on March 28, 1979, in the Three Mile Island Unit 2 (TMI-2) reactor, near Middletown, Pennsylvania. A combination of equipment malfunctions, design-related problems, and worker errors led to TMI-2 reactor core's partial meltdown and very small off-site release of radioactivity.

The Three Mile Island power plant had two reactors: TMI-1, a PWR of 880 MWe, in operation from 1974 to 2019 (one of the best-performing units in the USA), and TMI-2, a PWR of 959 MWe (almost brand new at the time of the accident).

The accident began in the secondary system, when, due to electrical or mechanical failure, the feedwater pumps stopped, preventing the removal of heat from steam generators. This caused the plant's turbine and the reactor to automatically shut down, and the pressure in the primary system to increase. In such circumstances the pilot-operated relief valve located at the top of the pressurizer automatically opened to control the pressure in the primary system. The valve should have then closed when the pressure fell to the proper level about 10 seconds later, but it became stuck open, leaking reactor coolant water into the reactor coolant tank. The plant operators, however, were unaware of this since the instruments in the control room indicated that a signal to close the valve was sent and they did not have an instrument indicating the valve's actual position.

With the relief valve still open, the pressure in the primary system dropped below a safety minimum level which triggered high-pressure injection pumps to push replacement water into the primary system. The water surged into the pressurizer, raising the water level in it. During normal operation the reactor pressure vessel is filled entirely with water and the pressurizer is filled only partly with water, the only instrument showing water level in the primary system was located in the pressurizer. Due to that reactor design flaw and limited training, operators believed that the primary system was full of water and responded by reducing the flow of water from the high-pressure injection pumps.

The main circulating pumps were still operating and pumping the remaining coolant through the core. However, due to the increasing content of steam in the reactor primary system, the coolant entering the pumps contained a mixture of steam and water, which caused the pumps to vibrate. To protect the pumps from damage, they were shut down by operators, which ended forced cooling of the reactor core. At this stage reactor coolant water boiled away and the reactor's fuel core was uncovered. Due to poor heat removal, the temperature of fuel rods significantly increased leading to their damage and release of radioactive material into the cooling water.

Only after 2 hours and 20 minutes of the accident progression, operators closed a block valve between the relief valve and the pressurizer. This action stopped the loss of coolant water through the relief valve. However, proper cooling of the reactor core was still not possible since superheated steam and non-condensable gases blocked the flow of water through the core. After more than 13 hours operators restored forced cooling of the reactor core when they were able to restart one of the reactor coolant pumps.

As a consequence of the accident, one-third of the fuel core was melted, but the reactor vessel itself maintained its integrity and contained the damaged fuel. Radioactive gases from the reactor cooling system were built up in the makeup tank in the auxiliary building and were moved to the waste gas decay tanks using compressors. A small amount of radiation was released from the plant since the compressors leaked. The leaking gases went through high-efficiency particulate air (HEPA) filters and charcoal filters which removed most of the radionuclides, except for noble gases. According to NRC, 1.6 PBq of krypton was released in July 1979. With a short half-life and being biologically inert, this release did not pose a health hazard.

The TMI-2 accident is described in numerous reports, publications, and journal articles that provide very deep and detailed analyses of root causes of the accident [190], reviews of identified operational errors and system misalignments, and lessons learned [170]. Brief descriptions of the accident, including illustrations presenting the core damage and the improvements in reactor designs to prevent such accidents in the future are also provided in textbooks [130, 194].

## 1.7.2 CHERNOBYL ACCIDENT

The Chernobyl accident that occurred on April 26, 1986, in northern Ukraine (then part of the USSR) was the result of major design deficiencies in the RBMK type of reactor, the violation of safety procedures, and the absence of a safety culture. Main features of the RBMK design that played a major role during the accident include:

1. The reactor core operated in an overmoderated regime and had a positive void coefficient of reactivity. As a result, an overheating of the coolant caused an increase in steam volume fraction and an increase in reactivity.
2. The large size of the RBMK active core region (11.8 meters in diameter by 7 meters in height) implied that different parts of the core were largely neutronically decoupled from one another. This made controlling the power level difficult, especially due to the small fraction of control rods inserted into the core.
3. The leading edge of the control rods contained graphite followers having no neutron absorbers, which, upon insertion into the core, added positive reactivity to the core. This was a design flaw and one of the main factors contributing to the accident.
4. Unlike Western reactors, the Chernobyl reactor did not have a containment structure, allowing for release of at least 5% of the total radioactive material in the reactor core into the atmosphere that was subsequently deposited as dust close by, and some were carried by wind over a wide area.

The accident was initiated during a turbogenerator coastdown test that was conducted with flagrant violation of operating procedures and willful disregard for safe operating practices.

The reactor power was reduced from 3200 to 200 MWt over 24 hours, resulting in a large reactivity defect due to the build-up of fission product $^{135}$Xe. To maintain criticality, the reactor operator withdrew control rods beyond the operating reactivity margin. Once the operator initiated the turbine isolation, the core coolant flow was reduced due to the reactor coolant pump rundown. As a result, the void fraction in the core increased, leading to the increase of the reactor power. The reactor operator then initiated a reactor scram, but due to the replacement of water by a control rod with graphite followers, the reactor became prompt critical.

The reactor was totally destroyed in the accident and the consequent reactor fire resulted in an unprecedented release of radioactive material. The accident had enormous adverse consequences for the public and the environment. To provide a balanced assessment of the environmental consequences and health effects of the accident, IAEA established Chernobyl Forum in 2003 and presented the findings and recommendations in a dedicated report [98].

### 1.7.3 FUKUSHIMA DAI-ICHI ACCIDENT

On March 11, 2011, the Fukushima Dai-ichi Nuclear Power Station operated by Tokyo Electric Power Company was damaged due to the Great East Japan Earthquake of magnitude 9.0 and the ensuing tsunami. A 15-meter tsunami disabled the power supply and cooling of unit 1–3 reactors. All three reactor cores largely melted in the first three days and unit 4 experienced hydrogen explosion after five days. The accident caused high radioactive releases of some 940 PBq ($^{131}$I eq.) and was rated level 7 on the International Nuclear and Radiological Event Scale. Units 1–4 with net capacity of 2719 MWe had to be written off due to damages caused by the accident.

The Fukushima Dai-ichi reactors were General Electric boiling water reactors of an early (1960s) design with what is known as a Mark I containment. Reactors came into commercial operation between 1971 and 1975 and had a capacity of 460 MWe for unit 1, 784 MWe for units 2–5, and 1100 MWe for unit 6.

The reactors proved robust seismically, but vulnerable to the tsunami. Units 1–3 were operating at the time and all shut down automatically when the earthquake hit with no significant damage to any from the earthquake. Units 4–6 were not operating at the time but were affected by the tsunami.

Almost one hour after the earthquake, when the entire Fukushima Dai-ichi power plant site was flooded by the tsunami, 12 of 13 backup generators onsite and also the heat exchangers for dumping reactor waste heat and decay heat into the sea were disabled. As a result, units 1–3 lost the ability to maintain proper reactor cooling and water circulation functions. At that time reactor cores still produced about 1.5% of their nominal thermal power from fission product decay, equivalent to 22 MW in unit 1 and 33 MW in units 2 and 3.

During the first day of the accident progression, the steam-driven reactor core isolation cooling systems and high-pressure coolant injection system in unit 3 provided cooling to all three units. During the second day of the accident, back-up battery supplies were depleted and the ability to cool the reactor cores of units 1–3 was significantly degraded or became unavailable.

Without heat removal from the cores, steam was generated in reactor pressure vessels. To control pressure in the reactor coolant system, the steam was discharged to suppression chambers designed for that purpose. However, this resulted in a pressure increase within primary containments and their venting became necessary. The venting was designed to be through external stacks, but in the absence of power, much of it apparently backflowed to the reactor buildings. The vented steam, noble gases and aerosols were accompanied by hydrogen. Accumulation of hydrogen in reactor buildings led to hydrogen explosions, first at unit 1 and later also at units 3 and 4. As for unit 2, a hydrogen explosion did not occur and therefore the building remained undamaged.

Due to insufficient cooling of reactor cores in units 1–3, major fuel melting occurred, though the fuel and fission products remained essentially contained. However, some volatile fission products were vented or released and some soluble ones were leaking with the water, especially from unit 2, where the containment was breached. A "cold shutdown condition" for units 1–3 was declared in mid-December 2011.

The spent fuel storage pools were not significantly damaged by the earthquake, tsunami, and hydrogen explosions. New cooling circuits were provided and analysis of water confirmed that most fuel rods were intact.

The Fukushima Dai-ichi accident revealed weaknesses in the plant design, emergency preparedness, response arrangement, and in planning for the management of a severe accident. In the design it was assumed that a loss of all electrical power at a nuclear power plant for a period longer than a few hours would never occur. Likewise, the possibility of several reactors at the same site experiencing accident conditions was never considered. Finally, insufficient provision was made for possibility of a nuclear accident occurring during ongoing major natural disaster.

In response to the accident and its management, Japanese Government has reformed its regulatory system and gave regulators clearer responsibilities and greater authority. Other countries responded with measures that included carrying out "stress tests" to reassess the design of nuclear power plants against site-specific extreme natural hazards, installing additional backup sources of electrical power and supplies of water, and strengthening the protection of plants against extreme external events [105].

# PROBLEMS

### PROBLEM 1.1

Assuming that the subject damage mechanism in a reactor core is embrittlement of the cladding material, explain which safety variables should be tracked.

### PROBLEM 1.2

A power pulse in a *reactivity-initiated accident* (RIA) is usually characterized by the maximum deposited fuel enthalpy in calories per gram (cal/g) and by the pulse half-width in milliseconds. Consider a pulse that deposits 200 cal/g in the uranium dioxide ($UO_2$) fuel pellets at room temperature. Assuming that the energy is deposited uniformly and momentarily (no heat is transferred from the pellet), estimate the temperature increase in the pellet.

### PROBLEM 1.3

For the same conditions as in Problem 1.2, assume that the linear expansion coefficient of $UO_2$ in the temperature range corresponding to the temperature increase of the pellet is 2.9%. In fresh fuel the as-fabricated gap between the pellets and the cladding is about 2% of the cladding diameter. Does the cladding accommodate the pellet expansion?

### PROBLEM 1.4

Explain which feature of the RBMK reactor in Chernobyl played the major role during the accident and led to the reactor prompt criticality.

# 2 Nuclear Power Reactors

The various topics in this book are concerned with selected power reactor types that have been evolving during the past several decades of nuclear power development. Three of these reactor types, namely, the pressurized water reactor (PWR), the boiling water reactor (BWR), and the pressurized heavy water reactor (PHWR) are dominating among the currently operating reactors. The remaining four types, namely, the light water-cooled graphite moderated reactor (LWGR), the gas-cooled reactor (GCR), the fast breeder reactor (FBR), and the high-temperature gas cooled reactor (HTGR) constitute only a small fraction of all currently operating power reactors worldwide.

The main features of the reactor types that are under operation are presented in §2.1. The configurations for PWRs, BWRs, PHWRs, GCRs, FBRs, and HTGRs are described. In §2.2, the new advanced generation III, III+, and IV reactors are discussed. These reactors are distinguished by improved safety and reliability features (mainly generation III and III+ reactors), and sustainability and non-proliferation features (generation IV reactors). The section also includes small modular reactors (SMRs), which are reactors with modular design and the output power of 300 MWe or less.

## 2.1 REACTOR TYPES

The common feature of all nuclear fission power reactors is that the fission power that is released in the reactor core has to be evacuated from the core to produce useful energy. There are many possible technology solutions for this process to be conducted. Even though these technologies are currently quite well established, there is ongoing research to develop new, safer, and more efficient methods to design and operate nuclear reactors. As a result of this research, several reactor types have been developed, as briefly described in this section.

### 2.1.1 PRESSURIZED WATER REACTOR

The pressurized water reactor (PWR), schematically shown in Fig. 2.1, is a light water-moderated and cooled nuclear reactor. The water coolant in the primary system is kept under high pressure (around 17.5 MPa) to prevent boiling in the reactor core. The coolant is circulating in the primary loop, removing heat from fuel elements in the reactor core and depositing it in the steam generator to produce steam. Depending on the reactor type, two to four independent primary loops are employed.

Water enters the reactor core at around 563 K (290°C) and its temperature at the core exit is approximately 35 K higher. Since the water pressure is high, the exit subcooling is high enough to prevent boiling. By keeping the coolant in liquid form, the control rod system is simplified and can be placed above the reactor core. In case

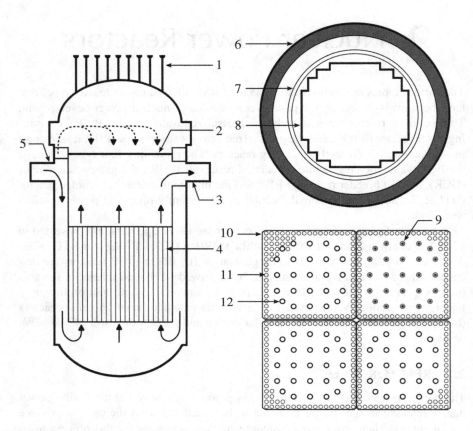

**Figure 2.1** Pressurized water reactor: 1–control rod drive mechanisms, 2–upper support plate, 3–outlet nozzle, 4–reactor core with fuel assemblies, 5–inlet nozzle, 6–reactor vessel, 7–core barel, 8–core baffle, 9–control cluster element, 10–fuel assembly, 11–fuel rod, 12–guide thimble.

of a loss of power in the plant, the electromagnetic system holding the rods will give out, and gravity will cause the rods to fall into the core, stopping the fission reaction.

The hot coolant water leaving the reactor pressure vessel flows through inverted U–tubes and heats up a secondary loop of water in steam generators. This secondary loop is at lower pressure, usually around 7 MPa, so the water boils and a saturated steam is generated. The steam then passes through turbines connected to generators that generate electricity.

PWRs must use enriched uranium as their nuclear fuel, because of their use of light water as a moderator. Enriched uranium, as ceramic uranium dioxide pellets, is packed into fuel rods which are bundled into fuel assemblies. There are 200–300 rods in each fuel assembly and up to several hundred fuel assemblies (typically 150 to 250) in the reactor core. This corresponds to five cubic meters of uranium or 80–100 tons of uranium.

The bundles are arranged vertically in fuel channels within the core. The water coolant, which is at the same time a moderator, flows vertically upwards between fuel rods, providing sufficient cooling of the fuel rod's outer surfaces. In case of insufficient cooling, for example resulting from a loss of coolant accident (LOCA), there will also be a loss of moderator causing the nuclear chain reaction to stop. Also if the coolant water evaporates and becomes water vapor inside a fuel assembly, there will be less moderator and therefor the chain reaction will stop.

## 2.1.2  BOILING WATER REACTOR

Similarly to PWRs, boiling water reactors (BWRs) are using light water as a moderator and coolant. A schematic of the boiling water reactor is shown in Fig. 2.2. The

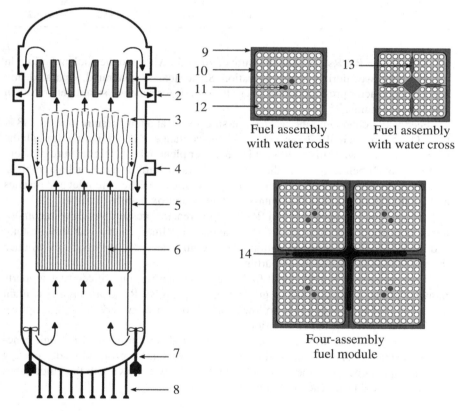

**Figure 2.2**  BWR pressure vessel and fuel assemblies: 1–steam dryer assembly, 2–steam outlet, 3–steam separator assembly, 4–feedwater inlet, 5–core shroud, 6–reactor core with fuel assemblies, 7–main circulation pump, 8–control rod drives, 9–core bypass water, 10–channel, 11–water rod, 12–fuel rod, 13–water cross, 14–control rod blade.

**TABLE 2.1**

**Fuel Core Characteristics for PWR and BWR Nuclear Power Plants [9]**

| Feature | PWR 900 | PWR 1400 | VVER-1000 | Typical BWR |
|---|---|---|---|---|
| Number of assemblies in the core | 157 | 214 | 163 | 700–900 |
| Type of assembly | $17 \times 17$ | $17 \times 17$ | Hexagonal | $10 \times 10$ |
| Number of water rods | 25 | 15 | 19 | |
| Number of fuel rods per assembly | 264 | 264 | 312 | 91–96 |
| Number of fuel rods in the core | 41 448 | 56 496 | 50 856 | 63 000–86 000 |
| Primary circuit coolant and moderator | $H_2O$ | $H_2O$ | $H_2O$ | $H_2O$ |
| Primary circuit pressure (MPa) | 15.5 | 15.5 | 15.7 | About 7 |
| Feed coolant temperature (°C) | 285 | 285 | 290 | 276 |
| Assembly length (m) | 4.0 | 4.50 | 4.57 | About 4.0 |
| Active fuel length (m) | 3.66 | 4.268 | 3.53–3.68 | About 3.7 |

water coolant in BWRs is kept at pressure of around 7 MPa and the coolant boiling in the core is allowed during normal operation. Since steam is generated, separated, and dried in the reactor pressure vessel, there is no need for separate steam generators in BWR power plants.

Feed water enters into the reactor pressure vessel at a temperature around 488 K (215°C) and mixes in the downcomer with the coolant that is circulating through the reactor core. The well-mixed coolant in the lower plenum has a temperature of about 549 K (276°C) before it enters the core and flows upward through fuel assemblies. The coolant starts boiling a few decimeters downstream of the core inlet and reaches at the core exit an average vapor mass content of approximately 11–12%.

The principal components of a BWR are the reactor pressure vessel with components, reactor water recirculation system, main steam lines, control rod drive system, and nuclear fuel and instrumentation. These components are necessary to produce the steam power required by the turbine.

The reactor core is made up of fuel assemblies that rest on orificed fuel supports mounted on top of the control rod guide tubes. Typical BWR fuel assemblies contain up to 100 fuel rods with a few hollow "water" rods or a "water" cross, to improve moderation within the bundle interior.

The insertion and withdrawal of the control rods are performed through penetration nozzles in the bottom head of the reactor vessel. Control rod blades occupy alternate spaces between fuel assemblies and can be withdrawn into the guide tubes below the core during reactor operation.

### 2.1.3 PRESSURIZED HEAVY WATER REACTOR

Pressurized heavy water reactors (PHWRs) are nuclear power reactors that use unenriched natural uranium as their fuel, and that use heavy water ($D_2O$) as their coolant and moderator. The heavy water is kept under pressure to avoid boiling, exactly as for PWRs. Low absorption of neutrons by heavy water greatly increases the neutron economy of the reactor, avoiding the need for enriched fuel.

CANDU[1] reactors are pressure tube, heavy-water moderated nuclear power reactors developed in Canada. The basic structure of the CANDU core is the fuel bundle, which contains natural uranium dioxide in zircaloy-clad fuel pins, separated with spacers. Fuel bundles are placed end to end in a pressure tube through which flows pressurized (10 MPa) heavy water. The reactor core consists of fixed calandria tubes in a vessel filled with a heavy-water moderator. A pressure tube is loaded into each calandria tube, which, together with a gas gap, insulates the pressure tube from the moderator. The moderator and the coolant system are therefore separate.

The reactor control and shutdown devices are located in the moderator and do not cross the coolant pressure boundary. They run between columns or rows of calandria tubes.

CANDU reactors employ three types of shutdown mechanisms. Draining the heavy water out of the calandria shuts down the chain reaction by removing the moderator which slows down the neutrons. Solid shutoff rods, falling in from the top under the action of gravity, effectively decrease the number of neutrons available for fission by absorption. The same effect can be achieved by an injection of liquid "poison" through tubes directly into the moderator water.

The emergency core cooling system (ECCS) is designed to re-establish fuel cooling following a loss-of-coolant accident. The ECCS water supply comes from either the heavy water moderator or an ordinary water storage tank. The injection points are at the reactor headers. These headers are large pipes to which every fuel channel is connected. They are at either end of the core, and above all the fuel channels.

The main advantage of PHWRs is the ability to use natural uranium as the fuel. Since the moderator can be kept at a relatively low temperature, the resulting thermal neutrons have lower energies, for which the neutron cross section for fission in $^{235}$U is higher. This feature means that a PHWR can use natural uranium and other fuels more efficiently than LWRs. CANDU reactors are claimed to be able to handle fuels including reprocessed uranium or even spent nuclear fuel from LWRs. A successful use of recovered uranium from LWRs in the Qinshan CANDU Unit 1 was announced by Atomic Energy of Canada Limited (AECL) on March 23, 2010.[2] Additional natural uranium equivalent fuel bundles were planned to be inserted into separate fuel channels at the Qinshan Unit 1 reactor in Haiyan, China.

## 2.1.4   GAS COOLED REACTOR

Gas-cooled reactors (GCRs) are nuclear power reactors that use graphite as a moderator and a gas (carbon dioxide, $CO_2$, or helium, He) as coolant. GCRs can provide efficient and cost-effective electricity and produce high-temperature process heat usable for various industrial applications. In early designs, such as Magnox reactors, the fuel consisted of natural uranium metal clad with an alloy of magnesium

---

[1]Canada Deuterium Uranium
[2]newswire.ca

known as Magnox. The newer advanced gas-cooled reactors (AGRs) use a slightly enriched uranium dioxide clad with stainless steel.

Current development of GCRs goes in two directions. On the one hand, high-temperature and very-high-temperature gas-cooled reactors are under investigation, permitting a very high outlet temperature in the order of 1273 K (1000°C). Such reactors can produce electricity at a very high thermodynamic efficiency. Alternatively, the reactors can provide high-temperature heat that can be used for hydrogen generation. On the other hand, a development of a gas-cooled fast reactor (GFR) is pursued within the framework of the generation IV reactors (see §2.2.2).

### 2.1.5 FAST BREEDER REACTOR

A **fast-neutron reactor**, called often in short a *fast reactor*, is designed to maintain its neutrons at high energies. Such a fast reactor needs no neutron moderator, and all moderating materials should be avoided at its core. As a result, the core of a fast reactor has a relatively small size. In order to reduce the required amount of fissile material in the core, it is desirable to extract as high power as possible from the fuel. Fast reactors have therefore a high power density in the core, and to avoid a high temperature in fuel rods, their diameter is rather small and is kept in a range of 6–8 mm.

The high power density in the fast reactor core causes a high heat flux on a surface of fuel rods. Therefore one of the requirements for a coolant in the fast reactor is to provide efficient cooling at such high heat fluxes.

A **fast breeder reactor** (FBR) is a nuclear reactor that uses fast neutrons to generate more nuclear fuels than it consumes while generating power. The design requirements for such reactors include additional accommodation of fertile material that is necessary for the breeding of new fuel. This is usually accomplished by placing a blanket containing fertile material around the fissile core.

### 2.1.6 HIGH-TEMPERATURE GAS-COOLED REACTOR

High-temperature gas-cooled reactors (HTGRs) are nuclear power reactors that use a graphite moderator and a gaseous coolant (usually helium) at high outlet temperature. The reactor core is made up of a number of hexagonal graphite blocks stacked in close-packed columns to form a cylindrical arrangement. Fuel rods are inserted in vertical holes drilled in the graphite. The fuel rods are made of graphite containing a mixture of particles of thorium and highly enriched uranium, as oxide or carbide, clad with multiple coatings of pyrolytic carbon or silicon carbide. The presence of these coatings eliminates the need for additional cladding on the fuel rod. The helium coolant flows through other vertical holes in the graphite. Large holes are provided for the control rods, which contain a mixture of boron carbide and graphite packed in metal tubes. The control rods are inserted into the core from above.

## 2.2   REACTOR GENERATIONS

Several generations of nuclear power reactors are commonly distinguished. Generation I reactors were developed in 1950s and 60s, and are now all shutdown.[3] This generation consisted of early prototype reactors with the main purpose to demonstrate the capability of nuclear power to contribute to electricity generation.

Generation II is a class of commercial reactors developed and built mainly during three decades starting from 1970s until 1990s. During this period prototypical and older versions of PWR, BWR, CANDU, AGR, VVER, and RBMK were developed. As of 2022, about 85% of electricity produced worldwide by nuclear power comes from the generation II reactors.

Current reactor development is addressing several challenges, including improved safety performance, more sustainable operation, and higher merchantability of nuclear power reactors. To this end generation III, III+, IV, and small modular reactors are under intensive research and development. The main directions of the current development are described below.

### 2.2.1   GENERATION III AND III+ REACTORS

Generation III and III+ power reactors are advanced reactors designed to succeed the reactors developed during 1970s until 1990s, commonly termed as the generation II reactors. These new reactors are significantly improved in terms of safety and reliability, as compared to the generation II reactors. Their core damage frequency is largely reduced thanks to enhanced safety systems, often based on passive safety principles. The generation III and III+ reactors are designed for 60 years of operation (compared to 40 years for generation II reactors), with a potential for lifetime extension to 100 or more years of operation (compared to 60 years for generation II reactors).

### 2.2.2   GENERATION IV REACTORS

Generation IV International Forum (GIF) was formally charted in 2001 and is now an international collective representing governments of 13 countries where nuclear energy is playing a significant role and is considered to be important for the future.[4] Late in 2002, after some two years' deliberation and review of about one hundred concepts, GIF announced the selection of six reactor technologies which are believed to represent the future shape of nuclear energy. The main criteria used in the selections include safety, cost-effectiveness, sustainability, security from terrorist attacks, and resistance to diversion of materials for weapons proliferation. The main features of the six generation IV technologies are summarized in Table 2.2.

---

[3]Magnox reactor Wylfa Unit I in the UK was the last generation I reactor shut down on December 30, 2015.

[4]www.gen-4.org/gif

**TABLE 2.2**

**Generation IV Reactor Technologies under Development by GIF [103]**

| Feature | GFR | LFR | MSR | SFR | SCWR | VHTR |
|---|---|---|---|---|---|---|
| Neutron spectrum | Fast | Fast | Fast/thermal | Fast | Thermal/fast | Thermal |
| Coolant | Helium | Lead or Pb-Bi | Fluoride Salts | Sodium | Water | Helium |
| Pressure | High | Low | Low | Low | Very high | High |
| Fuel cycle | Closed | Closed | Closed | Closed | Open/closed | Open |
| Output (MWe) | 1200 | 20–1000 | 1000 | 50–1500 | 300–1500 | 250–300 |
| Temperature (°C) | 850 | 480–570 | 700–1000 | 500–550 | 510–625 | 900–1000 |

## Gas-Cooled Fast Reactor (GFR)

The **gas-cooled fast reactor** (GFR) system, schematically shown in Fig. 2.3, is a helium-cooled fast-spectrum technology for long-term sustainability of uranium resources and waste reduction. With high outlet temperature of coolant, the system has high thermal cycle efficiency and is capable of generating heat for industrial uses, such as for hydrogen production.

The reference GFR is a 2400 MWt/1200 MWe unit with a core outlet temperature of 850°C, and with three 800 MWt loops. The core is contained in a steel pressure vessel and consists of an assembly of hexagonal fuel elements. A heat exchanger transfers the heat from the primary helium coolant to a secondary gas cycle containing a helium-nitrogen mixture which drives a closed-cycle gas turbine in a Bryton cycle. The waste heat from the gas turbine exhaust is used to raise steam in a steam generator which is then used to drive the steam turbine in a Rankine cycle. Such a combined cycle is common practice in a natural gas-fired power plant and thus is a very well-established technology.

The main challenges and research needs include fuel, materials, and thermal-hydraulics. The high core outlet temperature sets significant demands on the fuel to operate continuously with the high power density required for good neutron economy in the core.

## Lead-Cooled Fast Reactor (LFR)

The **lead-cooled fast reactor** (LFR), shown in Fig. 2.4, is a fast spectrum technology employing liquid lead or lead-bismuth eutectic as a coolant. The core cooling, at least for decay heat removal, is accomplished by the natural circulation of liquid metal coolant at atmospheric pressure. LFRs are expected to have multiple applications including the production of electricity, hydrogen, and process heat. Thanks to their design flexibility, LFRs can use depleted uranium or thorium fuel matrices, and burn actinides from LWR fuel.

A wide range of unit sizes is envisaged, from factory-built "battery" with 15–20 year life for small grids or for developing countries, through modular 300–400 MWe units, to large single units of 1400 MWe. An operating temperature of 550°C is

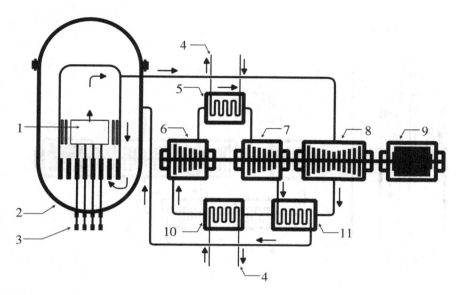

**Figure 2.3** Gas-cooled fast reactor operating in direct Brayton cycle: 1–reactor core, 2–reactor pressure vessel, 3–control rods, 4–heat sink, 5–intercooler, 6–compressor, 7–compressor, 8–turbine, 9–generator, 10–pre-cooler, 11–recuperator.

already achievable, but envisaged 800°C will require the development of advanced materials to provide lead corrosion resistance at high temperatures.

The main research needs include fuels and materials. Additional important challenges are created by some features of the liquid metal coolant such as a high melting temperature of lead (327°C), its opacity, and its high mass density (above 10500 kg m$^{-3}$). The high melting temperature requires that the primary coolant system be maintained at temperatures to prevent the solidification of the lead. The opacity of lead, in combination with its high melting temperature, presents challenges related to the inspection and monitoring of reactor in-core components as well as fuel handling. The high density and corresponding high mass of lead as a coolant result in the need for careful consideration of structural design to prevent seismic impacts to the reactor system. A single most important challenge results from the tendency of lead at high temperatures to be corrosive when in contact with structural steels.

## Molten Salt Reactor (MSR)

The **molten salt reactor** (MSR), shown in Fig. 2.5, has a fluid core consisting of fuel dissolved in molten fluoride salt. This technology is not new and was first studied in the 1960s for airplane propulsion applications. Modern interest is on fast reactor concept as a long-term alternative to solid-fuelled fast-spectrum reactors. The two variants of the system include a fast-spectrum reactor with fissile material dissolved

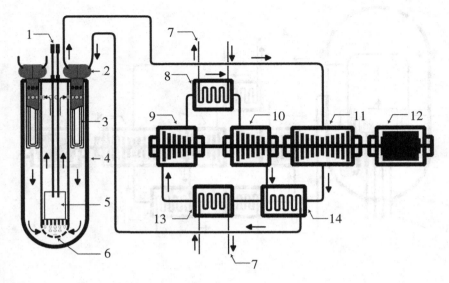

**Figure 2.4**  Lead-cooled fast reactor operating in indirect Brayton cycle: 1–control rods, 2–header, 3–U-tube heat exchanger, 4–reactor vessel, 5–reactor core, 6–inlet distributor, 7–heat sink, 8–intercooler, 9–compressor, 10–compressor, 11–turbine, 12–generator, 13–pre-cooler, 14–recuperator.

in the circulation fuel salt, and solid particle fuel in graphite and the salt functioning only as coolant. The former, termed as the molten salt fast neutron reactor (MSFR) is characterized by the thorium fuel cycle, recycling of actinides, closed Th/U fuel cycle with no enrichment, and with enhanced safety and minimal waste. The latter called the advanced high-temperature reactor (AHTR), employs the same graphite and solid fuel core structures as the very high-temperature gas reactor (VHTR) and molten salt as coolant instead of helium. This configuration will allow power densities 4 to 6 times greater than HTRs and power levels up to 4000 MWt with passive safety systems. Compared with the solid-fuel reactors, MSFR systems have lower fissile inventories, no radiation damage constraint on attainable fuel burnup, no requirement to fabricate and handle solid fuel, and a homogeneous isotopic composition of fuel in the core.

The main research needs include fuel treatment, materials, and system reliability. Still a lot of work must be done on salts before demonstration reactors will be operational.

### Sodium-Cooled Fast Reactor (SFR)

The **sodium-cooled fast reactor** (SFR), shown in Fig. 2.6, uses liquid sodium as the reactor coolant, allowing high-power density with low coolant volume and operation at low pressure. The SFR closed fuel cycle enables regeneration of fissile fuel and

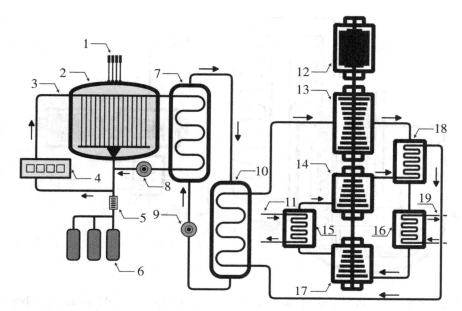

**Figure 2.5** Molten salt reactor: 1–control rods, 2–reactor, 3–purified salt, 4–chemical processing plant, 5–freeze plug, 6–emergency dump tanks, 7–heat exchanger, 8–fuel salt pump, 9–coolant salt pump, 10–heat exchanger, 11–heat sink, 12–generator, 13–turbine, 14–compressor, 15–intercooler, 16–pre-cooler, 17–compressor, 18–recuperator, 19–heat sink.

facilitates management of minor actinides. The outlet temperature is 550°C allowing the use of the materials developed and proven in prior fast reactor programs and making SFR suitable for electricity generating.

Three options with a pool layout or a compact loop layout are under consideration. The largest, with the electric power output of 600 to 1500 MWe, is the loop-type reactor with mixed uranium-plutonium oxide fuel and potentially minor actinides. The smallest in size is a modular-type reactor with an electric power output of 50 to 150 MWe, with uranium-plutonium-minor-actinide-zirconium metal alloy fuel. A pool-type units with oxide or metal fuel are envisaged in the intermediate-to-large size range with an electric power output of 300 to 1500 MWe.

The main research needs are fuels and advanced recycle options. Since sodium is chemically reacting with water and air, LFRs need a sealed coolant system.

### Supercritical Water-Cooled Reactor (SCWR)

The **supercritical water-cooled reactor** (SCWR), shown in Fig. 2.7, is a very high-pressure water-cooled reactor which operates above the thermodynamic critical point of water (22.1 MPa, 374°C) to eliminate the risk of the critical heat flux in the core. Supercritical water at a pressure of 25 MPa and temperature 510–550°C directly

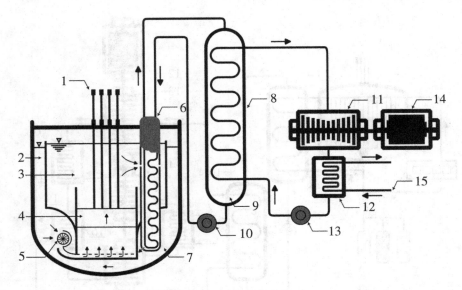

**Figure 2.6** Sodium-cooled fast reactor: 1–control rods, 2–cold plenum, 3–hot plenum, 4–reactor core, 5–pump, 6–heat exchanger, 7–primary sodium, 8–steam generator, 9–secondary sodium, 10–pump, 11–turbine, 12–condenser, 13–pump, 14–generator, 15–heat sink.

drives the turbine, using well-developed solutions employed in coal-fired plants. The thermodynamic efficiency of the system is expected to be one-third higher than today's LWRs.

Two design options are under consideration. The first one uses a pressure vessel and is expected to have similar operational and safety features to ABWRs. The second option employs pressure tubes and uses heavy water moderation, showing similarities to the earlier CANDU designs.

The main research needs include materials and thermal-hydraulics. Even though the critical heat flux is avoided in the core thanks to the high pressure well above the critical point of water, a local sudden heat transfer deterioration can occur in some fuel assemblies. More research is needed to better understand the reasons for the heat transfer deterioration, and guide the safe fuel assembly design.

### Very High-Temperature Gas Reactor (VHTR)

The **very high-temperature gas reactor** (VHTR), shown in Fig. 2.8, is primarily dedicated to the cogeneration of electricity and hydrogen. Its high outlet temperature of 1000°C makes the system attractive for other process heat applications, including the chemical, oil, and iron industries. Systems with direct high-efficiency driving of a gas turbine employing the Brayton cycle are considered. At low outlet temperatures, the Rankine steam cycle may be used for electricity generation. The technical

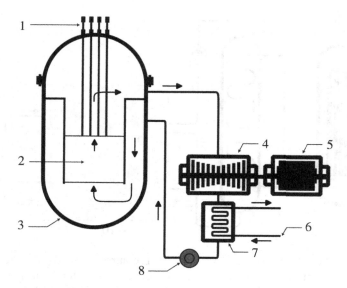

**Figure 2.7** Supercritical water-cooled reactor: 1–control rods, 2–reactor core, 3–reactor pressure vessel, 4–turbine, 5–generator, 6–heat sink, 7–condenser, 8–pump.

basis for VHTR is the TRISO particle-coated fuel, graphite as the core structure, and helium coolant. The VHTR has the potential for inherent safety, with decay heat removal by natural convection. More research is needed for fuels, materials, and hydrogen production. In particular, the capability to produce hydrogen from only heat and water is of interest. This can be achieved by using thermochemical processes (such as the sulfur-iodine (S-I) process or the hybrid sulfur process), high-temperature steam electrolysis, or from heat, water, and natural gas by applying the steam reforming technology.

### 2.2.3 SMALL MODULAR REACTORS

Small modular reactors (SMRs) are modern nuclear reactors that have a power capacity of up to 300 MWe per unit, whose systems and components are factory-assembled and transported as a unit to a location for installation. SMRs have many advantages compared to other types of reactors. Thanks to their smaller footprint, SMRs can be sited in locations not suitable for larger nuclear power plants. They can produce a large amount of low-carbon electricity at locations not suitable for other technologies, and where, so far, the only alternative has been to use fossil fuels.

Centralized prefabrication of systems and components will make SMRs more affordable to build than larger power reactors, which are often custom designed for a particular location. Shipment and installation of SMRs on site will offer savings in cost and construction time. This solution provides flexibility to deploy SMRs incrementally to match changing energy demand.

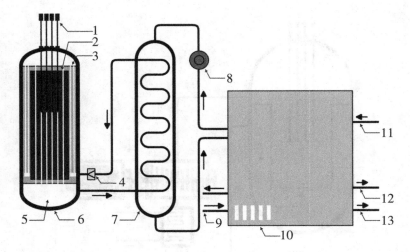

**Figure 2.8**  Very-high temperature reactor: 1–control rods, 2–graphite reactor core, 3–graphite reflector, 4–blower, 5–helium coolant, 6–reactor, 7–heat exchanger, 8–pump, 9–heat sink, 10–hydrogen production plant, 11–water, 12–oxygen, 13–hydrogen.

SMRs can be installed into an existing grid or remotely off-grid, providing power for local industry or the population. SMRs are particularly suitable for rural areas with limited grid coverage by meeting the requirement that a single power plant should contribute with no more than 10% of the total installed grid capacity. In some areas this requirement will be met by micro-reactors, which are a subset of SMRs designed to generate electrical power up to 10 MWe.

SMRs designs are in general simpler and use the passive safety principles in larger extent than the existing large power reactors. Several current solutions provide "walk-away" safety, since no human intervention or external power is required to safely shut down systems. Such phenomena and processes as natural circulation, natural convection, gravity-driven processes, and self-depressurization are commonly exploited. These features increase plant safety by elimination or significant reduction of the risks associated with releases of radioactivity to the environment in case of an accident.

SMRs are using proven elements and solutions to reduce the effort required to create a new design. For the GE Hitachi's BWRX-300 design, 90% of the components are already in use in the industry. That includes the fuel, the material in the control rods, and the control-rod drive mechanisms. The fuel requirements of SMRs will be significantly reduced and they may require less frequent refueling, every 3 to 7 years, in comparison to between 1 to 2 years for large power reactors. Some SMRs are designed to operate as long as 30 years without refueling.

## 2.3  NUCLEAR FUEL

The nuclear fuel that is used in nuclear reactors has to be manufactured from raw material, such as uranium, and has to undergo several steps of processing to get a form that allows an efficient and safe performance in the reactor core. All of the current generation of power reactors use a ceramic-grade uranium dioxide powder that is pressed, sintered at high temperatures, and milled to a very precise shape and size.

Fuel elements are the smallest, usually canned, construction elements containing nuclear fuel. They can have various shapes such as rods, plates, and pebbles. Most of the current power reactors employ fuel rods that are bundled together into a fuel assembly, with the pins arranged into a square or hexagonal lattice. The fuel assemblies must conform to the integral design of the reactor as well as the control rod mechanisms.

### 2.3.1  FUEL RODS

A fuel rod contains fuel pellets encapsulated in a cladding tube, sealed on both ends with plugs. Preserving the integrity of fuel rods is one of the primary goals of a fuel design, since it provides the first barrier preventing the fission products from escaping into the environment. As illustrated in Fig. 2.9, a fuel rod contains some extra space, called a plenum, to accommodate gaseous fission products. The rod is filled with helium at a pressure depending on the reactor type. A compression spring is used to keep the pellets axially in place during handling and transport. This arrangement also provides an adequate elastic stiffness to allow for axial thermal expansions of the rod. The fuel rod components are mainly made of zirconium alloys. A few

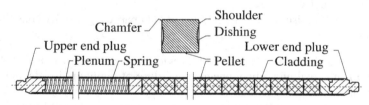

**Figure 2.9**   A fuel rod structure and its main components.

examples of cladding materials and their properties are presented in §3.1. The fuel cladding is a tube with length, diameter, and wall thickness depending on reactor type and assembly design. Typical values for light water reactors are given in Table 2.3. To improve the resistance of the cladding against the pellet-clad interaction (PCI) failures, graphite-coated cladding is used in CANDU reactors, and cladding with an inner liner in BWRs. The latter consists of a layer of material with better stress corrosion cracking (SCC) resistance and higher ductility than the cladding material. As an example, zirconium with 400 ppm Fe can be used for this purpose.

**TABLE 2.3**

**Fuel Rod Characteristics for PWR and BWR Nuclear Power Plants [9]**

| Feature | PWR 17x17 | VVER | BWR 8x8 | BWR 10x10 |
|---|---|---|---|---|
| Pellet material | $UO_2$ | $UO_2$ | $UO_2$ | $UO_2$ |
| Initial $^{235}U$ enrichment (%) | 3.2–5 | 3.8–5 | 3.2–5 | 3.2–5 |
| Theoretical density (kg m$^{-3}$) | 10 950 | 10 970 | 10 950 | 10 950 |
| Fuel initial porosity (%) | 3–7 | 2.2–5 | 4.5–7 | 3–5 |
| Possible neutron absorbent additives | Gd, Er, B | Gd | Gd | Gd |
| Pellet diameter (mm) | 8.19 | 7.6 | 10.4 | 8.2–8.5 |
| Pellet length (mm) | 10–12 | 9–12 | 10.4 | about 10 |
| Central hole diameter (mm) | - | 0/1.2 | - | - |
| Cladding material | Zirconium alloy | E110 | Zry–2 | Zry-2 |
| Cladding outer diameter (mm) | 9.45 | 9.1 | 12.30 | 9.6–9.8 |
| Cladding inner diameter (mm) | 8.347 | 7.8 | 10.68 | 9.0–9.2 |
| Cladding wall thickness (mm) | 0.57 | 0.65 | 0.81 | about 0.60 |

## 2.3.2 COATED PARTICLE (TRISO) FUEL

Coated particle fuel contains fuel microspheres with refractory coatings to contain fission products. This type of fuel experienced its first practical demonstrations starting in the 1960s in the Dragon reactor[5] in the UK, the AVR[6] in Germany, and Peach Bottom Unit I[7] in the US. The earliest versions of the fuel involved fuel kernels coated with only a single pyrolytic carbon layer intended to protect the kernel during fabrication. The design of coated particle fuel evolved later into a more complex and more effective fuel structures.

The tristructural isotropic (TRISO) coated fuel particle generally consists of the following elements:

- A spherical uranium-bearing kernel with a diameter in a range of 350–600 $\mu$m containing a heterogeneous mixture of uranium dioxide and uranium carbide ($UO_2$ or UCO).
- A spherical porous (approximately 50% of theoretical density) pyrocarbon buffer layer with a thickness of 95–100 $\mu$m, which provides a void space for the fission gases released from the kernel and accommodate fission recoils.
- A dense, highly isotropic inner pyrocarbon (IPyC) spherical layer with a thickness of 40 $\mu$m. This layer contributes to the retention of fission gases in the particle.
- A silicon-carbide (SiC) layer with a thickness of 35 $\mu$m provides the main structural strength of the particle and constitutes the primary barrier to the release of non-gaseous fission products not sufficiently retained by the kernel or the buffer layer.

---

[5]Prismatic-core HTGR with 20 MWth power.

[6]Arbaitsgemeinschaft Versuchsreaktor — first pebble-bed reactor with 46 MWth power.

[7]Prismatic-core HTGR with 115 MWth power.

- An outer pyrocarbon (OPyC) layer with a thickness of 40 $\mu$m protects SiC layer during handling, acts as a surface for bonding to the graphitic fuel matrix, and provides an additional barrier to fission product release.

Through several decades of development, fabrication, and quality control methods have been demonstrated that are capable of producing fuel with low-manufactured defect fractions and low residual contamination. High-quality fuel exhibits very low particle failure rates during irradiation, with failure fraction approximately $10^{-5}$ demonstrated for $UO_2$ fuel up to about 11% FIMA and UCO fuel up to about 20% FIMA. The particles exhibit remarkable duralibity under severe accident conditions, which can include temperatures up to 1600°C in modern modular HTGR designs [54].

## PROBLEMS

### PROBLEM 2.1

Compare steam cycles in BWRs and PWRs and explain why, despite their differences, the capital costs of both systems remain highly competitive.

### PROBLEM 2.2

Explain the main differences between generation IV reactors and previous reactor generations in terms of their expected operational safety.

### PROBLEM 2.3

Compare the six types of Generation IV reactors and provide an assessment as to which one presents the greatest challenges when it comes to estimating its thermal safety margins.

# 3 Thermophysical Properties of Reactor Core Materials

Safety and performance of nuclear reactors depend, to a large extent, on the physical properties of materials that are used in reactor cores. The major components of reactor cores are fuel, metal cladding, coolant, moderator, reflector, and structural materials that provide support. Additional materials are created as a result of a fission process. These materials, called fission products, stay within the fuel cladding, causing changes of the thermophysical properties of fuel elements.

The reactor core materials are selected to satisfy requirements from the neutronic and thermal-hydraulic points of view. Practically all materials, except for dedicated neutron absorbers, should have the macroscopic cross-section for neutron absorption as low as possible. It is required that the materials should demonstrate stable behavior in the reactor core environment with high neutron fluxes and high temperatures. In addition, the materials should preferably be relatively highly abundant and not prohibitively expensive. Additional particular requirements depend on the material application area, as described in the present chapter.

Clad and fuel pellet materials play a particularly important role since they constitute the first barrier for the release of fission products. The desirable properties of LWR fuel cladding and pellet depend on the conditions under consideration. During normal operation, high thermal conductivity in the pellet and cladding is desired to limit the fuel centerline temperature. High critical heat flux is desirable to avoid a sudden heat transfer coefficient drop at the coolant-cladding interface.

Under a power ramp transient, with a pellet expansion rate of the order of $10^{-5}$ $s^{-1}$, high thermal conductivity and a low coefficient of thermal expansion are desirable in the fuel pellet to minimize pellet–cladding mechanical interaction. The same properties are desirable for a sudden reactivity insertion accident scenario when a rapid pellet expansion rate of the order of 5 $s^{-1}$ can occur. However, for such rapid transients, in addition a low heat capacity in the fuel is desirable to limit the extent of energy deposition in the core. On the contrary, under a design basis loss of coolant accident, high heat capacity in the fuel can be beneficiary since it will reduce the rate of temperature rise as a result of decay heat. Finally, for a beyond design basis accident such as a station blackout, when a core temperature can exceed 1200°C, oxidation resistance at high steam temperatures, as well as overall chemical and physical stability, are of great importance.

Safety analyses require calculations with safety codes that need the appropriate thermophysical properties of various reactor core materials. The available open literature on these properties is very rich and an exhaustive review of the subject can be found in, e.g., [133]. A compilation of material property correlations with an extensive history of use for LWR safety analyses is provided in the MATPRO

DOI: 10.1201/9781003255000-3

database [83]. In this chapter we provide the basic thermophysical properties of reactor core materials relevant for safety analyses of current and nearest future nuclear reactor types.

## 3.1  CLADDING MATERIALS

Cladding materials should possess several important features, such as a low macroscopic neutron absorption cross section and high corrosion resistance. Such metals as beryllium, magnesium, and aluminum have the lowest macroscopic cross-section (0.001, 0.005, and 0.014 $cm^{-1}$, respectively), but they are not suitable for cladding applications due to various reasons. Beryllium is expensive, difficult to fabricate, and toxic. Magnesium has a low melting point (923 K), and has a poor resistance to hot water corrosion. Aluminum has a low melting point (933 K) and a poor high-temperature strength.

Austenitic stainless steels (type 304) have been used as cladding in BWRs, but they were prone to stress corrosion cracking (SCC) failures. Even though austenitic steel was successfully used in PWRs, the need for higher fuel burnup eventually led to the replacement of austenitic stainless steel with zirconium-based cladding.

Zirconium has a very low macroscopic neutron absorption cross-section (0.01 $cm^{-1}$), high hardness, ductility, and corrosion resistance. Commercial non-nuclear grade zirconium contains 1–5% of hafnium, whose absorption cross-section is very high and must therefore be almost entirely removed (to less than 0.02% of the alloy) for reactor applications. Zirconium alloys for cladding applications contain more than 95% of zirconium, less than 0.3% of iron and chromium, 0.1–0.14% of oxygen, and varying content of zinc and niobium. Due to the high content of zirconium, zirconium alloys have properties similar to pure zirconium. The most frequently used alloys include Zircaloy 2 (1.2–1.7% Sn and 0.0% Nb), Zircaloy 4 (1.2–1.7% Sn and 0.0% Nb), ZIRLO[1] (0.7–1.0% Sn and 1.0% Nb), and M5[2] (0.0% Sn and 0.8–1.2% Nb). Zirconium alloys suffer from bad creep rates and the phase transformation of zirconia[3] is of great concern.

### 3.1.1   ZIRCALOY 2 AND 4

Zircaloy 4 (Zry-4) is used in PWRs for the cladding tubes and guide tubes, whereas Zircaloy 2 (Zry-2) is used for the cladding tubes and channel boxes in BWRs.

### Zirconium Density

The density of zirconium varies with the temperature $T$ (K) as follows [103],

$$\rho = 6550 - 0.1685T, \tag{3.1}$$

where $\rho$ (kg m$^{-3}$) is the density.

---

[1] A trademark of Westinghouse that stands for **zir**conium low oxidation.
[2] A trademark of AREVA.
[3] Zirconium dioxide ($ZrO_2$)

## Thermal Conductivity

The recommended equations for the thermal conductivity of Zircaloy 2 and Zircaloy 4 are as follows [100]:

$$\lambda = 12.767 - 5.4348 \cdot 10^{-4}T + 8.9818 \cdot 10^{-6}T^2, \tag{3.2}$$

where $\lambda$ (W m$^{-1}$ K$^{-1}$) is the thermal conductivity and $T$ (K) is the temperature. The equation is valid for 300 K $< T <$ 1800 K with average uncertainty 7%.

## Specific Heat Capacity

The recommended equations for the specific heat capacity of Zircaloy 2 are as follows [100]:

For 273 K $< T <$ 1100 K ($\alpha$–phase),

$$c_p = 255.66 + 0.1024T. \tag{3.3}$$

For 1100 K $< T <$ 1214 K ($\alpha + \beta$–phase transition),

$$c_p = 255.66 + 0.1024T + 1058.4 \exp\left[\frac{(T - 1213.8)^2}{719.61}\right]. \tag{3.4}$$

For 1214 K $< T <$ 1320 K ($\alpha + \beta$–phase transition),

$$c_p = 597.1 - 0.4088T + 1.565 \cdot 10^{-4}T^2 + 1058.4 \exp\left[\frac{(T - 1213.8)^2}{719.61}\right]. \tag{3.5}$$

For 1320 K $< T <$ 2000 K ($\beta$–phase),

$$c_p = 597.1 - 0.4088T + 1.565 \cdot 10^{-4}T^2. \tag{3.6}$$

Here $c_p$ (J kg$^{-1}$ K$^{-1}$) is the specific heat capacity and $T$ (K) is the temperature. The two-standard deviation uncertainty for Zircaloy 2 $\alpha$–phase is 2–3% whereas for $\beta$–phase it increases from 10% (for temperatures in a range from 1300 K to 1600 K) to 20% (at 1700 K).

No equation for the specific heat capacity of Zircaloy 4 is available due to a lack of data. Until measurements of the heat capacity of Zircaloy 4 are available, the Zircaloy 2 equations are recommended with the caution that the actual heat capacity for Zircaloy 4 may be higher by 10–20% in the $\alpha$–phase and by 30% in the $\beta$–phase [100].

### 3.1.2  ZIRCONIUM ALLOY WITH 1% NOBIUM

The main application of Zr-1%Nb alloy is for the cladding tubes in VVERs.

**Thermal Conductivity**

The thermal conductivity $\lambda$ (W m$^{-1}$ K$^{-1}$) of Zr-1%Nb in the direction of the rod length is given as [100],

$$\lambda = \begin{cases} 23.48 - 1.92 \cdot 10^{-2}T + 1.68 \cdot 10^{-5}T^2 & \text{for} \quad 300 < T \leq 1150 \text{ K} \\ 1.51 + 0.020T & \text{for} \quad 1150 < T < 1600 \text{ K} \end{cases}, \quad (3.7)$$

where $T$ (K) is the temperature and the correlation uncertainty is 10%.

### 3.1.3   ZIRCONIUM ALLOY WITH 2.5% NOBIUM

Zr-2.5%Nb alloy is used for pressure tubes in PHWRs. Its density is $6.44 \cdot 10^3$ kg m$^{-3}$, the melting temperature is 1757 K, and the specific heat capacity (at 573 K) is 305.2 J kg$^{-1}$ K$^{-1}$.

**Thermal Conductivity**

Using the least square regression procedure, the following correlation for the thermal conductivity $\lambda$ (W m$^{-1}$ K$^{-1}$) of Zr-2.5%Nb was obtained from the available experimental data [100],

$$\lambda = 3.172 + 14.75\theta^2 - 2.435\theta^3 + \frac{4.831}{\theta} \quad (3.8)$$

where $\theta = T/1000$, $T$ (K) is the temperature, and the correlation uncertainty is less than 8%.

## 3.2   FUEL PELLET MATERIALS

Almost all current commercial nuclear power plants with Generation II and Generation III/III+ reactors utilize uranium dioxide ($UO_2$) fuel. Fuel design varies for different reactor types, including LWRs (both PWRs and BWRs), AGRs, VVERs, and CANDU reactors. Plutonium utilization and recycling have been demonstrated in light-water and heavy-water reactors. In some countries mixed oxide (MOX) fuel[4] is used on a commercial scales.

Oxide nuclear fuels such as $UO_2$ and $(U,Pu)O_2$ have fluorite structures in which the oxygen-to-metal (O/M) ratio equals 2.0. In advanced nuclear fuel, this ratio is adjusted to values lower than 2.0 in the sintering process to improve the compatibility between fuel and cladding materials. Basic properties of selected fuel materials at normal conditions are shown in Table 3.1.

A fuel pellet density is managed in the fuel production process. A theoretical fuel density is the standard value that can be determined from a lattice parameter, a composition, and an isotopic content. In general, the theoretical density of $(U,Pu)O_2$ increases with increasing Pu content and decreases with decreasing O/M ratio [119].

---

[4]The MOX fuel is obtained by separating the plutonium and mixing it with the depleted uranium.

**TABLE 3.1**

**Basic Properties of Selected Fuel Materials at Normal Conditions**

| Property | U | Pu | $UO_2$ | $PuO_2$ | $MOX^a$ | $UC^b$ | $UN^c$ |
|---|---|---|---|---|---|---|---|
| Atomic/Molecular mass (amu) | 238.03 | 244.06 | 270.3 | 276.045 | 271.2 | 250 | 252 |
| Density ($10^3$ kg m$^{-3}$) | 19.05 | 19.84 | 10.96 | 11.46 | 11.074 | 13.63 | 14.30 |
| Melting point (K) | 1405 | 913 | 3120 | 2663 | 3023 | 2638 | 3123 |
| Boiling point (K) | 4018 | 3500 | 3815 | 3600 | 3811 | 4691 | |
| Heat of fusion (kJ mol$^{-1}$) | 8.72 | 2.8 | 70 | 70.39 | 77.37 | 48.9 | |
| Heat of vaporization (kJ mol$^{-1}$) | 487 | 350 | 413 | 376.8 | 413.5 | 530 | |
| Heat capacity (J mol$^{-1}$ K$^{-1}$) | 27.67 | 31.2 | 63.7 | 66.25 | 65.09 | 50 | 48 |
| Thermal conductivity (W m$^{-1}$ K$^{-1}$) | 22.5 | 5.2 | 8.68 | 6.3 | $7.82^e$ | 25.3 | 13.0 |
| Linear expansion ($10^{-6}$ K$^{-1}$) | 13.9 | $46.85^d$ | 9.75 | 7.8 | 9.4 | 10.1 | 7.52 |

$a$ $U_{0.8}Pu_{0.2}O_2$ mixed oxide.
$b$ Uranium carbide.
$c$ Uranium mononitride.
$d$ For $\alpha$–phase.
$e$ For MOX of 95% density.
*Source:* [103]

During the early stages of reactor irradiation, sub-micron pores disappear from oxide fuels causing their *densification*. A maximum densification occurs at around 5–6 MWd kg$^{-1}$ burnup, which results in a peak fuel temperature. Therefore fuel densification is an important phenomenon that has to be considered in reactor safety evaluations. When a fuel burnup exceeds 5–6 MWd kg$^{-1}$, the oxide fuel density decreases due to fuel *swelling* caused by the accumulation of fission products.

A nuclear fuel is developed to allow high burnup, high operational flexibility, and high reliability. These goals are met by increasing the fuel density, improving the fission gas retention, improving the pellet-cladding interaction resistance, and improving the post-failure performance. The density and grain size of the fuel material can be increased by using various additions. As a result less fission gas bubbles migrate to the fuel surface, and the release of fission gases to the gap between the pellet and the cladding decreases. The fuel material composition determines the most important thermophysical properties such as the heat capacity, the thermal expansion coefficient, the melting temperature, and the thermal conductivity. In this section we present the typical equations and correlations that can be used to determine these properties for selected fuel materials.

### 3.2.1 URANIUM METAL

Metallic uranium was used as the fuel in most of the early reactors mainly because it provided the maximum number density of uranium atoms. Metallic fuels produce a very hard neutron spectrum and therefore they are suitable for fast reactors. Other benefits offered by metallic fuel include a high thermal conductivity and a relatively low heat capacity. The former reduces peak temperatures and local hot spots, whereas the latter limits the stored heat in the fuel, allowing the fuel to be cooled more readily.

In the past, metallic fuels had limited endurance due to excessive *swelling*. This drawback has been mitigated by incorporating space for swelling. Uranium-plutonium alloys with 10% zirconium (to raise the melting point) have been shown to be very reliable, and very high burnups are now routinely achieved [91].

The basic properties of the metallic uranium are provided in Table 3.1 for normal conditions only. The temperature dependence of the metallic uranium properties is described with correlations, as summarized below.

### Density of Uranium Metal

The density of uranium metal varies with the temperature $T$ (K) as follows [103],

$$\rho = \begin{cases} 19.36 \cdot 10^3 - 1.03347T & \text{for} \quad 273 \leq T \leq 942 \text{ K } (\alpha\text{-phase}) \\ 19.092 \cdot 10^3 - 0.9807T & \text{for} \quad 942 < T \leq 1049 \text{ K } (\beta\text{-phase}) \\ 18.447 \cdot 10^3 - 0.5166T & \text{for} \quad 1049 < T \leq 1405 \text{ K } (\gamma\text{-phase}) \end{cases} , \qquad (3.9)$$

where $\rho$ (kg m$^{-3}$) is the density.

### Heat Capacity of Uranium Metal

The uranium specific heat capacity $c_p$ (J kg$^{-1}$ K$^{-1}$) varies with the temperature $T$ (K) as follows [103],

$$c_p = \begin{cases} c_{p\alpha}(T) & \text{for} \quad 293 \leq T \leq 942 \text{ K } (\alpha\text{-phase}) \\ 176.4 & \text{for} \quad 942 < T \leq 1049 \text{ K } (\beta\text{-phase}) \\ 156.8 & \text{for} \quad 1049 < T \leq 1405 \text{ K } (\gamma\text{-phase}) \end{cases} , \qquad (3.10)$$

where

$$c_{p\alpha}(T) = 104.82 + 5.3686 \cdot 10^{-3}T + 10.1823 \cdot 10^{-5}T^2. \qquad (3.11)$$

### Thermal Conductivity of Uranium Metal

The thermal conductivity of uranium metal $\lambda$ (W m$^{-1}$ K$^{-1}$) varies with the temperature $T$ (K) as follows [103],

$$\lambda = 22.0 + 0.023(T - 273.15). \qquad (3.12)$$

The correlation is valid in the temperature range of 293–1405 K with accuracy of $\pm 10\%$.

### 3.2.2 URANIUM DIOXIDE

The ceramic uranium dioxide ($UO_2$) fuel is used in LWRs in the form of pellets. The pellets are enclosed in a metal cladding and are, in general, characterized by an excellent dimensional stability during fission. Since $UO_2$ is the most common type of fuel, its properties are widely investigated and well-known. The most common thermo-mechanical properties of $UO_2$ are given in this section.

**Density of Solid $UO_2$**

A density of solid $UO_2$ is given as,

$$\rho = \rho_0 \left[ \frac{L_0}{L(T)} \right]^3 , \tag{3.13}$$

where $\rho_0 = 1.0963 \cdot 10^4$ kg m$^{-3}$ is $UO_2$ density at temperature $T = 273$ K and $L(T)$ is the linear thermal expansion of solid $UO_2$ given as:

for $273$ K $\leq T \leq 923$ K,

$$L(T) = L_0 \left( 9.9734 \cdot 10^{-1} + 9.802 \cdot 10^{-6}T - 2.705 \cdot 10^{-10}T^2 \right); \tag{3.14}$$

for $923$ K $\leq T \leq 3120$ K,

$$L(T) = L_0 \left( 9.9672 \cdot 10^{-1} + 1.179 \cdot 10^{-5}T - 2.429 \cdot 10^{-9}T^2 + 1.219 \cdot 10^{-12}T^3 \right). \tag{3.15}$$

Here $L_0$ is the linear thermal expansion of solid $UO_2$ at temperature $T = 273$ K.

**Thermal Conductivity of Solid $UO_2$**

The thermal conductivity of solid $UO_2$ can be found from the following equation [100],

$$\lambda = \frac{1 - \alpha p}{1 - 0.95\alpha} \left[ \frac{100}{7.5408 + 17.692\theta + 3.6142\theta^2} + \frac{6400}{\theta^{5/2}} \exp\left( -\frac{16.35}{\theta} \right) \right], \tag{3.16}$$

where $\alpha = 2.6 - 0.5\theta$, $\theta = T/1000$, $T$ (K) is the temperature, $p$ is the porosity, and $\lambda$ (W m$^{-1}$ K$^{-1}$) is the thermal conductivity. The uncertainty of the formula varies with the temperature and is 10% for temperatures from 298 to 2000 K and increases to 20% for temperatures from 2000 to 3120 K.

**Specific Heat Capacity of Solid $UO_2$**

The molar heat capacity at constant pressure of solid $UO_2$ can be obtained as,

$$C_{Mp} = \frac{C_1 \theta^2 e^{\theta/T}}{T^2 \left( e^{\theta/T} - 1 \right)^2} + 2C_2 T + \frac{C_3 E_a e^{-E_a/T}}{T^2} , \tag{3.17}$$

where $C_{Mp}$ (J mol$^{-1}$ K$^{-1}$) is the molar heat capacity, $C_1 = 81.613$ J mol$^{-1}$ K$^{-1}$, $\theta = 548.68$ K, $C_2 = 2.285 \cdot 10^{-3}$ J mol$^{-1}$ K$^{-2}$, $C_3 = 2.360 \cdot 10^7$ J mol$^{-1}$, $E_a = 18531.7$ K, and $T$ (K) is the temperature. The uncertainty of the formula is $\pm 2\%$ from 298.15 to 1800 K and $\pm 13\%$ from 1800 to 3120 K.

The specific heat capacity at constant pressure in J kg$^{-1}$ K$^{-1}$ can be obtained as $c_p = M C_{Mp}$, where $M$ is the molar mass of UO$_2$.

### 3.2.3 MOX FUEL

Comparing to the UO$_2$ fuel, the MOX fuel has a lower thermal conductivity and a lower melting point. For both oxides, the thermal conductivity decreases with increasing temperature up to around 2000 K, and then it rises with the temperature. Moreover, the deviation of O/M from 2.00 reduces the thermal conductivity. Similar effect has an increasing burnup.

The thermal conductivity $\lambda$ (W m$^{-1}$ K$^{-1}$) of the solid MOX fuel at temperature $T$ (K), porosity $p$, and burnup $E$ (MWd kg$^{-1}$) is correlated to the fuel temperature and the burnup as [140]:

$$\lambda = 1.0789 \lambda_{95} \frac{1-p}{1+0.5p}, \tag{3.18}$$

where $\lambda_{95}$ (W m$^{-1}$ K$^{-1}$) is the thermal conductivity of MOX fuel with 5% porosity given as,

$$\lambda_{95} = \frac{1}{A(x) + B(x)T + 2.46 \cdot 10^{-4}T + h(E,T)} + \frac{1.5 \cdot 10^9 e^{-13520/T}}{T^2}, \tag{3.19}$$

and $h(E,T)$ is a burnup and temperature dependent factor:

$$h(E,T) = 1.87 \cdot 10^{-3}E + \frac{3.8 \cdot 10^{-2} \left(1 - 0.9e^{-0.04E}\right) E^{0.28}}{1 + 396e^{-6380/T}}. \tag{3.20}$$

The influence of the O/M ratio is taken into account by the following correlations,

$$A(x) = 2.85x + 0.035$$
$$B(x) = 2.86 \cdot 10^{-4} \qquad . \tag{3.21}$$
$$x = 2.00 - O/M$$

The porosity $p$ is defined in terms of the full dense fuel $\rho_{TD}$ and the actual density $\rho$ as,

$$p = \frac{\rho_{TD} - \rho}{\rho_{TD}}. \tag{3.22}$$

### 3.2.4 THORIUM METAL

Thorium is a chemical element with atomic number 90 and an atomic mass of 232.04 amu. Fertile $^{232}$Th can be converted to fissile $^{233}$U and thus is an important energy resource as a supplement to natural uranium. Thorium melts at 2023 K and boils at 5063 K.

## Thorium Density

The thorium density $\rho$ (kg m$^{-3}$) varies with the temperature $T$ (K) as follows [103],

$$\rho = 11836 - 0.4219T. \tag{3.23}$$

## Thorium Heat Capacity

The thorium specific heat capacity $c_p$ (J kg$^{-1}$ K$^{-1}$) varies with the temperature $T$ (K) as follows [103]:

for $T < 1623$ K ($\alpha$-phase)

$$c_p = 111.95 + 3.6384T - 0.0145T^2, \tag{3.24}$$

and for $1623 \leq T < 2023$ K ($\beta$-phase)

$$c_p = 145.77 + 5.7774T - 0.2032T^2. \tag{3.25}$$

## Thorium Thermal Conductivity

The thorium thermal conductivity $\lambda$ (W m$^{-1}$ K$^{-1}$) varies with the temperature $T$ (K) as follows [103],

$$\lambda = 34 + 0.0133T. \tag{3.26}$$

The correlation's temperature range is not indicated, but its accuracy is reported as $\pm 15\%$.

### 3.2.5  THORIUM-BASED FUEL

Thorium-based fuels, mainly as the oxides, are of interest due to high conversion ratio (theoretically exceeding unity) of $^{232}$Th to $^{233}$U. Various fuel compositions are considered, including ThO$_2$, (Th$_{1-y}$U$_y$)O$_2$, and (Th$_{1-y}$Pu$_y$)O$_2$. The thermophysical properties of thorium fuels are measured as a function of the composition (uranium and plutonium content determined by the $y$-factor) and the temperature.

## Density of Solid ThO$_2$ and (Th,U)O$_2$

The recommended equation for the theoretical density $\rho_{TD}$ (Mg m$^{-3}$) of ThO$_2$–UO$_2$ as a function of the temperature $T$ (K) and the content $x$ of UO$_2$ is as follows [100],

$$\rho_{TD} = 10.087 - 2.891 \cdot 10^{-4}T - x\left(6.354 \cdot 10^{-7}T + 9.279 \cdot 10^{-3}\right) \\ + 5.111 \cdot 10^{-6}x^2 \tag{3.27}$$

The equation is valid for the temperature range $298 < T < 1600$ and its uncertainty is $\pm 0.28\%$.

## Thermal Conductivity of Solid Thorium Fuels

The recommended correlation for the thermal conductivity $\lambda_{95}$ (W m$^{-1}$ K$^{-1}$) of $Th_yU_{1-y}O_2$ with 95% theoretical density is as follows [100],

$$\lambda_{95} = \frac{1}{-0.0464 + 0.0034y + (2.5185 \cdot 10^{-4} + 1.0733 \cdot 10^{-7}y)\,T}. \qquad (3.28)$$

The corresponding recommended correlation valid for the $Th_yPu_{1-y}O_2$ fuel is [100],

$$\lambda_{95} = \frac{1}{-0.08388 + 1.7378y + (2.62524 \cdot 10^{-4} + 1.7405 \cdot 10^{-4}y)\,T}, \qquad (3.29)$$

where $T$ (K) is the temperature. Both correlations are valid for the temperature range $873 < T < 1873$.

## Specific Heat Capacity of Solid $Th_yU_{1-y}O_2$

An equation for the specific heat capacity of the solid mixed oxide $Th_yU_{1-y}O_2$ has been derived from heat capacities of pure $ThO_2$ and $UO_2$. The following equation is recommended to calculate the molar heat capacity at constant pressure $C_{Mp}$ (J mol$^{-1}$ K$^{-1}$) [100],

$$C_{Mp} = yC_1 + (1-y)C_2, \qquad (3.30)$$

where,

$$C_1 = 55.962 + 0.05126T - 3.6802 \cdot 10^{-5}T^2 + 9.2245 \cdot 10^{-9}T^3$$
$$-\frac{5.74031 \cdot 10^5}{T^2}, \qquad (3.31)$$

and

$$C_2 = 52.1743 + 0.08795T - 8.4241 \cdot 10^{-5}T^2 + 3.1542 \cdot 10^{-8}T^3$$
$$-2.6334 \cdot 10^{-12}T^4 - \frac{7.1391 \cdot 10^5}{T^2}. \qquad (3.32)$$

The specific heat capacity at constant pressure, $c_p$ (J kg$^{-1}$ K$^{-1}$), can be obtained as $c_p = MC_{Mp}$, where $M$ is the molar mass of $Th_yU_{1-y}O_2$.

## 3.2.6 TRISO FUEL COMPACT

The TRISO fuel compact has a composite structure as described in §2.3.2. Numerous fuel performance codes with material properties have been developed for TRISO fuel analysis [115]. In this section we present the most important thermal physical properties of this type of fuel.

It is not practical to distinguish between the TRISO particle layers and the graphite matrix in full core models. Instead, effective properties of the TRISO fuel compact as a homogeneous material are used. Expressions and correlations used for the thermal conductivity, specific heat, and density are presented below.

## Thermal Conductivity

The effective thermal conductivity of the TRISO fuel compact, $\lambda_e(T)$, is a function the thermal conductivity of the graphite matrix, $\lambda_m(T)$, the thermal conductivity of the TRISO particle, $\lambda_p(T)$, the particle volume fraction, $\phi$, and usually the particle configuration, $c$. Thus, a general relationship for the effective conductivity of heterogeneous materials is as follows,

$$\lambda_e(T) = F\left(\lambda_m(T), \lambda_p(T), \phi, c\right). \tag{3.33}$$

Usually this relationship is expressed in terms of two parameters: the ratio of the thermal conductivity of particles to the continuous matrix,

$$\kappa = \frac{\lambda_p}{\lambda_m}, \tag{3.34}$$

and (in analogy with potential theory) reduced polarizability,

$$\beta = \frac{\kappa - 1}{\kappa + 2}. \tag{3.35}$$

Maxwell's model is one of the oldest approaches based on the potential theory, in which no assumptions are made as to the geometric configuration of the particles, and a low particle volume fraction is assumed. The model is given as,

$$\frac{\lambda_e}{\lambda_m} = \frac{1 + 2\beta\phi}{1 - \beta\phi}. \tag{3.36}$$

An improved model that agrees with experimental data for materials with $\kappa$ ranging from $10^{-3}$ to $10^4$ and $\phi$ from 0.15 to 0.85 is as follow [34, 78]:

$$\frac{\lambda_e}{\lambda_m} = \frac{1 + 2\beta\phi + (2\beta^3 - 0.1\beta)\phi^2 + 0.05\phi^3 \exp(4.5\beta)}{1 - \beta\phi}. \tag{3.37}$$

Thermal conductivity of the fuel kernel (for both UCO and UO$_2$ fuel) is taken as [160]:

$$\lambda = 0.0132e^{1.88 \cdot 10^{-3}(T - 273.15)} + \begin{cases} \frac{4040}{190.85 + T} & \text{for} \quad T < 1923.15 \text{ K} \\ 1.9 & \text{for} \quad 1923.15 \text{ K} \leq T \end{cases}, \tag{3.38}$$

where $\lambda$ (W m$^{-1}$ K$^{-1}$) is the thermal conductivity and $T$ (K) is the temperature. For the SiC layer, the thermal conductivity is given as,

$$\lambda = \frac{17885}{T} + 2, \tag{3.39}$$

where $\lambda$ (W m$^{-1}$ K$^{-1}$) is the thermal conductivity and $T$ (K) is the temperature. The thermal conductivity of the pyrocarbons is constant and equal to 4.0 W m$^{-1}$ K$^{-1}$ for IPyC and OPyC, and 0.5 W m$^{-1}$ K$^{-1}$ for the buffer layer.

**TABLE 3.2**

**Summary of Thermophysical Properties of TRISO Fuel Compact**

| Property | Fuel Kernel | Buffer | IPyC[a] | SiC | OPyC | Matrix |
|---|---|---|---|---|---|---|
| Density (kg m$^{-3}$) | 11040 | 1050 | 1900 | 3190 | 1900 | 1750 |
| Thermal conductivity (W m$^{-1}$ K$^{-1}$) | Eq. (3.38) | 0.5 | 4.0 | Eq. (3.39) | 4.0 | Eq. (3.68) |
| Specific heat capacity (J kg$^{-1}$ K$^{-1}$) | Eq. (3.40) | 720 | 720 | Eq. (3.41) | 720 | Eq. (3.65) |
| Radius/thickness ($\mu$m) | 500 | 95 | 40 | 35 | 40 | |

## Specific Heat Capacity

For both UO$_2$ and UCO fuel kernels, the specific heat capacity can be found as [66, 115],

$$Mc_p = 52.1743 + 87.951\theta - 84.2411\theta^2 + 31.542\theta^3 - 2.6334\theta^4 - \frac{0.71391}{\theta^2}, \quad (3.40)$$

where $c_p$ (J kg$^{-1}$ K$^{-1}$) is the specific heat capacity, $\theta = T/1000$ is the reduced temperature, $M$ (kg mol$^{-1}$) is the molar mass , and $T$ (K) is the temperature.

The correlation for specific heat capacity of SiC is given by [206],

$$c_p = 925.65 + 0.3772T - 7.9259 \cdot 10^{-5}T^2 - \frac{3.1946 \cdot 10^7}{T^2}, \quad (3.41)$$

where $c_p$ (J kg$^{-1}$ K$^{-1}$) is the specific heat capacity and $T$ (K) is temperature. The specific heat capacity for IPyC and OPyC is constant and equal to 720 J kg$^{-1}$ K$^{-1}$.

The effective specific heat capacity of the TRISO fuel compact, $c_{p,e}$, can be found as,

$$c_{p,e} = \sum_i^N r_i c_{p,i}, \quad (3.42)$$

where $r_i$ is the mass fraction of component $i$, $c_{p,i}$ is the component specific heat capacity, and $N$ is the number of components.

## 3.3 ACCIDENT TOLERANT FUEL

The concept of *accident-tolerant fuel* (ATF) was conceived following the Fukushima Daiichi accident, during which three reactor cores were destroyed. The ceramic fuel melted when its temperature exceeded about 3100 K, and significant amounts of hydrogen were generated from the steam-zirconium reaction. The objective of ATF is to make the fuel more resistant to high temperatures and the cladding less susceptible to chemical reactions with coolant at high temperature. In particular, less energy should be released in the chemical reaction and the kinetics of the reaction should be slower.

The initial ATF development focused on a near-term improvement to existing zirconium-based cladding and urania-based fuel coating and additives. The primary functions of the coating are to reduce corrosion during normal operation and to slow chemical reaction rates when high-temperature steam is present during accident conditions. The additives in urania fuel, in turn, are used to enhance the fuel thermal conductivity. This increased thermal conductivity leads to lower fuel centerline temperature and thus improves the safety margin.

As an interim solution, novel fuel and cladding materials are sought. These materials differ from the traditional $UO_2$–Zr fuel and cladding system but have the same fundamental design. Such alternative fuel materials as $U_3Si_2$ and cladding materials such as iron-chromium-aluminum (FeCrAl) alloys are considered.

The long-term ATF concepts will be drastically different from the previous and currently existing solutions. This includes ceramic micro-encapsulated fuel and silicon carbide fiber/silicon carbide matrix (SiC-f/SiC-m) composite concepts. These fuels will require a significant modification or alteration to reactor designs, but they will offer significant performance and safety gains [26].

## 3.4 COOLANTS

Heat generated in a fuel is transferred to a coolant and carried by it outside of the reactor core. The extent to which heat is transferred to a moving coolant depends upon the flow details along a *coolant channel* and upon the coolant thermodynamic properties. Thus, the primary requirement for a coolant is that it must efficiently remove heat from a high-power-density system. Other requirements for coolants include low parasitic neutron absorption, minimum (for fast reactors) or maximum (for thermal reactors) neutron moderation, and low activation by neutrons.

Water was among the first candidates investigated for cooling the thermal spectrum reactors due to its availability, high heat capacity, and high latent heat. A disadvantage is that water-moderated reactors require high-pressure primary systems to keep water in the liquid state at high temperatures, which is required for a high thermodynamic efficiency.

Fast spectrum reactors cannot use water or any organic coolant. For this reason liquid metals were the first candidates investigated for cooling such reactors. Sodium, sodium/potassium eutectic (NaK), mercury, lead, and lead/bismuth eutectic have all been considered. However, helium and steam were also given considerable thought, primarily in connection with the switch from metal to ceramic fuels, in which accompanying lower power densities allow the gas coolant to fulfill the basic heat transfer requirements [224].

Coolant properties vary with pressure and temperature, as shown in Figs. 3.1 through 3.4. The coolant density variation is shown in Fig. 3.1. The gaseous coolants have the lowest density, followed by the intermediate densities of water and sodium, and the highest density of the lead-bismuth eutectic. Water exhibits a significant density drop due to a phase change. Even supercritical water at 25 MPa pressure suffers from a density drop by an order of magnitude when its temperature is increased from 300–500 K to above 700 K.

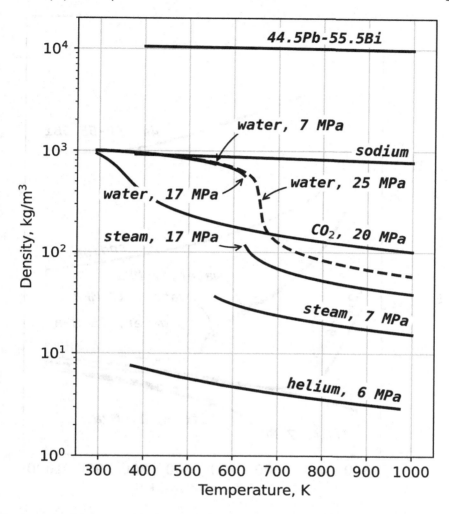

**Figure 3.1**   Density as a function of temperature for selected reactor coolants.

In Fig. 3.2 a plot is shown of the temperature influence on the dynamic viscosities of selected coolants. This chart shows that the viscosities of gases are lower than the viscosities of liquids. Moreover, the viscosities of gases increase with increasing temperature, whereas the viscosities of liquids decrease with increasing temperature. The viscosity of supercritical water at 25 MPa drops significantly when the temperature increases from about 600 to 700 K.

Figure 3.3 is a plot of the temperature influence on the thermal conductivities of selected coolants. Similarly as for the dynamic viscosities, when the temperature is high enough above the saturation temperature, the thermal conductivities of gases

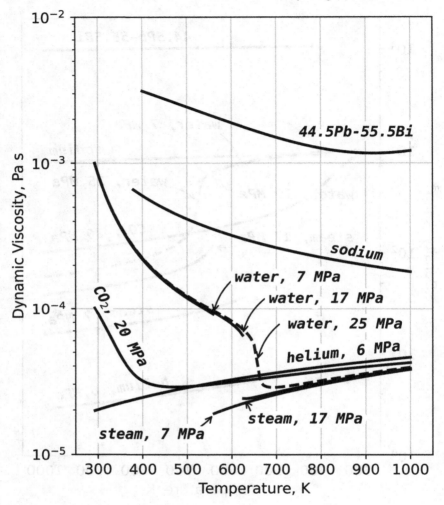

**Figure 3.2** Dynamic viscosity as a function of temperature for selected reactor coolants.

increase with increasing temperature. A similar trend can be observed for the lead-bismuth eutectic in the indicated range of temperatures and for water at temperatures below about 400 K. For sodium, $CO_2$ at temperatures below 400 K, and water at temperatures above 400 K the thermal conductivities decrease with increasing temperature.

Variations of the specific isobaric heat capacity with the temperature for selected coolants are shown in Fig. 3.4. Only a weak dependence of the heat capacities on the temperature variations can be observed for helium, $CO_2$ at temperatures above

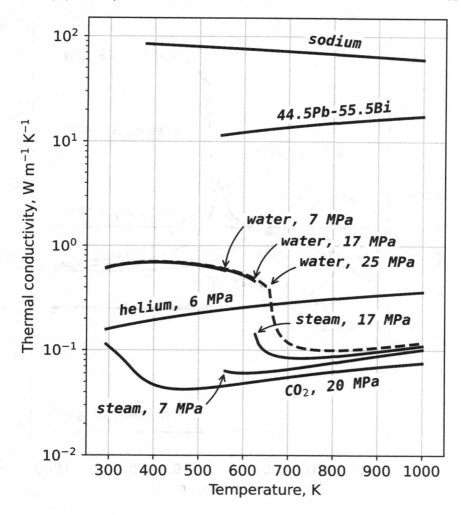

**Figure 3.3** Thermal conductivity as a function of temperature for selected reactor coolants.

500 K, sodium, and the lead-bismuth eutectic. The most significant variations of the heat capacities can be observed for water. In particular, the specific heat of the supercritical water at 25 MPa and at the pseudocritical temperature reaches a value that is more than an order of magnitude higher than the typical values of the specific heat for water and steam at subcritical pressures.

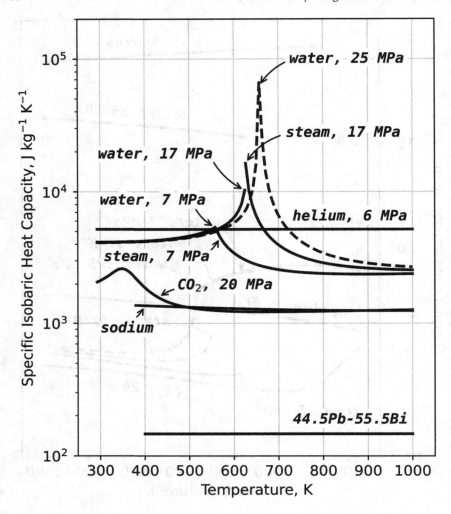

**Figure 3.4**   Specific isobaric heat capacity as a function of temperature for selected reactor coolants.

### 3.4.1   WATER COOLANTS

Water and steam are commonly used in various engineering applications and thanks to this their properties are very well known. Traditionally water and steam properties have been tabulated in so-called "Steam Tables". Currently a very comprehensive set of water property functions is proposed by the International Association for the Properties of Water and Steam (IAPWS)[5]. The functions are valid in the following

---

[5]www.iapws.org

**TABLE 3.3**

**Reference Constants of Ordinary Water used in the IAPWS Formulation**

| Constant | Symbol | Value | Unit |
|---|---|---|---|
| Specific gas constant | $R_{sp}$ | 0.461562 | kJ kg$^{-1}$ K$^{-1}$ |
| Critical temperature | $T_c$ | 647.096 | K |
| Critical pressure | $p_c$ | 22.064 | MPa |
| Critical density | $\rho_c$ | 322 | kg m$^{-3}$ |

range of pressure $p$ and temperature $T$:

$$273.15 \text{ K} \leq T \leq 1073.15 \text{ K} \quad \text{and} \quad p \leq 100 \text{ MPa},$$

$$1073.15 \text{ K} \leq T \leq 2273.15 \text{ K} \quad \text{and} \quad p \leq 50 \text{ MPa}.$$

Reference constants of ordinary water used in the IAPWS formulation are shown in Table 3.3.

## 3.4.2 GASEOUS COOLANTS

The main features of gas cooling, making it particularly useful for reactor applications, are as follows [202]:

- Pure single-phase operation providing freedom from all concern about boiling and phase change.
- Low neutron absorption and moderation.
- Inertness, both chemical and radioactive, including, in the case of helium, compatibility with water, air, and fuel.
- Impossibility of total coolant loss (only depressurization) and low total stored energy.

Early gas-cooled reactors, such as Magnox reactors and AGRs, used $CO_2$ as a coolant. The succeeding generation of reactors, called the high-temperature gas reactors (HTGRs), were designed with helium as the coolant. Currently there is a revived interest in developing gas-cooled fast reactors (GFRs) that use $CO_2$ as the reactor coolant. Thus both $CO_2$ and helium can be expected as the coolant in future gas-cooled reactors. Since the two gases have significantly different thermophysical properties, the reactor designs must be different as well. In particular, it has been shown that helium requires less pumping power than $CO_2$ and produces substantially lower lift forces in the core, but it requires somewhat more core flow area. However, $CO_2$ is considered to be a better convective coolant than helium for decay heat removal [152].

## Equation of State

Equations of state are useful in describing fluid properties such as density or specific volume. In general, an equation of state has the following form,

$$f(\rho, p, T) = 0, \tag{3.43}$$

where $\rho$–density, $p$–pressure, and $T$–temperature.

There is no single equation of state that is valid for all fluids. For gases, the simplest form of the equation of state is given by the ideal gas law,

$$\rho = \frac{p}{RT}, \tag{3.44}$$

where $\rho$–gas density, $p$–pressure, $R$–specific gas constant, and $T$–absolute temperature. Sometime the equation is written as,

$$\rho = \frac{p}{(\kappa - 1)u}, \tag{3.45}$$

where $\kappa = c_p/c_v$ is the specific heat ratio, $c_p$–specific heat at constant pressure, $c_v$–specific heat at constant volume, and u–specific thermal energy: $u = c_v T$, where $T$ is the absolute temperature.

For real gases, a more complex, gas-specific, equation of state is required. Such equations are provided in standard thermodynamics textbooks, e.g., [31]. The following equation is based on five experimentally determined constants and was proposed by Beattie and Bridgeman for several common gases [11],

$$p = \frac{R_u T}{v_M^2} \left( 1 - \frac{c}{v_M T^3} \right) (v_M + B) - \frac{A}{v_M^2},$$

$$\text{where} \quad A = A_0 \left( 1 - \frac{a}{v_M} \right), \ B = B_0 \left( 1 - \frac{b}{v_M} \right). \tag{3.46}$$

Here $p$–pressure, $T$–absolute temperature, $v_M$–specific molar volume (m$^3$ kmol$^{-1}$), and $R_u$–universal gas constant (kPa m$^3$/(kmol K)). Since the equation is implicit in specific volume, thus density, iterations are needed to find a gas density in function of pressure and temperature.

## Helium

*Helium* (He) is a chemical element with atomic number $Z = 2$, relative atomic mass $A_r = 4.002602$, crustal average abundance 0.008 mg kg$^{-1}$, and ocean abundance $7 \cdot 10^{-6}$ mg L$^{-1}$. Helium coolant is particularly suitable for reactors operating at high-temperatures. The very high-temperature reactor (VHTR), one of the six candidates of the Gen IV nuclear systems, is designed to have core outlet temperatures between 970 and 1220 K, or more than 1270 K in the future. The VHTR can operate in the direct cycle during which a helium gas turbine system can be directly set in the primary coolant loop. The combination of high thermal conductivity and specific heat together with chemical inertness gives helium unique advantages over any other gas as a reactor coolant. The temperature dependencies of the most common physical properties of helium at 6 MPa pressure are shown in Figs. 3.1–3.4.

**TABLE 3.4**

**Constants in Eq. (3.46) for Selected Gases**

| Gas | $A_0$ | $a$ | $B_0$ | $b$ | $c$ |
|---|---|---|---|---|---|
| Air | 131.8441 | 0.01931 | 0.04611 | -0.001101 | $4.34 \times 10^4$ |
| Argon, Ar | 130.7802 | 0.02328 | 0.03931 | 0.0 | $5.99 \times 10^4$ |
| Carbon dioxide, $CO_2$ | 507.2836 | 0.07132 | 0.10476 | 0.07235 | $6.60 \times 10^5$ |
| Helium, He | 2.1886 | 0.05984 | 0.01400 | 0.0 | 40 |
| Hydrogen, $H_2$ | 20.0117 | -0.00506 | 0.02096 | -0.04359 | 504 |
| Nitrogen, $N_2$ | 136.2315 | 0.02617 | 0.05046 | -0.00691 | $4.20 \times 10^4$ |
| Oxygen, $O_2$ | 151.0857 | 0.02562 | 0.04624 | 0.004208 | $4.80 \times 10^4$ |

## Carbon Dioxide

*Carbon dioxide* ($CO_2$) is a chemical compound with relative molecular weight $M_r = 44.01$. It is a colorless and non-flammable gas present in the air with a concentration of over 400 ppm. The temperature and pressure of the $CO_2$ coolant in Magnox reactors were about 673 K and 1.6 MPa, respectively. To improve the conversion efficiency, Generation II advanced gas-cooled reactors (AGRs) were designed to operate at significantly higher temperatures and pressures, to match the inlet steam conditions of prevailing coal-fired power plants. This resulted in a maximum $CO_2$ temperatures of ~920 K and pressures of ~4.1 MPa. Generation IV reactors are expected to operate at even higher pressures, where supercritical $CO_2$ ($sCO_2$) power cycles with pressure in the range of 20–30 MPa are under consideration. The temperature dependencies of the most common physical properties of $sCO_2$ at 20 MPa pressure are shown in Figs. 3.1–3.4.

### 3.4.3 LIQUID METAL COOLANTS

Liquid metal coolants have been considered as first candidates for fast spectrum reactors since they provide efficient cooling without significantly interacting with neutrons. They have relatively high melting and boiling temperatures at atmospheric pressure, allowing for a single-phase operation in a wide range of temperatures. The temperature dependencies of the most common physical properties of sodium and lead-bismuth eutectic are shown in Figs. 3.1–3.4.

## Sodium

*Sodium* (Na) is a chemical element with atomic number $Z = 11$, relative atomic mass $A_r = 22.98977$, crustal average abundance $2.36 \cdot 10^4$ mg kg$^{-1}$, and ocean abundance $1.08 \cdot 10^4$ mg L$^{-1}$. Since sodium is one of the most electropositive[6] metals, it reacts exothermically with water yielding sodium hydroxide (NaOH) and hydrogen gas.

---

[6]An electropositive element is one that has a tendency to lose electrons and form a positively charged ion.

**TABLE 3.5**

**Basic Thermophysical Properties of Liquid Metal Coolants**

| Property | Na | Pb | Pb-Bi[a] |
|---|---|---|---|
| Melting temperature (K (°C)) | 371.15(98) | 606.6(327.45) | 398.15(125) |
| Boiling temperature (K (°C)) | 1156.15(883) | 2021(1747.85) | 1911.15(1638) |
| Density at 293 K (solid) (kg m$^{-3}$) | 966 | 11340 | 10474 |
| Density at 723 K (liquid) (kg m$^{-3}$) | 844 | 10514 | 10150 |
| Dynamic viscosity at 723 K (Pa s) | $2.532 \cdot 10^{-4}$ | $2.0 \cdot 10^{-3}$ | $1.421 \cdot 10^{-3}$ |
| Thermal conductivity at 293 K (solid) (W m$^{-1}$ K$^{-1}$) | 130 | 35 | 10 |
| Thermal conductivity at 723 K (liquid) (W m$^{-1}$ K$^{-1}$) | 71.2 | 17.1 | 14.2 |
| Heat capacity at 293 K (solid) (kJ kg$^{-1}$ K$^{-1}$) | 1.230 | 0.127 | 0.128 |
| Heat capacity at 723 K (liquid) (kJ kg$^{-1}$ K$^{-1}$) | 1.272 | 0.1473 | 0.146 |
| Prandtl number at 723 K (-) | 0.0045 | 0.0174 | 0.0147 |

a An alloy/eutectic with 44.5 wt% Pb + 55.5 wt% Bi composition.
*Source:* [103, 104, 207]

Basic thermophysical properties of sodium are given in Table 3.5. Analytic expressions for density, dynamic viscosity, thermal conductivity, and specific heat capacity as a function of temperature are given below.

*Density*

$$\rho = 896.6 + 516.1\theta - 1829.7\theta^2 + 2201.6\theta^3 \\ - 1397.6\theta^4 + 448.7\theta^5 - 57.96\theta^6 , \tag{3.47}$$

where $\rho$ (kg m$^{-3}$) is the density and $\theta = T/1000$, with $T$ (K) is the temperature [103].

*Dynamic Viscosity*

$$\mu = \exp\left(-6.4406 - 0.3958\ln T + \frac{556.8}{T}\right), \tag{3.48}$$

where $\mu$ (Pa·s) is the dynamic viscosity and $T$ (K) is the temperature [103].

*Thermal Conductivity*

$$\lambda = 99.5 - 39.1 \cdot 10^{-3}T, \tag{3.49}$$

where $\lambda$ (W m$^{-1}$ K$^{-1}$) is the thermal conductivity and $T$ (K) is the temperature [103].

*Specific Isobaric Heat Capacity*

$$c_p = 43.497\left(38.12 - \frac{0.069 \cdot 10^6}{T^2} - 19.493 \cdot 10^{-3}T + 10.24 \cdot 10^{-6}T^2\right), \tag{3.50}$$

where $c_p$ (J kg$^{-1}$ K$^{-1}$) is the specific isobaric heat capacity and $T$ (K) is the temperature [103].

## Lead

*Lead* (Pb) is a chemical element with atomic number $Z = 82$, relative atomic mass $A_r = 207.2$, crustal average abundance 14 mg kg$^{-1}$, and ocean abundance $3 \cdot 10^{-5}$ mg L$^{-1}$. Solid lead is practically not oxidizing in dry air; however, in humid air it is coated with an oxide film (PbO). Air causes molten lead oxidization initially to $Pb_2O$ and then to PbO oxide. In the temperature range 673–773 K lead interacts with water producing hydroxide Pb(OH)$_2$. The choice of structural materials resistant to erosion-corrosion effects is probably the main issue in lead-cooled reactors.

Basic thermophysical properties of lead are given in Table 3.5. Analytic expressions for density, dynamic viscosity, thermal conductivity, and specific heat capacity as a function of temperature are given below.

### Density

The following linear correlation can be used to describe the temperature dependence of the density at normal atmospheric pressure [207],

$$\rho = 10671 - 1.2795\,(T - 600.6), \tag{3.51}$$

where $\rho$ (kg m$^{-3}$) is the density and $T$ (K) is the temperature. The uncertainty of the correlation at the 95% confidence level is 0.5–1% for the temperature range 601 K $< T <$ 2000 K.

### Dynamic Viscosity

The dynamic viscosity $\mu$ (Pa·s) depends on the temperature $T$ (K) as follows [207],

$$\mu = 0.479 \cdot 10^{-3} \cdot \exp\left(\frac{8600}{R_u T}\right), \tag{3.52}$$

where $R_u \approx 8.3145$ J mol$^{-1}$ K$^{-1}$ is the *universal gas constant*. The uncertainty of the correlation is $\pm 6\%$ for the temperature range 601 K $< T <$ 1473 K.

### Thermal Conductivity

The following linear correlation for the thermal conductivity is proposed [207],

$$\lambda = 16.093 + 0.0078462\,(T - 600.6), \tag{3.53}$$

where $\lambda$ (W m$^{-1}$ K$^{-1}$) is the thermal conductivity and $T$ (K) is the temperature. The correlation is valid in the temperature range 602–1150 K at standard pressure with uncertainty $\pm 16.9\%$ at the 95% confidence level.

*Specific Isobaric Heat Capacity*

The specific isobaric heat capacity of liquid lead can be found from the following correlation [207],

$$c_p = 176.2 - 4.923 \cdot 10^{-2}T + 1.544 \cdot 10^{-5}T^2 - 1.524 \cdot 10^6 T^{-2}, \qquad (3.54)$$

where $c_p$ (J kg$^{-1}$ K$^{-1}$) is the specific isobaric heat capacity and $T$ (K) is the temperature.

## Lead-Bismuth Eutectic

The *lead-bismuth alloy* phase diagram has a eutectic point at 55.5 wt% Bi with a melting temperature 398.15 K and the relative molecular weight $M_r = 208.188$. Basic thermophysical properties of the lead-bismuth 44.5Pb-55.5Bi alloy[7] are given in Table 3.5. Analytic expressions for density, dynamic viscosity, thermal conductivity, and specific heat capacity as a function of temperature are given below.

*Density*

$$\rho = 1.105 \cdot 10^4 - 1.249T, \qquad (3.55)$$

where $\rho$ (kg m$^{-3}$) is the density and $T$ (K) is the temperature [103].

*Dynamic Viscosity*

$$\mu = \rho \left( 68.9 - 0.126T + 6.95 \cdot 10^{-5}T^2 \right) \cdot 10^{-8}, \qquad (3.56)$$

where $\mu$ (Pa·s) is the dynamic viscosity, $\rho$ (kg m$^{-3}$) is the density, and $T$ (K) is the temperature [103].

*Thermal Conductivity*

The thermal conductivity of lead-bismuth eutectic at normal atmospheric pressure is given as [207],

$$\lambda = 9.35 + 0.01434(T - 398) - 2.305 \cdot 10^{-6}(T - 398)^2, \qquad (3.57)$$

where $\lambda$ (W m$^{-1}$ K$^{-1}$) is the thermal conductivity and $T$ (K) is the temperature. The correlation is valid for temperature range: 405 K $< T <$ 1073 K and the uncertainty is about 10–15%.

*Specific Isobaric Heat Capacity*

The specific isobaric heat capacity of lead-bismuth eutectic can be found from the following correlation [207],

$$c_p = 164.8 - 3.939 \cdot 10^{-2}T + 1.249 \cdot 10^{-5}T^2 - 4.563 \cdot 10^5 T^{-2}, \qquad (3.58)$$

---

[7]Properties of the lead-bismuth alloy depend on its composition. At present most of the researchers fix this composition at 44.5 wt% Pb + 55.5 wt% Bi, corresponding to 44.3 at% Pb + 55.7 at% Bi.

where $c_p$ (J kg$^{-1}$ K$^{-1}$) is the specific isobaric heat capacity and $T$ (K) is the temperature. The uncertainty of the correlation is $\pm 7\%$.

## 3.5  OTHER MATERIALS

Wide range of additional materials are essential for the safe and reliable operation of the reactor core. Structural components within the core, commonly 304 or 316 stainless steel, are often the most critical for the core integrity. Materials such as graphite and various molten salts are important for gas-cooled and molten salt reactors.

### 3.5.1  AUSTENITIC STAINLESS STEELS

Austenitic stainless steel is one of the classes of stainless steel by crystalline structure, with the austenite as the primary crystalline structure that prevents steel from being hardenable by heat treatment and makes them essentially non-magnetic. Austenitic stainless steel contains chromium to give a high corrosion resistance, nickel, carbon (which is rather impurity than an alloying element), and many other elements in solution. Austenitic stainless steels have a good performance on corrosion resistance and radiation resistance, which makes them suitable for nuclear power applications.

#### Austenitic Stainless Steel Type 304

The austenitic stainless steel type 304 is characterized by the following composition (%): 0.08C, 18–20Cr, 8–12Ni, and $\leq$ 2Mn. The austenitic stainless steel type 304 is general stainless steel with good corrosion resistance and inter-granular corrosion resistance. Its variant denoted 304L stainless steel has a lower carbon content, making it more suitable for being welded. The resistance to the inter-granular corrosion of 304L stainless steel is better than that of 304 stainless steel. The density of 304 stainless steel is 7960 kg m$^{-3}$ and its solidus and liquidus phase transition temperatures are 1671 K and 1727 K, respectively. The recommended expressions for the thermophysical properties of 304 stainless steel are as follows [95]:

*Specific Heat Capacity*

$$c_p = \begin{cases} 326 + 0.298T - 9.56 \cdot 10^{-3}T^2 & \text{for} \quad 300 \leq T \leq 1558 \text{ K} \\ 558.228 & \text{for} \quad 1558 \text{ K} < T \end{cases}, \quad (3.59)$$

where $c_p$ (J kg$^{-1}$ K$^{-1}$) is the specific heat capacity and $T$ (K) is the temperature.

*Thermal Conductivity*

$$\lambda = \begin{cases} 7.58 + 0.0189T & \text{for} \quad 300 \leq T < 1671 \text{ K} \\ 610.9393 - 0.3421767T & \text{for} \quad 1671 \leq T < 1727 \text{ K} \\ 20.0 & \text{for} \quad 1727 \text{ K} \leq T \end{cases}, \quad (3.60)$$

where $\lambda$ (W m$^{-1}$ K$^{-1}$) is the thermal conductivity and $T$ (K) is the temperature.

*Thermal Strain*

$$\varepsilon = \begin{cases} 1.57 \cdot 10^{-5}T + 1.69 \cdot 10^{-9}T^2 & \text{for} \quad 300 \leq T < 1671 \text{ K} \\ -2.986634 \cdot 10^{-1} + 1.972573 \cdot 10^{-4}T & \text{for} \quad 1671 \leq T < 1727 \text{ K} , \quad (3.61) \\ 0.042 & \text{for} \quad 1727 \text{ K} \leq T \end{cases}$$

where $\varepsilon$ (mm/mm) is the thermal expansion in terms of thermal strain and $T$ (K) is the temperature.

## Austenitic Stainless Steel Type 316

The austenitic stainless steel type 316 is characterized by the following composition (%): 0.08C, 16–18Cr, 10–14Ni, $\leq$ 2Mn, 2–3Mo, $\leq$ 1Si, < 0.045P, and < 0.03S. 316L stainless steel is a low-carbon (0.04%C) variant of 316 stainless steel with great pitting corrosion resistance, inter-granular corrosion resistance, and good chloride resistance. The density of 316 stainless steel is 7960 kg m$^{-3}$ and its solidus and liquidus phase transition temperatures are 1683 K and 1708 K, respectively. The recommended expressions for the thermophysical properties of 316 stainless steel are as follows [95]:

*Specific Heat Capacity*

$$c_p = \begin{cases} 407.8 + 0.2111T & \text{for} \quad 300 \leq T \leq 1683 \text{ K} \\ 760.0 & \text{for} \quad 1683 \text{ K} < T \end{cases}, \quad (3.62)$$

where $c_p$ (J kg$^{-1}$ K$^{-1}$) is the specific heat capacity and $T$ (K) is the temperature.

*Thermal Conductivity*

$$\lambda = \begin{cases} 9.735 + 0.01434T & \text{for} \quad 300 \leq T < 1683 \text{ K} \\ 10.981 + 0.003214T & \text{for} \quad 1683 \leq T < 2073 \text{ K (liquid)} \end{cases}, \quad (3.63)$$

where $\lambda$ (W m$^{-1}$ K$^{-1}$) is the thermal conductivity and $T$ (K) is the temperature.

*Thermal Expansion Coefficient*

$$\alpha = 1.417 \cdot 10^{-5} + 4.381 \cdot 10^{-9}T, \quad (3.64)$$

where $\alpha$ (K$^{-1}$) is the mean linear expansion coefficient and $T$ (K) is the temperature. The expression is valid for the temperature range $293 < T \leq 1683$ K.

### 3.5.2   GRAPHITE

Graphite is a naturally occurring mineral that is mined in many places worldwide. Graphite can also be manufactured for use as a moderator or reflector within nuclear reactors and is then called an artificial, synthetic, or nuclear graphite. Isotropic behavior is a very desirable property in nuclear graphite and is achieved in modern nuclear graphite through the use of cokes with an isotropic structure in the initial formulation.

Graphite is a modification of carbon with atomic number $Z = 6$, relative atomic mass $A_r = 12.011$, and theoretical density of 2266 kg m$^{-3}$. Thermophysical properties of graphite can greatly differ depending on its production method and owing to its anisotropic structure. It is therefore difficult to determine an exact value of the melting point of graphite, which according to early data is about 3700–4000 K, and according to more recent data exceeds 4530 K [103].

The bulk density of synthetic graphite varies according to the manufacturing process and it can be as low as 1600 kg m$^{-3}$ for the coarse-grain graphite and can exceed 1850 kg m$^{-3}$ for the fine-grained isotropic graphite. In general such properties as strength, stiffness, and thermal conductivity increase with increasing density.

Isobaric specific heat capacity of graphite $c_p$ (kJ kg$^{-1}$ K$^{-1}$) varies with the temperature $T$ (K) and for $298 \leq T \leq 1273$ K is given as [103]:

$$c_p = 2.031 + 7.8645 \cdot 10^{-5} T - 4.2671 \cdot 10^5 T^{-2}$$
$$+ 1.3203 \cdot 10^8 T^{-3} - 1.199 \cdot 10^{10} T^{-4} \qquad (3.65)$$

for $1273 < T \leq 3273$ K

$$c_p = 1.131 + 6.62 \cdot 10^{-4} T - 9.969 \cdot 10^{-8} T^2, \qquad (3.66)$$

and for $3273 < T \leq 5000$ K

$$c_p = 6.12 \cdot 10^{-5} T^{1.3}. \qquad (3.67)$$

The uncertainty of the correlation is $\pm 10\%$.

Thermal conductivity of graphite depends on temperature, fast neutron fluence, and density. Assuming an isotropic type of graphite, the conductivity can be found as,

$$\lambda = \lambda_u \lambda_\phi \frac{\rho}{1700}, \qquad (3.68)$$

where,

$$\lambda_u = \lambda_0 \left[ 1 - \alpha (T - 100) e^{\delta T} \right], \qquad (3.69)$$

$$\lambda_\phi = 1.0 - (0.94 - 0.604\theta) \{ 1 - \exp[-(2.96 - 1.955\theta)\phi_0] \}$$
$$- (0.043\theta - 0.008\theta^8) \phi_0 \qquad (3.70)$$

Here $\phi_0 = \phi/(1.52 \cdot 10^{25})$, $\phi$–fluence in n/m$^2$, $\theta = T/1000$, $T$–temperature in K, $\rho$–graphite density in kg m$^{-3}$, $\lambda_0$–graphite thermal conductivity at 373.15 K in W m$^{-1}$ K$^{-1}$, $\alpha$, $\delta$–non-dimensional empirical constants. The coefficients $\alpha$, $\delta$, and $\lambda_0$ depend on the type of graphite. For graphite A3-27 with heat treatment at 2073.15 K they are as follows: $\lambda_0 = 47.4$ W m$^{-1}$ K$^{-1}$, $\alpha = 9.7556 \cdot 10^{-4}$, and $\delta = -6.0360 \cdot 10^{-4}$. Various models to calculate the thermophysical properties of graphite are provided in the literature [176].

**TABLE 3.6**
**Selected Properties of the Coolant Salts**

| Property | LiF–BeF$_2$ [a] | NaF–NaBF$_4$ [b] | LiF–NaF–KF [c] |
|---|---|---|---|
| Melting point (K) | 728 | 657±1 | 727 |
| $\rho$ (kg m$^{-3}$); T (K) | $2146.3 - 0.4884T$ | $2446.3 - 0.7110T$ | $2579.3 - 0.6240T$ |
| $\mu$ (mPa s); T (K) | $1.81\exp(1912.2/T)$ | $0.0877\exp(2240/T)$ | $0.0248\exp(4477/T)$ |
| $\lambda$ (W m$^{-1}$ K$^{-1}$); T (K) | 1.1 | $0.66 - 2.37\cdot10^{-4}T$ | $0.36 + 5.6\cdot10^{-4}T$ |
| $c_p$ (kJ kg$^{-1}$ K$^{-1}$) | 2.39 | 1.506 | 1.88 |

[a] Molar fraction composition 0.66–0.34
[b] Molar fraction composition 0.08–0.92
[c] Molar fraction composition 0.465–0.115–0.42
*Source:* [14]

### 3.5.3 MOLTEN SALTS

Molten salt is salt which is solid at room temperature and atmospheric pressure but it melts due to elevated temperature. The fuel of a molten salt reactor (MSR) is based on the dissolution of the fissile material in an inorganic liquid. The operation temperature of the MSR is between 800 and 1000 K, where the lower limit is determined by the melting temperature of the salt and the upper one by the corrosion rate of the structural material.

The fuel in the MSR must fulfill several requirements with respect to its physico-chemical properties and nowadays is commonly recognized that fluoride and chloride systems are the best candidates to meet these requirements. Although more data exist on fluoride salts, there are gaps in physico-chemical properties data for both system. In general chloride salts melt at somewhat lower temperatures compared to fluorides, but metal halides of both groups melt well above the room temperature. Even though boiling points are generally lower for chlorides compared to fluorides, both halides offer sufficient margin between melting points and boiling points.

The density, viscosity, thermal conductivity, and specific heat for typical coolant and fuel compositions are given in Tables 3.6 and 3.7.

### 3.6 GAS GAP

Fuel rod designs provide for a small annular gap between the fuel pellet and the cladding. During fabrication, the gap is filled with helium, but during irradiation, gaseous fission products migrate into the gap and mix with the helium. The gas mixture thermal conductivity, viscosity, density, and specific heat capacity need to be determined as a function of the mixture composition, pressure, and temperature. In addition, the swelling and cracking of the pellet tend to close the gap and increase the surface roughness. For safety evaluations it is necessary to use semi-empirical

**TABLE 3.7**

**Selected Properties of the Fuel Salts**

| Property | LiF–ThF$_4$ [a] | LiF–BeF$_2$–ThF$_4$ [b] | LiF–NaF–BeF$_2$–PuF$_3$ [c] | NaCl–UCl$_3$–PuCl$_3$ [d] |
|---|---|---|---|---|
| Melting point (K) | 841 | 771 | 775 | 873 |
| $\rho$ (kg m$^{-3}$); T (K) | $5543 - 1.25T$ | $4124.3 - 0.869T$ | $2759.9 - 0.573T$ | 3600[e] |
| $\mu$ (mPa s); T (K) | $0.365\exp(2735/T)$ | $0.062\exp(4636/T)$ | $0.100\exp(3724/T)$ | - |
| $\lambda$ (W m$^{-1}$ K$^{-1}$); T (K) | $\sim 1.5$[f] | 1.5 | $0.402 + 0.5 \cdot 10^{-3}/T$ | - |
| $c_p$ (kJ kg$^{-1}$ K$^{-1}$) | 1.0 | 1.55 | 2.15 | 0.908[e] |

[a] Molar fraction composition 0.78–0.22
[b] Molar fraction composition 0.717–0.16–0.123
[c] Molar fraction composition 0.203–0.571–0.212–0.013
[d] Molar fraction composition 0.55–0.294–0.156
[e] Value for $T = 963$ K.
[f] Value for $T = 1023$ K.
*Source:* [14]

computational models based on the thermal conductance of a gas layer between two rough surfaces in partial contact [204].

# PROBLEMS

### PROBLEM 3.1

Compare the thermal conductivity of various zirconium alloys within a temperature range of 550 to 850 K to that of stainless steel type 304, and discuss which material is more suitable as a cladding material from the point of view of maximum fuel temperature.

### PROBLEM 3.2

Estimate the range of variation of the thermal conductivity of the MOX fuel assuming that the fuel temperature changes in a range from 750 to 1600 K. Investigate the influence of the porosity on thermal conductivity knowing that the mean porosity is equal to 8% and it varies within $\pm 1\%$.

### PROBLEM 3.3

Calculate the change of the thermal conductivity of graphite A3-27 when the fluence increases from $1.52 \cdot 10^{25}$ to $5.5 \cdot 10^{25}$ n/m$^2$. Assume a constant temperature of the graphite equal to 2073.15 K.

# Section II

## Reactor Core Thermal-Hydraulics

# 4 Conservation Equations for Single-Phase Flow

Power reactors need a coolant to evacuate heat from a core. In light water reactors, ordinary water under pressure is used as a coolant. In pressurized water reactors, the coolant is kept at high enough pressure and at a sufficiently low temperature to prevent phase change. Thus, a coolant flow in the core of a pressurized water reactor can be treated as a single-phase liquid flow. Similar conditions are kept in cores of liquid-metal-cooled reactors, in which boiling is not allowed under normal operating conditions. In gas-cooled reactors single-phase flow regime is prevailing under all conceivable conditions, and high pressure (such as around 10 MPa in high-temperature gas-cooled reactors) is kept to improve the gas heat transfer properties. Even in molten salt cooled reactors, where molten salt material properties vary significantly, the boiling point is high enough to ensure a single-phase flow condition in the core.

In this chapter we formulate generic conservation equations applicable to a single-phase flow. The equations are first derived in an instantaneous form, for an arbitrary and non-stationary open system. Next, several special cases are analyzed, including time-averaged equations, equations for flow in porous media, and equations for flow in channels.

## 4.1 PRELIMINARIES

A coolant flow theory belongs to *fluid mechanics*, which is a branch of mechanics dealing with the movement of liquid and gaseous media, and their interactions with surrounding or submerged solids. In its most basic formulation, the purpose of solving fluid mechanics problems is to find key fluid properties as a function of spatial location and time. The properties of interest can be divided into kinematic properties, transport properties and thermodynamic properties, as described below.

### 4.1.1 KINEMATIC PROPERTIES

**Kinematic properties** are flow field properties and include such quantities as linear velocity, defined as,

$$\mathbf{v}(\mathbf{r},t) = \mathbf{v}(x,y,z,t) = u(x,y,z,t)\mathbf{e}_x + v(x,y,z,t)\mathbf{e}_y + w(x,y,z,t)\mathbf{e}_z, \qquad (4.1)$$

where $x,y,z$ are Cartesian coordinates, $\mathbf{e}_x$, $\mathbf{e}_y$, $\mathbf{e}_z$ are unit vectors of the Cartesian coordinates, and $u(x,y,z,t), v(x,y,z,t), w(x,y,z,t)$ are Cartesian velocity components in the $x$, $y$, and $z$ directions, respectively. The position vector $\mathbf{r}$ is defined as,

$$\mathbf{r}(x,y,z) = x\mathbf{e}_x + y\mathbf{e}_y + z\mathbf{e}_z. \qquad (4.2)$$

DOI: 10.1201/9781003255000-4

Other kinematic properties, which can be derived from the linear velocity include linear acceleration,

$$\mathbf{a}(\mathbf{r},t) = \frac{d\mathbf{v}}{dt}, \tag{4.3}$$

and vorticity,

$$\omega(\mathbf{r},t) = \nabla \times \mathbf{v}(\mathbf{r},t). \tag{4.4}$$

In the kinematic description of the theoretical fluid mechanics we investigate separate fluid particles, filling the volume of the considered region in a continuous manner (Lagrangian approach), or as a velocity field in a region occupied by a fluid in motion (Eulerian approach). In the *Lagrangian approach*, coordinates $x, y, z$ determine the instantaneous location of the fluid particle, considering them as functions of the Lagrange variables, that is the time $t$ and certain numbers $a, b, c$, representing generalized coordinates of the particle at the initial time instant. In the *Eulerian approach*, the flow properties such as, e.g., velocity $\mathbf{v}(x,y,z)$, pressure $p(x,y,z)$ or density $\rho(x,y,z)$, are described at a fixed point in the space given by coordinates $x, y, z$ or by a position vector $\mathbf{r}(x,y,z)$.

One of the important consequences of the two formulations is the existence of two derivatives with respect to time. The time derivative when keeping constant position $\mathbf{r}$ is defined as,

$$\frac{\partial}{\partial t} \equiv \left(\frac{\partial}{\partial t}\right)_{\mathbf{r} \ fixed}, \tag{4.5}$$

and the time derivative when following a fixed material particle, called *substantive derivative* or *material derivative*, is defined as,

$$\frac{D}{Dt} \equiv \left(\frac{\partial}{\partial t}\right)_{particle \ fixed}. \tag{4.6}$$

The two derivatives are related by the following relationship,

$$\frac{D\Phi}{Dt} = \frac{\partial\Phi}{\partial t} + \mathbf{v} \cdot \nabla\Phi, \tag{4.7}$$

where $\Phi$ is some fluid property.

Substituting fluid velocity vector $\mathbf{v}$ into Eq. (4.7) yields,

$$\mathbf{a}(\mathbf{r},t) \equiv \frac{D\mathbf{v}}{Dt} = \frac{\partial\mathbf{v}}{\partial t} + \mathbf{v} \cdot \nabla\mathbf{v}, \tag{4.8}$$

where $\mathbf{a}(\mathbf{r},t)$ is the fluid acceleration. The first term on the right-hand side of Eq. (4.8) represents the local acceleration, whereas the second term describes the convective part of the acceleration. For stationary flows the first term is equal to zero and the fluid acceleration is,

$$\mathbf{a}(\mathbf{r}) \equiv \frac{D\mathbf{v}}{Dt} = \mathbf{v} \cdot \nabla\mathbf{v}. \tag{4.9}$$

As can be seen, the fluid acceleration for stationary flows, expressed with the total derivative $\mathbf{v}$ in Eq. (4.9), is equal to zero only in case when the dot product $\mathbf{v} \cdot \nabla\mathbf{v}$ is equal to zero.

## 4.1.2 TRANSPORT PROPERTIES

*Dynamic viscosity* $\mu$, or just *viscosity*, belongs to the most important **transport properties** in fluid dynamics. In conservation equations describing a fluid motion, viscosity $\mu$ is often divided by the density of the fluid $\rho$. Thus, it is convenient to introduce another quantity called the *kinematic viscosity* $\nu$, defined as,

$$\nu = \frac{\mu}{\rho}. \tag{4.10}$$

Fluid viscosity varies over many orders of magnitudes with varying pressure and temperature conditions. Selected viscosity data for main coolant materials are provided in §3.

If a fluid is present between two parallel plates separated by a distance $d$ and one of the plates is set in motion with constant velocity $U$, momentum is transported between the plates through the fluid by a viscous process. To maintain the process at a steady state, the following force must be applied to the plate in motion,

$$F = \mu \frac{U}{d} A, \tag{4.11}$$

where $A$ is the plate area. The validity of this empirical relationship has been confirmed for laminar flow of all gases and liquids with a molecular weight of less than 5000. Due to this universality, Eq. (4.11) is expressed in terms of a local *shear stress* $\tau_{yx}$ and a local velocity gradient $\frac{du}{dy}$ as follows,[1]

$$\tau_{yx} = \mu \frac{du}{dy}. \tag{4.12}$$

This equation is sometimes called *Newton's law of viscosity* and all fluids that can be described by the equation are referred to as *Newtonian fluids*. The *shear stress tensor* for fluids can be in general described with the following function,

$$\boldsymbol{\tau} = \phi(\mathbf{v}, \nabla\mathbf{v}, \mathbf{D}), \tag{4.13}$$

where $\mathbf{D}$ is the **deformation tensor** (also called the *rate-of-strain tensor*),

$$\mathbf{D} = \frac{1}{2} \begin{bmatrix} 2\dfrac{\partial u}{\partial x} & \left(\dfrac{\partial u}{\partial y} + \dfrac{\partial v}{\partial x}\right) & \left(\dfrac{\partial u}{\partial z} + \dfrac{\partial w}{\partial x}\right) \\[2mm] \left(\dfrac{\partial v}{\partial x} + \dfrac{\partial u}{\partial y}\right) & 2\dfrac{\partial v}{\partial y} & \left(\dfrac{\partial v}{\partial z} + \dfrac{\partial w}{\partial y}\right) \\[2mm] \left(\dfrac{\partial w}{\partial x} + \dfrac{\partial u}{\partial z}\right) & \left(\dfrac{\partial w}{\partial y} + \dfrac{\partial v}{\partial z}\right) & 2\dfrac{\partial w}{\partial z} \end{bmatrix}, \tag{4.14}$$

---

[1] By convention, the first subscript in the shear stress indicates the surface on which the stress acts (here constant y-plane) and the second subscript denotes the direction of the stress (here x-direction). Some authors use a minus sign in the equation in analogy to transport of heat; however, we adopt here a convention that is more common in the fluid dynamics literature.

and $\nabla\mathbf{v}$ is the **velocity gradient tensor** given as follows,

$$\nabla\mathbf{v} = \begin{bmatrix} \dfrac{\partial u}{\partial x} & \dfrac{\partial v}{\partial x} & \dfrac{\partial w}{\partial x} \\[2mm] \dfrac{\partial u}{\partial y} & \dfrac{\partial v}{\partial y} & \dfrac{\partial w}{\partial y} \\[2mm] \dfrac{\partial u}{\partial z} & \dfrac{\partial v}{\partial z} & \dfrac{\partial w}{\partial z} \end{bmatrix}. \tag{4.15}$$

It can be shown that

$$\nabla\mathbf{v} = \mathbf{D} + \mathbf{S}, \tag{4.16}$$

where $\mathbf{S}$ is the anti-symmetric *vorticity tensor*,

$$\mathbf{S} = \frac{1}{2}\begin{bmatrix} 0 & -\left(\dfrac{\partial v}{\partial x} - \dfrac{\partial u}{\partial y}\right) & \left(\dfrac{\partial u}{\partial z} - \dfrac{\partial w}{\partial x}\right) \\[3mm] \left(\dfrac{\partial v}{\partial x} - \dfrac{\partial u}{\partial y}\right) & 0 & -\left(\dfrac{\partial w}{\partial y} - \dfrac{\partial v}{\partial z}\right) \\[3mm] -\left(\dfrac{\partial u}{\partial z} - \dfrac{\partial w}{\partial x}\right) & \left(\dfrac{\partial w}{\partial y} - \dfrac{\partial v}{\partial z}\right) & 0 \end{bmatrix}. \tag{4.17}$$

For constant-density Newtonian fluids, the local shear stress is entirely due to the local deformation tensor and the *shear stress tensor* is well-approximated with the following equation:

$$\boldsymbol{\tau} = 2\mu\mathbf{D} = \mu\left[\nabla\mathbf{v} + (\nabla\mathbf{v})^T\right]. \tag{4.18}$$

For variable-density fluids, this expression is generalized as proposed by Stokes,

$$\boldsymbol{\tau} = \mu\left[\nabla\mathbf{v} + (\nabla\mathbf{v})^T\right] + \left(\mu' - \frac{2}{3}\mu\right)(\nabla\cdot\mathbf{v})\mathbf{I}, \tag{4.19}$$

where $\mu'$ is called the *bulk coefficient of viscosity* or *dilatational viscosity*. As can be easily noticed, for constant-density flows $\nabla\cdot\mathbf{v} = 0$ and the second term on the right-hand side of Eq. (4.19) is zero. Thus, Eqs. (4.18) and (4.19) take the same form.

The shear stress tensor $\boldsymbol{\tau}$ consists of the tangential and viscous shear stresses. In addition, the fluid pressure $p$, which is compressive, acts in the opposite direction to that of the convention, and needs to be added to the shear stress tensor to give the **total stress tensor** as follows,[2]

$$\mathbf{T} = -p\mathbf{I} + \boldsymbol{\tau}. \tag{4.20}$$

Here $-p\mathbf{I}$ represents the isotropic pressure stress and $\boldsymbol{\tau}$ is the anisotropic viscous stress.

Defining the mean pressure as $\overline{p} = -\frac{1}{3}(\mathbf{I} : \mathbf{T})$ we obtain,

$$p - \overline{p} = \mu'(\nabla\cdot\mathbf{v}). \tag{4.21}$$

---

[2] Also known as the complete stress tensor.

This equation shows that $p = \overline{p}$ when either $\mu' = 0$ or $\nabla \cdot \mathbf{v} = 0$. The former is valid only for monoatomic gases, whereas the latter for constant-density flows.

In fluid dynamics the pressure $p$ is postulated to be equivalent to the *thermodynamic pressure*. An immediate consequence of this assumption is that pressure $p$ should depend only on the local instantaneous values of the mass density $\rho$ and the specific thermal energy u and not on $\nabla \cdot \mathbf{v}$.

The consequence of $p \neq \overline{p}$ can be seen when we multiply both sides of Eq. (4.21) by $\nabla \cdot \mathbf{v}$:

$$-\overline{p}\nabla \cdot \mathbf{v} = -p\nabla \cdot \mathbf{v} + \mu'(\nabla \cdot \mathbf{v})^2. \tag{4.22}$$

Here the term $\mu'(\nabla \cdot \mathbf{v})^2 > 0$ represents the dissipation of energy (in W m$^{-3}$) and $-p\nabla \cdot \mathbf{v}$ is the reversible contribution to the work done by the isotropic part of the stress tensor. When $\nabla \cdot \mathbf{v}$ is not very large, it is acceptable to assume $\mu' = 0$ for all fluids, even though it might not be strictly true. Then the pressure is the negative of the mean of the three normal stresses [230].

According to the **Stokes hypothesis** the assumption of $\mu' = 0$ is adopted for fluid flows with $\nabla \cdot \mathbf{v} \neq 0$ and the *shear stress tensor* becomes,

$$\boldsymbol{\tau} = \mu \left[ \nabla \mathbf{v} + (\nabla \mathbf{v})^T - \frac{2}{3}(\nabla \cdot \mathbf{v})\mathbf{I} \right] = \mu \left[ 2\mathbf{D} - \frac{2}{3}(\nabla \cdot \mathbf{v})\mathbf{I} \right]. \tag{4.23}$$

*Thermal conductivity* $\lambda$ plays a similar role in energy transport as viscosity in a momentum transport. The important difference is that, unlike viscosity, the thermal conductivity is a transport property of both fluids and solids.

Similarly as in the derivation of Eq. (4.11), we can consider a stagnant fluid or a solid between two parallel plates with area $A$, which are separated by a distance $d$. The plates are maintained at constant and uniform temperatures $T_1$ and $T_2$. When a steady-state condition is attained, a constant rate of a heat flow will exist between the plates, given as,

$$q_{1\to 2} = \lambda \frac{T_1 - T_2}{d}A = -\lambda \frac{T_2 - T_1}{d}A, \tag{4.24}$$

where $q_{1\to 2}$ is the heat flowing from plate 1 to 2 per unit time and $\lambda$ is the thermal conductivity of the substance between the plates. We can note that $q_{1\to 2}$ is positive when $T_1 > T_2$. Dividing both sides of Eq. (4.24) by area $A$ and expressing the equation in terms of a local temperature gradient $\nabla T$, the following **Fourier's law of heat conduction** is obtained,

$$\mathbf{q}'' = -\lambda \nabla T, \tag{4.25}$$

where $\mathbf{q}''$ is a heat flow rate per unit time and unit area called a **heat flux vector**.

Thermal conductivity belongs to the most important properties characterizing materials used in nuclear reactor cores. It can vary over five orders of magnitude from about $10^{-2}$ W m$^{-1}$ K$^{-1}$ for gases to about $10^3$ W m$^{-1}$ K$^{-1}$ for pure metals. Selected thermal conductivity data for the main coolant, construction materials, and fuel materials are provided in §3.

### 4.1.3 THERMODYNAMIC PROPERTIES

Materials used in nuclear reactor cores are exposed to significant heat generation and transport, and, as a result, to spatial and temporal temperature variations. The state of these materials can be described by their characteristic features called **thermodynamic properties**, which can be divided into *intensive properties*, such as pressure and temperature, and *extensive properties*, such as mass, enthalpy, thermal energy, heat capacity, and entropy. Mass, when divided by a volume, becomes an intensive property called density. All other extensive properties can be turned into intensive properties when expressed on a per-mass basis, and are then referred to as *specific properties*. Selected thermodynamic property data for the main coolant, construction materials, and fuel materials are provided in §3.

The **internal energy** is defined in thermodynamics as the sum of all microscopic energies such as the motion energy of molecules and the potential energy of particles in various bindings, such as nucleon bindings in nucleons, electron bindings in atoms, atom bindings in molecules, and bindings between molecules. The part of internal energy resulting from the relative motion of particles in a macroscopic system is called the **thermal energy**. Thus,

$$E_I = \mathscr{U} + \text{binding energies} + \cdots, \tag{4.26}$$

where $E_I$ is the internal energy and $\mathscr{U}$ is the thermal energy. When the binding energy is not changed during a certain thermodynamic process, the change of the internal energy during that process is equal to the change of the thermal energy: $\Delta E_I = \Delta \mathscr{U}$. Similarly, for specific values of the internal energy $e_I$ and the thermal energy u we have $\Delta e_I = \Delta u$. Thus, in the analysis of thermodynamic processes, where only differences between different states are of interest, we can consider the thermal energy changes to describe the process.

In the energy conservation equation the *combined energy flux* vector on an arbitrary differential surface $dS$ is,

$$\mathbf{e}'' = \mathbf{q}'' + \left(\frac{1}{2}\rho v^2 + \rho u\right)\mathbf{v} - \mathbf{T} \cdot \mathbf{v}, \tag{4.27}$$

where $\mathbf{v}$ is the velocity with which the energy is transported by convection across the area $dS$, $v = \|\mathbf{v}\|$, $\mathbf{q}''$ is the molecular heat flux vector on that area, given by Eq. (4.25), and $\mathbf{T}$ is the total stress tensor. Using Eq. (4.20) we have $-\mathbf{T} \cdot \mathbf{v} = p\mathbf{v} - \boldsymbol{\tau} \cdot \mathbf{v}$. The term $p\mathbf{v}$ can then be combined with the internal energy term $\rho u\mathbf{v}$ to give an enthalpy term $\rho u\mathbf{v} + p\mathbf{v} = \rho(u + p/\rho)\mathbf{v} = \rho i\mathbf{v}$, where $i = u + p/\rho$ is the **specific enthalpy**. Thus, the combined energy flux vector becomes,

$$\mathbf{e}'' = \mathbf{q}'' + \left(\frac{1}{2}\rho v^2 + \rho i\right)\mathbf{v} - \boldsymbol{\tau} \cdot \mathbf{v}. \tag{4.28}$$

For a surface element $dS$ of orientation $\mathbf{n}$, the quantity $(\mathbf{n} \cdot \mathbf{e}'')dS$ gives the convective energy, heat, and work passing across the surface element from its negative to its positive side.

We make use of the standard equilibrium thermodynamics formulas when expressing relationships between the internal energy, enthalpy, pressure, and temperature. For example, to evaluate the specific enthalpy in Eq. (4.28), we have,

$$\mathrm{di} = \left(\frac{\partial \mathrm{i}}{\partial T}\right)_p \mathrm{d}T + \left(\frac{\partial \mathrm{i}}{\partial p}\right)_T \mathrm{d}p = c_p \mathrm{d}T + \left[v - T\left(\frac{\partial v}{\partial T}\right)_p\right]\mathrm{d}p. \qquad (4.29)$$

For an ideal gas the term in the square brackets is zero and we have $\mathrm{di} = c_p \mathrm{d}T$. For fluids with a constant specific volume (and thus with a constant density) the term in the square bracket is $v$, which gives $\mathrm{di} = c_p \mathrm{d}T + v \mathrm{d}p$.

## 4.2   INSTANTANEOUS CONSERVATION EQUATIONS

The *instantaneous conservation equations* are describing fundamental conservation principles for mass, momentum, and energy, at any given point in time. Using the concept of continuum mechanics, the equations of fluid motion can be expressed in terms of differentiable functions well defined at any point in space. Thanks to this, computational fluid mechanics can leverage on a wide range of useful and powerful mathematical tools [8]. The most useful tools are provided in Appendix B.

Conservation principles play a fundamental role in fluid mechanics. They lead to a system of governing equations, which after solution provide the required flow properties, such as velocity, pressure or fluid temperature. The most important conserved quantities in fluid mechanics include mass, linear and angular momentum, and total energy.

There are several different approaches which are used to derive the conservation equations. In all cases, however, the concept of a control volume is used. A control volume is an arbitrary volume in space through which fluid flows. The control volume is surrounded by a control surface, which can be either real (that is a solid wall) or imaginary. The control surface may be at rest or in motion. The control volume can be infinitesimally small or it can be big enough to contain a whole component (e.g. a steam separator) or a system (e.g. primary circuit of a nuclear reactor).

The concept of the *material control volume* surrounded by a material surface is particularly useful in fluid mechanics. The material volume is an arbitrary volume in space that contains the same fluid particles. This means that the volume deforms and follows particles as they move in space. As a result there is no mass exchange between the material volume and the surrounding. However, the fluid contained in the material volume exchanges momentum and energy with surroundings through diffusion, radiation or gravity field. Using the material volume concept and applying several integral theorems it is possible to derive local conservation equations valid at any point in space containing fluid. This process is elucidated in the following subsections.

### 4.2.1 THE MASS CONSERVATION

*Conservation of mass*, $m$, contained in a time-dependent material volume, $V_m(t)$, can be written as,

$$\frac{Dm}{Dt} = 0. \tag{4.30}$$

This equation expresses the fact that in frames of classical mechanics, the mass cannot be created nor destroyed. The material derivative is used since mass $m$ is contained in a material volume, which follows distinct fluid particles. Assuming that the mass density in volume $V_m(t)$ is $\rho$, Eq. (4.30) becomes,

$$\frac{D}{Dt} \left( \iiint_{V_m(t)} \rho \, dV \right) = 0. \tag{4.31}$$

Using *Reynolds' transport theorem*, the *mass conservation equation* in the integral form can be written as,

$$\iiint_{V_m(t)} \frac{\partial \rho}{\partial t} dV + \iint_{S_m(t)} \rho \mathbf{v} \cdot \mathbf{n} dS = 0. \tag{4.32}$$

Here $S_m(t)$ is the surface of $V_m(t)$ and $\mathbf{n}$ is a unit vector, normal to surface $S_m(t)$, pointing outwards from volume $V_m(t)$. Equation (4.32) can be further transformed by applying the *divergence theorem* to the surface integral,

$$\iiint_{V_m(t)} \left[ \frac{\partial \rho}{\partial t} + \nabla \cdot (\rho \mathbf{v}) \right] dV = 0. \tag{4.33}$$

Since the above equation has to be satisfied for any material volume $V_m(t)$, the term under the integral symbol must be equal to zero. This leads to the following *mass conservation equation* in the differential form,

$$\frac{\partial \rho}{\partial t} + \nabla \cdot (\rho \mathbf{v}) = 0. \tag{4.34}$$

### 4.2.2 THE LINEAR MOMENTUM CONSERVATION

The *linear momentum conservation* for a material volume $V_m(t)$ is given by *Newton's second law of motion*,

$$\frac{D\mathbf{p}}{Dt} = \mathbf{F}, \tag{4.35}$$

where $\mathbf{p}$ is the total momentum of the fluid contained in the volume and $\mathbf{F}$ is the sum of all forces acting on the fluid in the volume.

The linear momentum of a differential fluid element with mass $dm$, volume $dV$, and velocity $\mathbf{v}$ is $d\mathbf{p} = dm\mathbf{v} = \mathbf{v}\rho dV$. The total linear momentum of fluid in volume $V_m(t)$ is thus,

$$\mathbf{p} = \iiint_{V_m(t)} \rho \mathbf{v} dV, \tag{4.36}$$

where $\rho$ is the fluid mass density. Substituting Eq. (4.36) into the right-hand side of Eq. (4.35) and applying *Reynolds' transport theorem* yields,

$$\frac{D}{Dt}\left(\iiint_{V_m(t)} \rho v dV\right) = \iiint_{V_m(t)} \frac{\partial(\rho v)}{\partial t} dV + \iint_{S_m(t)} v\rho(\mathbf{v}\cdot\mathbf{n})dS. \qquad (4.37)$$

The total force $\mathbf{F}$ on the right-hand side of Eq. (4.35) can be partitioned into two components: a force $\mathbf{F}_S$ acting on the fluid through surface $S_m(t)$ and a body force $\mathbf{F}_B$ acting on the fluid in the whole volume $V_m(t)$.

The surface force $\mathbf{F}_S$ results from the normal and tangential stresses applied to surface $S_m(t)$ and can be obtained as,

$$\mathbf{F}_S = \iint_{S_m(t)} \mathbf{n}\cdot\mathbf{T}dS, \qquad (4.38)$$

where $\mathbf{T}$ is the total stress tensor, containing both the viscous stresses and the normal stresses resulting from the pressure. The body force $\mathbf{F}_B$ can be obtained as,

$$\mathbf{F}_B = \iiint_{V_m(t)} \rho\mathbf{b}dV, \qquad (4.39)$$

where $\mathbf{b}$ is a force per unit mass. This force can result from gravity or an electromagnetic field. If only gravity is present, $\mathbf{b} = \mathbf{g}$, where $\mathbf{g}$ is a vector of the gravitational acceleration.

Substituting Eqs. (4.37)–(4.39) into (4.35) leads to the following *linear momentum conservation equation* in the integral form,

$$\iiint_{V_m(t)} \frac{\partial(\rho v)}{\partial t} dV + \iint_{S_m(t)} \rho\mathbf{v}(\mathbf{v}\cdot\mathbf{n})dS = \\ \iint_{S_m(t)} \mathbf{n}\cdot\mathbf{T}dS + \iiint_{V_m(t)} \rho\mathbf{g}dV. \qquad (4.40)$$

This equation can be turned into an integral equation over volume $V_m(t)$ by applying the divergence theorem to the surface integrals,

$$\iiint_{V_m(t)} \left[\frac{\partial(\rho v)}{\partial t} + \nabla\cdot(\rho\mathbf{v}\mathbf{v}) - \nabla\cdot\mathbf{T} - \rho\mathbf{g}\right]dV = 0. \qquad (4.41)$$

The above equation can be satisfied for an arbitrary volume $V_m(t)$ only if the integrand is equal to zero, thus, the following *linear momentum conservation equation in the differential form* is obtained,

$$\frac{\partial(\rho v)}{\partial t} + \nabla\cdot(\rho\mathbf{v}\mathbf{v}) = \nabla\cdot\mathbf{T} + \rho\mathbf{g}. \qquad (4.42)$$

As can be seen, the linear momentum conservation equation contains terms which all are vectors. In three-dimensional space the equation can be represented as three scalar equations, each expressing the linear momentum conservation principle along one spacial direction.

### 4.2.3 THE TOTAL ENERGY CONSERVATION

The principle of *total energy conservation* is using the *first law of thermodynamics*, which states that the increase of the total energy $E_T$ in the material volume is equal to the amount of heat $Q$ added to the volume, minus the amount of energy loss as a result of work $W$ done by the volume against its surroundings. Just as in conservation of mass and momentum, the energy equation is written as a time rate of change, thus,[3]

$$\frac{DE_T}{Dt} = \frac{dQ}{dt} - \frac{dW}{dt}.$$  (4.43)

Here, by a convention adopted in thermodynamics, work done by the system is assumed to be positive.

The total energy $E_T$ consists of three components: the internal energy $E_I$,

$$E_I = \iiint_{V_m(t)} \rho e_I dV,$$  (4.44)

the kinetic energy $E_K$,

$$E_K = \frac{1}{2} \iiint_{V_m(t)} \rho \mathbf{v} \cdot \mathbf{v} dV,$$  (4.45)

the potential energy $E_P$,

$$E_P = - \iiint_{V_m(t)} \rho \mathbf{b} \cdot \mathbf{r} dV,$$  (4.46)

where $e_I$ in Eq. (4.44) is the internal energy per unit mass and $\mathbf{r}$ and $\mathbf{b}$ in Eq. (4.46) is the displacement vector of a fluid particle and the total body force consisting of gravity and other fields acting on the particle, respectively. The minus sign in the above expression for the potential energy results from the fact that if $\mathbf{g} \cdot \mathbf{r} < 0$, the particle is displaced against gravity and its potential energy increases.

It is assumed that the time rate of heat $Q$ added to the volume consists of two terms: a surface term given as,

$$\frac{dQ_S}{dt} = - \iint_{S_m(t)} \mathbf{q}'' \cdot \mathbf{n} dS,$$  (4.47)

and a bulk (volume) term, given as,

$$\frac{dQ_B}{dt} = \iiint_{V_m(t)} q''' dV,$$  (4.48)

where $\mathbf{q}''$ is the *heat flux vector* and $q'''$ is the heat source per unit volume and unit time.

The net rate of work done by the material volume against the surroundings is due to the total stresses $\mathbf{T}$ on the surface, which during unit time is displaced by $\mathbf{v}$, thus,

$$\frac{dW}{dt} = - \iint_{S_m(t)} (\mathbf{T} \cdot \mathbf{v}) \cdot \mathbf{n} dS.$$  (4.49)

---

[3] We use the *inexact differential*, đ, since heat and work are path-dependent quantities.

Substituting Eqs. (4.44)–(4.49) into (4.43) yields the following *total energy conservation equation* in the integral form,

$$\frac{D}{Dt}\left[\iiint_{V_m(t)} \rho \left(e_I + \frac{1}{2}\mathbf{v}\cdot\mathbf{v} - \mathbf{g}\cdot\mathbf{r}\right) dV\right] =$$
$$-\iint_{S_m(t)} \mathbf{q}'' \cdot \mathbf{n}dS + \iiint_{V_m(t)} q''' dV + \iint_{S_m(t)} (\mathbf{T}\cdot\mathbf{v})\cdot\mathbf{n}dS. \tag{4.50}$$

Using Reynolds' transport theorem and performing some additional rearrangements, the following *total energy conservation equation* in the integral form is obtained,

$$\iiint_{V_m(t)} \frac{\partial\left[\rho\left(e_I + \frac{1}{2}\mathbf{v}\cdot\mathbf{v}\right)\right]}{\partial t} dV =$$
$$-\iint_{S_m(t)}\left[\rho\left(e_I + \frac{1}{2}\mathbf{v}\cdot\mathbf{v}\right)\mathbf{v} + \mathbf{q}'' - \mathbf{T}\cdot\mathbf{v}\right]\cdot\mathbf{n}dS. \tag{4.51}$$
$$+\iiint_{V_m(t)} \rho\left(\mathbf{b}\cdot\mathbf{v} + \frac{q'''}{\rho}\right) dV$$

Applying the divergence theorem we obtain an integral over the material volume which can be equal to zero only when the integrand is zero. This leads to the following *total energy conservation equation* in the differential form:

$$\frac{\partial\left[\rho\left(e_I + \frac{1}{2}\mathbf{v}\cdot\mathbf{v}\right)\right]}{\partial t} + \nabla\cdot\left[\rho\left(e_I + \frac{1}{2}\mathbf{v}\cdot\mathbf{v}\right)\mathbf{v}\right] =$$
$$-\nabla\cdot\mathbf{q}'' + \nabla\cdot(\mathbf{T}\cdot\mathbf{v}) + \rho(\mathbf{v}\cdot\mathbf{b}) + q'''. \tag{4.52}$$

As can be seen, the derivation of the instantaneous conservation equations of mass, linear momentum, and total energy follows a similar chain of mathematical operations. Starting from fundamental conservation principles, the equations are expressed in various forms, including the integral form over a material volume $V_m(t)$, and the partial differential form, valid at an arbitrary point in space and time. We could also derive various useful special forms of equations valid for typical situations, such as a stationary fixed control volume or a moving control volume. Finally, we could perform time and space averaging of the equations to make them suitable for numerical calculations. Since all these operations present the above-mentioned similarities, we will first introduce a generic conservation equation as a model equation for all these derivations.

## 4.3  THE GENERIC CONSERVATION EQUATION

Conservation equations of mass, linear momentum and total energy constitute the basis for fluid mechanics and heat transfer. Usually the equations are derived and analyzed separately, since they concern quantities that have quite distinct physical interpretations. However, as already mentioned in the previous section, the equations have many common aspects that can be treated in a *generic* way.

Let $\Psi$ be any *extensive property* contained in an open system with mass, $m$, and time-dependent volume, $V(t)$, bounded by a time-dependent surface, $S(t)$, shown in Fig. 4.1, for which,

$$m = \iiint_{V(t)} \rho \, dV, \tag{4.53}$$

where $\rho$ is a local mass density in the volume $V(t)$. The conservation law for $\Psi$ contained in any open system (that is a system which is exchanging mass with the surroundings) can be formulated as follows,

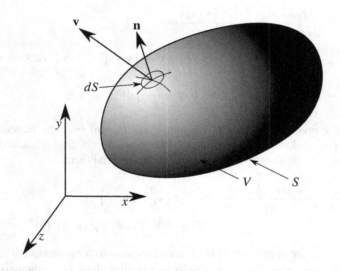

**Figure 4.1**  A macroscopic volume, $V$, bounded by a surface, $S$. The fluid flows with a velocity, $\mathbf{v}$, through a surface area element, $dS$, which has an outward normal vector, $\mathbf{n}$.

$$\frac{d\Psi}{dt} = \Gamma, \tag{4.54}$$

where $\Gamma$ represents all sources and sinks per unit time of the property $\Psi$ in the system under consideration. We do not use here the material derivative since $\Psi$ is contained in an arbitrary moving volume $V(t)$, which is not necessarily a material volume. Since $\Psi$ is an extensive property, it can be represented in an integral form as follows,

$$\Psi = \iiint_{V(t)} \rho \psi \, dV, \tag{4.55}$$

where $\psi$ is property $\Psi$ per unit mass. The source term $\Gamma$ can be partitioned into two parts,

$$\Gamma = \Gamma_v + \Gamma_s, \tag{4.56}$$

where $\Gamma_v$ results from volumetric sources distributed in volume $V(t)$ and $\Gamma_s$ results from sources distributed over surface $S(t)$. Thus,

$$\Gamma_v = \iiint_{V(t)} \rho \phi dV, \tag{4.57}$$

where $\phi$ is a source or sink of $\Psi$ per unit mass and unit time. The surface source $\Gamma_s$, in turn, can result from convection and diffusion of the property $\Psi$ through surface $S$. Thus,

$$\Gamma_s = \Gamma_{sc} + \Gamma_{sd}, \tag{4.58}$$

where the convective part is given as,

$$\Gamma_{sc} = -\iint_{S(t)} \rho \psi (\mathbf{v} \cdot \mathbf{n}) dS, \tag{4.59}$$

and the diffusive part is given as,

$$\Gamma_{sd} = -\iint_{S(t)} \mathbf{n} \cdot \mathbf{J} dS. \tag{4.60}$$

Here $\mathbf{n}$ is a unit vector normal to the surface, $S(t)$, and pointing outward from the volume, $V(t)$, $\mathbf{v}$ is the fluid velocity vector, and $\mathbf{J}$ is a diffusive flux of $\Psi$ per unit area and unit time. The minus signs in Eqs. (4.59) and (4.60) are used since the positive surface sources are assumed when the property $\Psi$ is convected into volume $V(t)$, that is in the direction of vector $-\mathbf{n}$. Using terms given by Eqs. (4.55)–(4.60), the conservation equation for the property $\Psi$ in an open system with volume $V(t)$ bounded by surface $S(t)$ can be written as follows,

$$\frac{d}{dt} \left( \iiint_{V(t)} \rho \psi dV \right) = -\iint_{S(t)} [\rho \psi (\mathbf{v} \cdot \mathbf{n}) + \mathbf{n} \cdot \mathbf{J}] dS + \iiint_{V(t)} \rho \phi dV. \tag{4.61}$$

This equation is valid for any additive (extensive) property $\Psi$, which in general can be either a scalar (for mass and total energy conservation) or a vector (for linear and angular momentum conservation). Since the surface flux term, $\mathbf{J}$ is dot-multiplied with the unit vector $\mathbf{n}$, its tensor rank must be higher by 1 compared to the tensor rank of $\Psi$.

The term on the left-hand side of Eq. (4.61) represents a time change of property $\Psi$ due to its changes inside the volume $V(t)$ and due to changes of the volume. Since the volume $V(t)$ is a function of time, the differentiation operator $d/dt$ cannot be moved under the integral symbol.

The right-hand side of Eq. (4.61) describes sources and sinks of property $\Psi$ due to convection and diffusion through the surface $S(t)$, and the volumetric sources, $\phi$.

The diffusive flux of mass is equal to zero, $\mathbf{J} = \mathbf{0}$, since net mass is not transported through a diffusive process in a single species. However, this type of process plays an important role in the transport of momentum and energy. The momentum is transported by the total stress in flowing fluids, $\mathbf{T}$. When operating on a surface,

**TABLE 4.1**

**Definitions of Terms in the Generic Conservation Equation (4.61)**

| Conserved Quantity | $\psi$ | J | $\phi$ |
|---|---|---|---|
| Mass | 1 | 0 | 0 |
| Linear momentum | $\mathbf{v}$ | $-\mathbf{T}$ | $\mathbf{b}$ |
| Angular momentum | $\mathbf{r} \times \mathbf{v}$ | $-(\mathbf{r} \times \mathbf{T})^T$ | $\mathbf{r} \times \mathbf{b}$ |
| Total energy | $e_I + \frac{1}{2}v^2$ | $\mathbf{q}'' - \mathbf{T} \cdot \mathbf{v}$ | $\mathbf{b} \cdot \mathbf{v} + \frac{1}{\rho}q'''$ |
| Entropy | $s$ | $\frac{1}{T}\mathbf{q}''$ | $\frac{1}{\rho}\Delta_s$ |

the total stress tensor produces traction $\mathbf{T} \cdot \mathbf{n}$. Thus, for the momentum equation, we have $\mathbf{J} = -\mathbf{T}$.

In a similar manner we obtain the terms $\psi$, $\mathbf{J}$, and $\phi$ for the remaining conservation equations, such as the angular momentum conservation and entropy conservation, as summarized in Table 4.1. The vector $\mathbf{r}$ in Table 4.1 describes the position of a point in the system with respect to a selected origin of the coordinate system. The form of diffusion vector, $\mathbf{J}$, for the angular momentum equation, contains the assumption that the shear stress tensor, $\boldsymbol{\tau}$, is symmetric, as for Newtonian fluids. A symbol $()^T$ denotes a transpose of a tensor. In the total energy conservation equation, $e_I$ is the internal specific energy, $\mathbf{q}''$ is the heat flux vector and $q'''$ is the volumetric heat source. In the entropy conservation equation, $s$ is the specific entropy, $T$ is the absolute temperature and $\Delta_s$ is the local entropy source per unit volume and time.

Several special cases of the generic conservation equation (4.61) can be considered. For many engineering applications, a stationary and a moving volume can be considered. For local flow predictions, the material volume and the differential formulation are convenient. All these special formulations Eq. (4.61) are presented in the following sub-sections.

### 4.3.1 A STATIONARY CONTROL VOLUME

For stationary control volume, when both the volume, $V$, and the surface, $S$, are time-independent, the integral conservation equation given by Eq. (4.61) is as follows,

$$\frac{d}{dt}\left(\iiint_V \rho\psi dV\right) = -\iint_S [\rho\psi(\mathbf{v} \cdot \mathbf{n}) + \mathbf{n} \cdot \mathbf{J}]\,dS + \iiint_V \rho\phi dV. \tag{4.62}$$

Since now $V = const$, the differentiation on the left-hand-side can be moved through the integral operator and the following *generic conservation equation* in the integral form for a stationary control volume is obtained,

$$\iiint_V \frac{\partial(\rho\psi)}{\partial t}dV = -\iint_S [\rho\psi(\mathbf{v} \cdot \mathbf{n}) + \mathbf{n} \cdot \mathbf{J}]\,dS + \iiint_V \rho\phi dV. \tag{4.63}$$

Partial differentiation is introduced in Eq. (4.63) since the product $\rho\psi$ is in general a function of time and coordinates.

## 4.3.2 A MOVING CONTROL VOLUME

When the control volume is moving, it is necessary to describe the nature of the movement, that is, to specify the velocity (for inertial systems) and the acceleration (for non-inertial systems) of the control volume in some reference inertial coordinate system.

For a control volume moving with a constant velocity along a straight line Eq. (4.61) is valid, however, it should be remembered that all velocities and time derivatives are specified relative to the control volume.

Applying *Leibniz's rule* to the left-hand-side of Eq. (4.61) yields,

$$\iiint_{V(t)} \frac{\partial (\rho \psi)}{\partial t} dV + \iint_{S(t)} \rho \psi \mathbf{v}_s \cdot \mathbf{n} dS = \\ - \iint_{S(t)} [\rho \psi (\mathbf{v} \cdot \mathbf{n}) + \mathbf{n} \cdot \mathbf{J}] dS + \iiint_{V(t)} \rho \phi dV \tag{4.64}$$

The equation can now be written as the following *generic conservation equation* in the integral form for a moving control volume,

$$\iiint_{V(t)} \frac{\partial (\rho \psi)}{\partial t} dV = - \iint_{S(t)} [\rho \psi (\mathbf{v}_r \cdot \mathbf{n}) + \mathbf{n} \cdot \mathbf{J}] dS + \iiint_{V(t)} \rho \phi dV. \tag{4.65}$$

where $\mathbf{v}_r = \mathbf{v} - \mathbf{v}_s$ is the relative velocity of fluid at the surface $S(t)$ and $\mathbf{v}_s$ is the surface velocity.

## 4.3.3 THE MATERIAL VOLUME

As already mentioned in §4.2, a volume that contains the same fluid particles in all time instants is called a **material volume**. The material volume can alter its shape, move in space, and deform. The closed surface that surrounds the material volume is called a **material surface**. Assuming that the volume, $V(t)$, and the surface, $S(t)$, in Eq. (4.61) are the material volume and the material surface, respectively, the generic conservation equation is as follows,

$$\frac{D}{Dt} \left( \iiint_{V_m(t)} \rho \psi dV \right) = - \iint_{S_m(t)} \mathbf{n} \cdot \mathbf{J} dS + \iiint_{V_m(t)} \rho \phi dV, \tag{4.66}$$

where notations $V_m(t)$ and $S_m(t)$ are used to indicate that integration is over time-dependent material volume and material surface. To indicate that the time derivative on the left-hand side of the equation concerns the material volume, the *material derivative*, $D/Dt$, is used. It should be noted that for a material surface, there is no convective component in the surface source term on the right-hand-side of Eq. (4.66), since the relative velocity between the fluid and the surface is equal to zero.

Applying Reynolds' transport theorem, Eq. (4.66) can be written as,

$$\iiint_{V_m(t)} \frac{\partial (\rho \psi)}{\partial t} dV + \iint_{S_m(t)} \rho \psi (\mathbf{v} \cdot \mathbf{n}) dS \\ = - \iint_{S_m(t)} \mathbf{n} \cdot \mathbf{J} dS + \iiint_{V_m(t)} \rho \phi dV \tag{4.67}$$

After rearrangements, the following *generic conservation equation* in the integral form for material volume is obtained,

$$
\iiint_{V_m(t)} \frac{\partial(\rho\psi)}{\partial t} dV =
$$
$$
- \iint_{S_m(t)} [\rho\psi(\mathbf{v}\cdot\mathbf{n}) + \mathbf{n}\cdot\mathbf{J}] dS + \iiint_{V_m(t)} \rho\phi dV
\tag{4.68}
$$

This integral conservation equation is useful for cases when the material boundaries of a system, which in general are part of the solution, are determined. In particular, the equation can be used to derive a local differential form of a generic conservation equation, as shown in the next subsection.

### 4.3.4 THE LOCAL DIFFERENTIAL FORMULATION

The surface integrals in Eq. (4.68) can be replaced with the volume integrals using the *Gauss integral theorem* and the following equation is obtained,

$$
\iiint_{V_m(t)} \frac{\partial(\rho\psi)}{\partial t} dV + \iiint_{V_m(t)} \nabla\cdot(\rho\psi\mathbf{v} + \mathbf{J}) dV - \iiint_{V_m(t)} \rho\phi dV = 0.
\tag{4.69}
$$

Since all integrals in Eq. (4.69) are over the same arbitrary volume, they can be combined as a single integral, which can be zero only when the expression under the integral operator is zero. Thus, the following *generic conservation equation* in the differential form is obtained,

$$
\frac{\partial(\rho\psi)}{\partial t} + \nabla\cdot(\rho\psi\mathbf{v} + \mathbf{J}) - \rho\phi = 0.
\tag{4.70}
$$

---

### BOX 4.1   DIFFERENTIAL CONSERVATION EQUATIONS

Substituting variables from Table 4.1 into Eq. (4.70), the following differential equations for conservation of mass, linear momentum, and total energy are obtained:

$$
\frac{\partial\rho}{\partial t} + \nabla\cdot(\rho\mathbf{v}) = 0,
\tag{4.71}
$$

$$
\frac{\partial(\rho\mathbf{v})}{\partial t} + \nabla\cdot(\rho\mathbf{v}\mathbf{v}) = -\nabla p + \nabla\cdot\boldsymbol{\tau} + \rho\mathbf{b},
\tag{4.72}
$$

$$
\frac{\partial(\rho e_{IK})}{\partial t} + \nabla\cdot(\rho e_{IK}\mathbf{v}) = -\nabla\cdot\mathbf{q}'' - \nabla p\cdot\mathbf{v} + \nabla\cdot\boldsymbol{\tau}\cdot\mathbf{v}
$$
$$
+ \rho\mathbf{v}\cdot\mathbf{b} + q'''
\tag{4.73}
$$

where $e_{IK} \equiv e_I + \frac{1}{2}v^2$.

## 4.4 THE TIME-AVERAGED CONSTANT-PROPERTY EQUATION

Due to the presence of turbulence, a required computational effort to resolve all flow details, including the smallest eddies, would be prohibitively high. A remedy is to express the conservation equations in terms of the time-averaged variables. This is a very useful approach, since in many applications detailed solutions of instantaneous conservation equations in time and space are not needed. Instead time or space averaged values of main flow parameters can be used. In such situations significant savings of the computational time are possible when solving averaged conservation equations.

Both time and space averaging of the conservation equations can be performed. In this section we discuss the *Reynolds averaged equations*, which are useful time-averaged conservation equations in studying flows of fluids with constant properties. For flows of fluids with variable properties, other averaging techniques should be used, as discussed in §4.5. The space averaging methods, including averaging of flows in porous media, are presented in §4.6.

### 4.4.1 REYNOLDS AVERAGING RULES

The *Reynolds averaging* is primarily used to derive time-averaged equations of motion to describe turbulent flows. According to an idea first proposed by Osborne Reynolds, the instantaneous quantities can be decomposed into their time-averaged and fluctuating components. This technique referred to as the *Reynolds decomposition*, is the basic tool required for the derivation of averaged balance equations from their instantaneous formulations.

If $f$ is any instantaneous and local flow variable, its time average at location $\mathbf{r}$ is given as:

$$\overline{f}(\mathbf{r},t) \equiv \frac{1}{\Delta t} \int_{t-\Delta t}^{t} f(\mathbf{r},\tau) \, d\tau, \qquad (4.74)$$

where the integral is taken over a sufficiently long time interval $\Delta t$ so that the average is not influenced by its length. A difference between the instantaneous and the time-average value is called a *fluctuation of the flow variable*:

$$f'(\mathbf{r},t) = f(\mathbf{r},t) - \overline{f}(\mathbf{r},t). \qquad (4.75)$$

Substituting Eq. (4.75) into Eq. (4.74) shows that a time average of the fluctuation is equal to zero. Dropping arguments of function $f$, we get:

$$\overline{f'} = \frac{1}{\Delta t} \int_{t-\Delta t}^{t} \left( f - \overline{f} \right) d\tau = \overline{f} - \overline{f} = 0. \qquad (4.76)$$

Similarly, for two independent variables $f$ and $g$, the following relationships can be derived:

$$\overline{\overline{f}} = \overline{f}, \quad \overline{f+g} = \overline{f}+\overline{g}, \quad \overline{\overline{f} \cdot g} = \overline{f} \cdot \overline{g}, \quad \overline{\int f d\xi} = \int \overline{f} d\xi, \quad \overline{\partial_\xi f} = \partial_\xi \overline{f}. \quad (4.77)$$

The time average of a product of two variables is given as:

$$\overline{f \cdot g} = \overline{(\overline{f} + f') \cdot (\overline{g} + g')} = \overline{\overline{f}\,\overline{g}} + \overline{\overline{f}g'} + \overline{\overline{g}f'} + \overline{f'g'} = \overline{f}\,\overline{g} + \overline{f'g'}. \tag{4.78}$$

This important result indicates that a time average of a product of two fluctuating quantities is equal to a sum of a product of averages $\overline{f}\,\overline{g}$ and a time average of a product of the fluctuations $\overline{f'g'}$. The latter term is a measure of correlation between fluctuations. The quantities $f'$ and $g'$ are said to be correlated if $\overline{f'g'} \neq 0$. For turbulent flows the flow variables are correlated and, as a result, the mean value of the product is different from the product of the mean values, $\overline{f \cdot g} \neq \overline{f} \cdot \overline{g}$.

### 4.4.2 THE REYNOLDS-AVERAGED EQUATION

The Reynolds decomposition approach can be applied to the generic instantaneous differential conservation equation (4.70). Performing time averaging and using the averaging rules presented in the previous section yields,

$$\frac{\partial \left[ \rho \left( \overline{\psi} + \psi' \right) \right]}{\partial t} + \nabla \cdot \left[ \rho \overline{\left( \overline{\psi} + \psi' \right) \left( \overline{\mathbf{v}} + \mathbf{v}' \right)} + \mathbf{J} \right] - \rho \overline{\phi} = 0. \tag{4.79}$$

Assuming constant density and knowing that $\overline{\psi'} = 0$ and $\overline{\mathbf{v}'} = 0$ we get the following *Reynolds averaged generic conservation equation* in the differential form:

$$\frac{\partial \overline{\psi}}{\partial t} + \nabla \cdot \left( \overline{\psi}\,\overline{\mathbf{v}} + \overline{\psi'\mathbf{v}'} + \frac{1}{\rho}\mathbf{J} \right) - \overline{\phi} = 0. \tag{4.80}$$

---

**BOX 4.2   REYNOLDS AVERAGED CONSERVATION EQUATIONS IN THE DIFFERENTIAL FORM**

Substituting variables from Table 4.1 into Eq. (4.80), the following Reynolds averaged differential equations for conservation of mass, linear momentum, and total energy are obtained:

$$\nabla \cdot \overline{\mathbf{v}} = 0, \tag{4.81}$$

$$\frac{\partial \overline{\mathbf{v}}}{\partial t} + \nabla \cdot (\overline{\mathbf{v}}\,\overline{\mathbf{v}}) = -\frac{1}{\rho}\nabla \overline{p} - \nabla \cdot \left( \overline{\mathbf{v}'\mathbf{v}'} \right) + \frac{1}{\rho}\nabla \cdot \overline{\boldsymbol{\tau}} + \overline{\mathbf{b}}, \tag{4.82}$$

$$\rho \left[ \frac{\partial \overline{e_{IK}}}{\partial t} + \nabla \cdot \left( \overline{e_{IK}}\,\overline{\mathbf{v}} \right) \right] = -\nabla \cdot \overline{\mathbf{q}''} - \rho \nabla \cdot \overline{e'_{IK}\mathbf{v}'} - \overline{\nabla p \cdot \mathbf{v}} + \nabla \cdot \overline{\boldsymbol{\tau} \cdot \mathbf{v}}$$
$$+ \rho \overline{\mathbf{v} \cdot \mathbf{b}} + \overline{q'''} \tag{4.83}$$

where $\overline{e_{IK}} \equiv \overline{e_I} + \frac{1}{2}\overline{v^2}$.

To close the system of equations, the time average of products of variables in Eqs. (4.82) and (4.83) have to be expressed in terms of the averaged variables, as discussed in §4.7.

Reynolds averaging of the generic integral conservation equation (4.63) yields the following *Reynolds averaged generic conservation equation* in the integral form:

$$\iiint_V \frac{\partial \overline{\psi}}{\partial t} dV = - \iint_S \left[ \overline{\psi} \, (\overline{\mathbf{v}} \cdot \mathbf{n}) + \overline{\psi'(\mathbf{v'} \cdot \mathbf{n})} + \frac{1}{\rho} \mathbf{n} \cdot \overline{\mathbf{J}} \right] dS + \iiint_V \overline{\phi} dV. \quad (4.84)$$

As can be seen, the Reynolds average equation contains an additional flux term $\overline{\psi' \mathbf{v'}}$ on the surface $S$. This flux term results from a non-zero correlation of $\psi'$ and $\mathbf{v'}$ on the surface $S$.

---

**BOX 4.3   REYNOLDS AVERAGE CONSERVATION EQUATIONS IN THE INTEGRAL FORM**

Substituting variables from Table 4.1 into Eq. (4.84), the following Reynolds average integral equations for conservation of mass, linear momentum, and total energy are obtained:

$$\iint_S \overline{\mathbf{v}} \cdot \mathbf{n} dS = 0, \quad (4.85)$$

$$\iiint_V \frac{\partial \overline{\mathbf{v}}}{\partial t} dV = - \iint_S \overline{\mathbf{v}}(\overline{\mathbf{v}} \cdot \mathbf{n}) dS - \iint_S \overline{\mathbf{v'}(\mathbf{v'} \cdot \mathbf{n})} dS \\ - \frac{1}{\rho} \iint_S \overline{p} \mathbf{n} dS + \frac{1}{\rho} \iint_S \mathbf{n} \cdot \overline{\tau} dS + \iiint_V \overline{\mathbf{b}} dV, \quad (4.86)$$

$$\rho \iiint_V \frac{\partial \overline{e_{IK}}}{\partial t} dV = -\rho \iint_S \overline{e_{IK}} \, \overline{\mathbf{v}} \cdot \mathbf{n} dS - \rho \iint_S \overline{e'_{IK} \mathbf{v'}} \cdot \mathbf{n} dS \\ - \iint_S \overline{\mathbf{q''}} \cdot \mathbf{n} dS - \iint_S \overline{p\mathbf{v}} \cdot \mathbf{n} dS + \iint_S \overline{\mathbf{n} \cdot \tau \cdot \mathbf{v}} dS, \quad (4.87) \\ + \rho \iiint_V \overline{\mathbf{v} \cdot \mathbf{b}} dV + \iiint_V \overline{q'''} dV$$

where $\overline{e_{IK}} \equiv \overline{e_I} + \frac{1}{2} \overline{v^2}$.

---

## 4.5   THE TIME-AVERAGED VARIABLE-PROPERTY EQUATION

Variable-property flows are very common in nuclear applications. In LWRs the water coolant entering the core has a lower temperature than at the core exit by 10 to

40 K. Even though water coolant can be treated as an incompressible fluid, its density variation due to the temperature change can be quite significant. Even more severe property variations can be experienced in SCWRs, since no boiling is present in the core, and the coolant temperature can increase several hundred kelvins once passing through the core. Similar conditions are present in other non-boiling reactors, such as liquid-metal or gas-cooled reactors. In all these applications the assumption of constant density of coolant will not be valid and the Reynolds averaging approach appears to be questionable. A better choice is then a density-weighted averaging, as discussed in the present section.

### 4.5.1   FAVRE AVERAGING RULES

When fluid density is not constant, time averaging of the product of the density and any other flow variable would give rise to an additional term, representing a time average of a product of fluctuations of the density and the variable (see Eq. (4.78)). For such conditions, an averaged mass conservation equation would contain a source term and no stream function of the mean flow could be formed. To remove this obvious annoyance, the mass-weighted time averaging is used. This averaging, known also as the **Favre averaging**, is defined as follows:

$$\widetilde{f}(\mathbf{r},t) \equiv \frac{\int_{t-\Delta t}^{t} \rho(\mathbf{r},\tau) f(\mathbf{r},\tau)\,d\tau}{\int_{t-\Delta t}^{t} \rho(\mathbf{r},\tau)\,d\tau} = \frac{\overline{\rho f}}{\overline{\rho}}. \tag{4.88}$$

We can notice that the Favre-averaged and the time-averaged variables are not equal, since combining Eqs. (4.88) and (4.78) gives:

$$\widetilde{f} = \frac{\overline{\rho f}}{\overline{\rho}} = \frac{\overline{\rho}\overline{f} + \overline{f'\rho'}}{\overline{\rho}} = \overline{f} + \frac{\overline{f'\rho'}}{\overline{\rho}}. \tag{4.89}$$

Since $\overline{f} = f - f'$, we get:

$$f = \widetilde{f} + f' - \frac{\overline{f'\rho'}}{\overline{\rho}} \equiv \widetilde{f} + f'', \tag{4.90}$$

where we introduced the **Favre fluctuation** of the variable $f$ as:

$$f'' = f' - \frac{\overline{f'\rho'}}{\overline{\rho}}. \tag{4.91}$$

Multiplying both sides of the above equation by $\rho$ and taking the time average of the resulting terms yields:

$$\overline{\rho f''} = \overline{\rho f'} - \overline{\rho}\frac{\overline{f'\rho'}}{\overline{\rho}} = 0. \tag{4.92}$$

Thus the time average of a product of a density and a Favre fluctuation is equal to zero.

## 4.5.2  FAVRE-AVERAGED EQUATIONS

Time averaging of the instantaneous generic conservation equation given by Eq. (4.70) gives,

$$\overline{\frac{\partial (\rho \psi)}{\partial t}} + \overline{\nabla \cdot (\rho \psi \mathbf{v} + \mathbf{J})} - \overline{\rho \phi} = 0. \tag{4.93}$$

Assuming a variable density and employing the averaging rules given by Eq. (4.77), the equation can be written as,

$$\frac{\partial (\overline{\rho} \widetilde{\psi})}{\partial t} + \nabla \cdot (\overline{\rho \psi \mathbf{v}}) + \nabla \cdot \overline{\mathbf{J}} - \overline{\rho} \widetilde{\phi} = 0. \tag{4.94}$$

The second term of the equation contains a time averaging of a product of three variables, $\rho$, $\psi$, and $\mathbf{v}$. It can be expressed in terms of average variables as follows,

$$\begin{aligned} \overline{\rho \psi \mathbf{v}} &= \overline{\rho \left( \widetilde{\psi} + \psi'' \right) \left( \widetilde{\mathbf{v}} + \mathbf{v}'' \right)} = \overline{\rho} \widetilde{\psi} \widetilde{\mathbf{v}} + \overline{\rho \widetilde{\psi} \mathbf{v}''} + \overline{\rho \psi'' \widetilde{\mathbf{v}}} + \overline{\rho \psi'' \mathbf{v}''} \\ &= \overline{\rho} \widetilde{\psi} \widetilde{\mathbf{v}} + \overline{\rho \psi'' \mathbf{v}''} \end{aligned}, \tag{4.95}$$

where we used a relationship given by Eq. (4.92). As can be seen, time averaging of the triple product of variables gives rise to a covariance term resulting from fluctuations of the variables. This term can be written in two equivalent forms,

$$\overline{\rho \psi'' \mathbf{v}''} = \overline{\rho} \widetilde{\psi'' \mathbf{v}''}. \tag{4.96}$$

Substituting Eq. (4.95) and Eq. (4.96) into Eq. (4.94) yields,

$$\frac{\partial (\overline{\rho} \widetilde{\psi})}{\partial t} + \nabla \cdot \left( \overline{\rho} \widetilde{\psi} \widetilde{\mathbf{v}} + \overline{\rho} \widetilde{\psi'' \mathbf{v}''} \right) + \nabla \cdot \overline{\mathbf{J}} - \overline{\rho} \widetilde{\phi} = 0. \tag{4.97}$$

---

**BOX 4.4   FAVRE AVERAGE CONSERVATION EQUATIONS IN THE DIFFERENTIAL FORM**

Substituting variables from Table 4.1 into Eq. (4.97), the following Favre average differential equations for conservation of mass, linear momentum, and total energy are obtained:

$$\frac{\partial \overline{\rho}}{\partial t} + \nabla \cdot (\overline{\rho} \widetilde{\mathbf{v}}) = 0, \tag{4.98}$$

$$\frac{\partial (\overline{\rho} \widetilde{\mathbf{v}})}{\partial t} + \nabla \cdot (\overline{\rho} \widetilde{\mathbf{v} \mathbf{v}}) = -\nabla \overline{p} + \nabla \cdot \overline{\boldsymbol{\tau}} - \nabla \cdot \overline{\rho} \widetilde{\mathbf{v}'' \mathbf{v}''} + \overline{\rho} \mathbf{b}, \tag{4.99}$$

$$\begin{aligned} \frac{\partial (\overline{\rho} \widetilde{e_{IK}})}{\partial t} + \nabla \cdot (\overline{\rho} \widetilde{e_{IK}} \widetilde{\mathbf{v}}) &= -\nabla \cdot \left( \overline{\rho} \widetilde{e_{IK}'' \mathbf{v}''} \right) - \nabla \cdot \overline{\mathbf{q}''} \\ &\quad - \overline{\nabla p \cdot \mathbf{v}} + \overline{\nabla \cdot \boldsymbol{\tau} \cdot \mathbf{v}} + \overline{\rho} \widetilde{\mathbf{v}} \cdot \mathbf{b} + \overline{q'''} \end{aligned}, \tag{4.100}$$

where $\widetilde{e_{IK}} \equiv \widetilde{e_I} + \frac{1}{2} \widetilde{v^2}$.

In a similar way, the integral conservation equations can be time averaged. Using the averaged terms derived above, the time average conservation equation for a stationary control volume given by Eq. (4.63) becomes,

$$\iiint_V \frac{\partial (\overline{\rho} \widetilde{\psi})}{\partial t} dV = - \iint_S \left[ \overline{\rho} \widetilde{\psi} (\widetilde{\mathbf{v}} \cdot \mathbf{n}) + \overline{\rho \psi'' (\mathbf{v}'' \cdot \mathbf{n})} + \mathbf{n} \cdot \overline{\mathbf{J}} \right] dS \\ + \iiint_V \overline{\rho} \widetilde{\phi} dV \tag{4.101}$$

---

**BOX 4.5   FAVRE AVERAGE CONSERVATION EQUATIONS IN THE INTEGRAL FORM**

Substituting variables from Table 4.1 into Eq. (4.101), the following Favre average integral equations for conservation of mass, linear momentum, and total energy are obtained:

$$\iiint_V \frac{\partial \overline{\rho}}{\partial t} dV = - \iint_S \overline{\rho} \widetilde{\mathbf{v}} \cdot \mathbf{n} dS, \tag{4.102}$$

$$\iiint_V \frac{\partial (\overline{\rho} \widetilde{\mathbf{v}})}{\partial t} dV = - \iint_S \overline{\rho} \widetilde{\mathbf{v}} (\widetilde{\mathbf{v}} \cdot \mathbf{n}) dS - \iint_S \overline{\rho} \widetilde{\mathbf{v}'' (\mathbf{v}'' \cdot \mathbf{n})} dS \\ - \iint_S \overline{p} \mathbf{n} dS + \iint_S \mathbf{n} \cdot \overline{\boldsymbol{\tau}} dS + \iiint_V \overline{\rho} \mathbf{b} dV \tag{4.103}$$

$$\iiint_V \frac{\partial (\overline{\rho} \widetilde{e_{IK}})}{\partial t} dV = - \iint_S \overline{\rho} \widetilde{e_{IK}} \widetilde{\mathbf{v}} \cdot \mathbf{n} dS - \iint_S \overline{\rho} \widetilde{e_{IK}'' \mathbf{v}''} \cdot \mathbf{n} dS \\ - \iint_S \overline{\mathbf{q}''} \cdot \mathbf{n} dS - \iint_S \overline{p \mathbf{v}} \cdot \mathbf{n} dS + \iint_S \overline{\mathbf{n} \cdot \boldsymbol{\tau} \cdot \mathbf{v}} dS , \tag{4.104} \\ + \iiint_V \overline{\rho} \widetilde{\mathbf{v}} \cdot \mathbf{b} dV + \iiint_V \overline{q'''} dV$$

where $\widetilde{e_{IK}} \equiv \widetilde{e_I} + \frac{1}{2} \widetilde{v^2}$.

---

## 4.6   SPACE-AVERAGED CONSERVATION EQUATIONS

The *spatial averaging* of conservation equations is used when some flow details in a complex geometry can be ignored. This approach is particularly useful to describe flows in porous media, where the exact geometry of flow paths is not known. For fluid flows within conduits, averaging over the cross-sectional area frequently yields a satisfactory one-dimensional model. This model can be utilized to predict the axial distribution of mean values for variables such as pressure or temperature.

The spatial averaging can be performed in a volume, over an area, or along a segment. The *volume averaging* leads to the simplest conservation equations, since flow details are lost in all three spatial dimensions. This type of averaging can be applied to a certain well-defined volume with several inflow and outflow ports connected to it. The volume averaging will then lead to a "point" model of the volume that describes the state of a conserved quantity by a single value. We can consider either stationary or non-stationary volumes for this type of volume-averaging.

Another type of volume averaging is applied to "smooth out" local flow variations, such as those present in porous media flows. The averaging volume is then assumed to be large enough to contain representative features of the porous body, but small enough to allow for the description of the flow in terms of local partial differential equations.

The *area averaging* is a natural choice to simplify conservation equations for flows in conduits. Such averaging leads to a transient, one-dimensional set of conservation equations. The distributions of flow parameters in a cross-section are then lost and have to be re-introduced as a closure relationship. One of the possible solutions is to use several parallel one-dimensional channels that exchange mass, momentum, and energy in the lateral directions. This technique is referred to as the *subchannel analysis* approach and is often used to find the lateral distribution of coolant enthalpy within a rod bundle.

The *segment averaging* reduces a three-dimensional flow to a two-dimensional flow, since distributions of flow parameters in one direction, along the segment, are replaced with single values. The equations can be used to describe the propagation of a thin liquid film on a surface, where the distribution of flow parameters is either known, can be assumed, or can be ignored. The various modes of spatial averaging are discussed in this section.

## 4.6.1 VOLUME AVERAGING METHODS

The primary purpose of the time averaging presented in the previous section is smoothing out of turbulent fluctuations. As a consequence, the conservation equations can be expressed in terms of time-averaged values. However, as a side effect, a covariance term $\overline{\rho \, \widetilde{\psi'' \mathbf{v}''}}$ appears, and an additional relationship to determine this term is required. We will return to this topic in section §4.7 that is dealing with the closure relationships for the conservation equations. In this section we discuss methods of averaging of conservation equations in space.

A volume averaging is useful when flow details in a certain space can be ignored. In such situations each flow variable can be represented by a single value, averaged in the space around an arbitrary point. For example, in a point-dynamic model of a nuclear reactor, the averaging is performed over the whole core volume, and the core state is represented by single values of the average coolant and fuel temperature. In a *subchannel analysis* a more detailed distribution of flow variables in the core can be obtained by averaging conservation equations in spaces between neighboring fuel rods.

For any volume $V(\mathbf{r}_C)$ with a centroid located at $\mathbf{r}_C$, the volume average value of a function $f(\mathbf{r},t)$ is defined as follows,

$$\langle f(\mathbf{r},t)\rangle_3(\mathbf{r}_C,t) \equiv \frac{1}{V(\mathbf{r}_C)} \iiint_{V(\mathbf{r}_C)} f(\mathbf{r}_C+\mathbf{s},t)\,dV(\mathbf{s}), \qquad (4.105)$$

where $\mathbf{s}$ is a relative position vector of the integration differential volume $dV(\mathbf{s})$. The symbol $\langle\rangle_3$ is used to indicated the averaging of function $f(\mathbf{r},t)$ in a three-dimensional space. As a result, a volume-averaged function of $\mathbf{r}_C$ and $t$ is obtained. In the continuation, we skip independent variables once writing volume-averaged equations and, for brevity, the volume average over $V$ is simply represented as:

$$\langle f\rangle_3 \equiv \frac{1}{V} \iiint_V f\,dV. \qquad (4.106)$$

In a similar manner, an area average over a surface area, $S$, is defined as:

$$\langle f\rangle_2 \equiv \frac{1}{S} \iint_S f\,dS, \qquad (4.107)$$

and the definition of a line average along a straight or curved line segment, $L$, is

$$\langle f\rangle_1 \equiv \frac{1}{L} \int_L f\,dL. \qquad (4.108)$$

We could apply the various averaging operators into Eq. (4.101), but this would lead to different representations of the function $f$. Instead we can apply spatial decomposition of the function as follows,

$$f = \langle f\rangle_3 + f^{\star 3}, \qquad (4.109)$$

where $f^{\star 3}$ is a **spatial deviation** of $f$. Substituting Eq. (4.109) into Eq. (4.106) yields:

$$\langle f\rangle_3 = \frac{1}{V} \iiint_V \left(\langle f\rangle_3 + f^{\star 3}\right)dV = \langle f\rangle_3 + \langle f^{\star 3}\rangle_3. \qquad (4.110)$$

Thus, as expected,

$$\langle f^{\star 3}\rangle_3 = 0. \qquad (4.111)$$

This result is similar to the results given by Eqs. (4.76) and (4.92) for the time and Favre averaging, respectively. For flows with significant variation of the density, it is convenient to introduce a **density-weighted spatial average** variable as follows,

$$\langle f\rangle_{3\rho} \equiv \frac{\iiint_V \rho f\,dV}{\iiint_V \rho\,dV} = \frac{\langle \rho f\rangle_3}{\langle \rho\rangle_3}, \qquad (4.112)$$

where now the volume location is properly determined by its *center of mass* position vector, defined as,

$$\mathbf{r}_G \equiv \frac{\iiint_V \rho\mathbf{r}\,dV}{\iiint_V \rho\,dV} = \frac{\langle \rho\mathbf{r}\rangle_3}{\langle \rho\rangle_3}, \qquad (4.113)$$

where $\mathbf{r} = \mathbf{r}_G + \mathbf{s}$, and $\mathbf{s}$ is the relative position vector of the differential volume of integration $dV(\mathbf{s})$.

In analogy to Eq. (4.109), the following decomposition of function $f$ against the density-weighted spatial average value can be done,

$$f = \langle f \rangle_{3\rho} + f^{\star 3\rho}, \tag{4.114}$$

with the density-weighted deviation satisfying the following condition,

$$\langle \rho f^{\star 3\rho} \rangle_3 = 0. \tag{4.115}$$

In general, we can write for $n$–dimensional space the following relationships,

$$f = \langle f \rangle_n + f^{\star n}, \quad \langle f^{\star n} \rangle_n = 0, \tag{4.116}$$

and

$$f = \langle f \rangle_{n\rho} + f^{\star n\rho}, \quad \langle \rho f^{\star n\rho} \rangle_n = 0. \tag{4.117}$$

These relationships are similar to the expressions given by Eqs. (4.76) and (4.92), derived for the time-averaged variables, and indicate that the space-averaged variable deviations are equal to zero.

The volume averaging of conservation equations can be performed in different ways, depending on the purpose for which the averaging is performed. As an example, let us consider coolant flow through a spacer grid in a fuel rod assembly. A typical spacer grid has dimensions comparable with the rod assembly cross-section, say of the order of magnitude of $10^{-1}$ m, but its geometry features, like grid wall thickness, can be less than $10^{-3}$ m. A rigorous analysis of flow around such structures requires a detailed description of the spacer grid geometry, with a computational mesh containing millions of cells, and a solution of conservation equations in the part of the volume of the spacer grid filled with fluid. Such analyses are feasible for a single fuel assembly, where the estimated number of computational cells would be about $10^8$, but in many situations, they will provide too complex and unnecessary information. A full resolution of flow within grid spacers for the whole reactor core would be prohibitively expensive.

Rather than trying to solve the flow equations for the spacer region filled with coolant in terms of local differential equations with boundary conditions on the solid walls, we can use the geometry details of the spacer grid to derive local volume-averaged equations that are valid everywhere in the spacer volume, including the part filled with solids. In that way spatially smoothed equations can be derived that can be solved for the whole assembly, using traditional approaches derived from the continuum mechanics principles. During the smoothing process, we separated length scales that are characteristic of the spacer grid from the length scales of the fuel rod assembly. The obtained conservation equation for the larger scales includes then the effect of boundary conditions on the smaller scales through proper closure relationships.

In addition to the space smoothing described above, spatial averaging can be used to derive conservation equations that are applicable to certain well-defined regions.

For example, in system codes used for thermal-hydraulic analyses of whole nuclear power plants, it is enough to determine average flow variables in cross sections of pipes or vessels. Such simplification can be obtained by area averaging of the generic three-dimensional conservation equations. This type of space averaging is used in §5.2 to derive a one-dimensional approximation of flow in a channel.

## 4.6.2 VOLUME-AVERAGED CONSERVATION EQUATIONS

The instantaneous conservation equation given by Eq. (4.61) is valid for an arbitrary volume, $V(t)$, bounded with a surface, $S(t)$, and for fluids with constant or variable properties. A derivation of a more specific conservation equation can be performed using additional simplifications. In this section we consider several special cases, including averaging of the instantaneous conservation equation in a stationary volume, $V$, bounded with a stationary surface, $S$, and in a time-dependent $V(t)$ and $S(t)$. The former leads to a set of equations valid for rigid flow systems, whereas the latter gives the equations for flow in elastic, time-dependent systems, such as a waterhammer flow in an elastic conduit.

A considerable importance has a **composite averaging** of the conservation equation, in which time and space averaging is employed. The composite averaging can be obtained in two different ways. In the first approach the time-averaged local variables are subsequently averaged over a volume. In the second approach, time averaging is applied to volume-averaged variables. It can be shown that both these approaches are equivalent and lead to the same results.

### Averaging of Instantaneous Conservation Equations in Stationary Volumes

The instantaneous conservation equation given by Eq. (4.63) is valid for an arbitrary volume, $V$, bounded with a surface, $S$. We can express the equation in terms of volume-averaged parameters as follows. First we divide both sides of the conservation equation with the time-independent averaging volume $V$,

$$\frac{1}{V} \iiint_V \frac{\partial (\rho \psi)}{\partial t} \, dV = -\frac{1}{V} \iint_S [\rho \psi (\mathbf{v} \cdot \mathbf{n}) + \mathbf{n} \cdot \mathbf{J}] \, dS + \frac{1}{V} \iiint_V \rho \phi \, dV. \quad (4.118)$$

The term on the left-hand side can be transformed in the following way,

$$\frac{1}{V} \iiint_V \frac{\partial (\rho \psi)}{\partial t} \, dV = \frac{\partial}{\partial t} \left( \frac{1}{V} \iiint_V \rho \psi \, dV \right) = \frac{\partial \langle \rho \psi \rangle_3}{\partial t}. \quad (4.119)$$

A partial differentiation of the term $\langle \rho \psi \rangle_3$ is necessary since the term depends both on the time and the location of the averaging volume. The first term on the right-hand side can be written as,

$$\frac{1}{V} \iint_S [\rho \psi (\mathbf{v} \cdot \mathbf{n}) + \mathbf{n} \cdot \mathbf{J}] \, dS = \frac{S}{V} \langle \rho \psi (\mathbf{v} \cdot \mathbf{n}) + \mathbf{n} \cdot \mathbf{J} \rangle_2, \quad (4.120)$$

where now averaging over a surface $S$ is used. Combining Eqs. (4.118)–(4.120) and expressing the second term on the right-hand side of Eq. (4.118) in terms of the spacial average value, the volume average, instantaneous local conservation equation becomes,

$$\frac{\partial \langle \rho \psi \rangle_3}{\partial t} = -\frac{S}{V} \langle \rho \psi (\mathbf{v} \cdot \mathbf{n}) + \mathbf{n} \cdot \mathbf{J} \rangle_2 + \langle \rho \phi \rangle_3. \tag{4.121}$$

For cases with a significant variation of a density, the density-weighted spatial averaging should be used. According to Eq. (4.112) we have:

$$\langle \rho \psi \rangle_3 = \langle \rho \rangle_3 \langle \psi \rangle_{3\rho}, \tag{4.122}$$

and,

$$\langle \rho \phi \rangle_3 = \langle \rho \rangle_3 \langle \phi \rangle_{3\rho}. \tag{4.123}$$

The triple product average $\langle \rho \psi \mathbf{v} \rangle_2$ can be decomposed as follows,

$$\begin{aligned} \langle \rho \psi \mathbf{v} \rangle_2 &= \langle \rho \left( \langle \psi \rangle_{2\rho} + \psi^{\star 2\rho} \right) \left( \langle \mathbf{v} \rangle_{2\rho} + \mathbf{v}^{\star 2\rho} \right) \rangle_2 \\ &= \langle \rho \langle \psi \rangle_{2\rho} \langle \mathbf{v} \rangle_{2\rho} \rangle_2 + \langle \rho \psi^{\star 2\rho} \langle \mathbf{v} \rangle_{2\rho} \rangle_2 \\ &\quad + \langle \rho \langle \psi \rangle_{2\rho} \mathbf{v}^{\star 2\rho} \rangle_2 + \langle \rho \psi^{\star 2\rho} \mathbf{v}^{\star 2\rho} \rangle_2 \\ &= \langle \rho \rangle_2 \langle \psi \rangle_{2\rho} \langle \mathbf{v} \rangle_{2\rho} + \langle \rho \rangle_2 \langle \psi^{\star 2\rho} \mathbf{v}^{\star 2\rho} \rangle_{2\rho} \end{aligned} \tag{4.124}$$

Substituting Eqs. (4.122)–(4.124) into Eq. (4.121) yields the following *volume-averaged conservation equation*,

$$\begin{aligned} \frac{\partial \left( \langle \rho \rangle_3 \langle \psi \rangle_{3\rho} \right)}{\partial t} &= -\frac{S}{V} \left( \langle \rho \rangle_2 \langle \psi \rangle_{2\rho} \langle \mathbf{v} \cdot \mathbf{n} \rangle_{2\rho} + \langle \rho \rangle_2 \langle \psi^{\star 2\rho} (\mathbf{v}^{\star 2\rho} \cdot \mathbf{n}) \rangle_{2\rho} \right) \\ &\quad - \frac{S}{V} \langle \mathbf{n} \cdot \mathbf{J} \rangle_2 + \langle \rho \rangle_3 \langle \phi \rangle_{3\rho} \end{aligned} \tag{4.125}$$

It should be noted that this equation is a non-local formulation of the transport equation since the volume average variables $\langle \rangle_3$ and $\langle \rangle_{3\rho}$ are defined at the center of mass of the volume $V$, whereas area average variables $\langle \rangle_2$ and $\langle \rangle_{2\rho}$ are defined at the center of mass of the surface $S$. The analysis of non-local phenomena is very complex and beyond the scope of this text. As a remedy, a Taylor series expansion can be used to express the surface average in terms of the volume average. More details on this topic can be found in studies performed by Cushman [49], Koch and Brady [132], Quintard and Whitaker [180, 181], and many others.

### Averaging of Instantaneous Conservation Equations in Non-Stationary Volumes

The instantaneous conservation equation given by Eq. (4.61) is valid for an arbitrary volume $V(t)$ bounded with a surface $S(t)$. For a non-stationary volume $V(t)$, the integral on the left-hand side of Eq. (4.61) can be written as,

$$\iiint_{V(t)} \rho \psi dV = \langle \rho \rangle_3 \langle \psi \rangle_{3\rho} V(t). \tag{4.126}$$

Similarly, for a non-stationary bounding surface $S(t)$, the first term on the right-hand side of Eq. (4.61) can be expressed in terms of area-averaged variables as,

$$\iint_{S(t)} [\rho\psi(\mathbf{v}\cdot\mathbf{n}) + \mathbf{n}\cdot\mathbf{J}]\,\mathrm{d}S = \langle\rho\rangle_2\langle\psi\rangle_{2\rho}\langle\mathbf{v}\cdot\mathbf{n}\rangle_{2\rho}S(t)$$
$$+ \langle\rho\rangle_2\langle\psi^{\star 2\rho}(\mathbf{v}^{\star 2\rho}\cdot\mathbf{n})\rangle_{2\rho}S(t) + \langle\mathbf{n}\cdot\mathbf{J}\rangle_2 S(t) \qquad (4.127)$$

Finally, the second term on the right-hand side of Eq. (4.61) is given as,

$$\iiint_{V(t)} \rho\phi\,\mathrm{d}V = \langle\rho\rangle_3\langle\phi\rangle_{3\rho}V(t). \qquad (4.128)$$

Thus, the instantaneous conservation equation can be expressed in terms of volume- and area-averaged values as follows,

$$\frac{\partial}{\partial t}\left[\langle\rho\rangle_3\langle\psi\rangle_{3\rho}V(t)\right] + \langle\rho\rangle_2\langle\psi\rangle_{2\rho}\langle\mathbf{v}\cdot\mathbf{n}\rangle_{2\rho}S(t)$$
$$+ \langle\rho\rangle_2\langle\psi^{\star 2\rho}(\mathbf{v}^{\star 2\rho}\cdot\mathbf{n})\rangle_{2\rho}S(t) + \langle\mathbf{n}\cdot\mathbf{J}\rangle_2 S(t) \qquad (4.129)$$
$$= \langle\rho\rangle_3\langle\phi\rangle_{3\rho}V(t)$$

A partial differential is used in the first term on the left-hand side of the equation since the term in the square parentheses can, in general, be a function of space and time.

## Composite-Averaged Conservation Equations in Non-Stationary Volumes

A *composite averaging* can be used if a time-averaged local variable has to be further averaged in space. This particular situation can take place when a three-dimensional time-averaged system of conservation equations should be replaced at some locations with a system of reduced order in space, such as one- or two-dimensional system of equations. For example, when pipes carrying various fluids are connected to a vessel, it can be desirable to use a space-averaged approximation for flow in the pipes, whereas a three-dimensional, time-averaged approach would be required to capture the mixing phenomena in the vessel.

Composite averaging can be, in principle, performed in two ways determined by the order in which the variables are averaged. Here we first time average the instantaneous equations, and in the next step, the equations are space averaged. This order of averaging has a much higher practical importance than the reversed order, since the time-averaged variables can easily be obtained from numerical predictions or experiments. The reversed order of averaging would require a knowledge of an instantaneous distribution of variables in space, which would require a significant experimental or numerical effort.

To obtain a composite-averaged conservation equation we first time average the instantaneous form of the equation, valid in an arbitrary time-dependent volume. Assuming variable-property fluid and using the Favre averaging in Eq. (4.61), we get,

$$\frac{\partial}{\partial t}\left(\iiint_{V(\mathbf{r},t)}\overline{\rho}\widetilde{\psi}dV\right) = -\iint_{S(\mathbf{r},t)}\left[\overline{\rho}\widetilde{\psi}(\widetilde{\mathbf{v}}\cdot\mathbf{n})+\overline{\rho}\widetilde{\psi''(\mathbf{v''}\cdot\mathbf{n})}+\mathbf{n}\cdot\overline{\mathbf{J}}\right]dS$$
$$+\iiint_{V(\mathbf{r},t)}\overline{\rho}\widetilde{\phi}dV \qquad (4.130)$$

A partial derivative on the left-hand side of the equation is used since we assumed that the integration volume and its bounding surface are functions of space and time. In the next step, we introduce the volume and surface averaged variables to the equation. By doing this, the volume integrals in the equation become,

$$\iiint_{V(\mathbf{r},t)}\overline{\rho}\widetilde{\psi}dV = \langle\overline{\rho}\rangle_3\langle\widetilde{\psi}\rangle_{3\rho}V(\mathbf{r},t), \qquad (4.131)$$

$$\iiint_{V(\mathbf{r},t)}\overline{\rho}\widetilde{\phi}dV = \langle\overline{\rho}\rangle_3\langle\widetilde{\phi}\rangle_{3\rho}V(\mathbf{r},t), \qquad (4.132)$$

and the surface integral is as follows,

$$\iint_{S(\mathbf{r},t)}\left[\overline{\rho}\widetilde{\psi}(\widetilde{\mathbf{v}}\cdot\mathbf{n})+\overline{\rho}\widetilde{\psi''(\mathbf{v''}\cdot\mathbf{n})}+\mathbf{n}\cdot\overline{\mathbf{J}}\right]dS =$$
$$\langle\overline{\rho}\rangle_2\langle\widetilde{\psi}\rangle_{2\rho}\langle\widetilde{\mathbf{v}}\cdot\mathbf{n}\rangle_{2\rho}S(\mathbf{r},t)+\langle\overline{\rho}\rangle_2\langle\widetilde{\psi}^{\star2\rho}(\widetilde{\mathbf{v}}^{\star2\rho}\cdot\mathbf{n})\rangle_{2\rho}S(\mathbf{r},t)\cdot \qquad (4.133)$$
$$+\langle\overline{\rho}\widetilde{\psi''(\mathbf{v''}\cdot\mathbf{n})}\rangle_2 S(\mathbf{r},t)+\langle\mathbf{n}\cdot\overline{\mathbf{J}}\rangle_2 S(\mathbf{r},t)$$

Substituting Eqs. (4.131)–(4.133) into Eq. (4.130) yields the following *composite-averaged conservation equation*,

$$\frac{\partial}{\partial t}\left[\langle\overline{\rho}\rangle_3\langle\widetilde{\psi}\rangle_{3\rho}V(\mathbf{r},t)\right]+\langle\overline{\rho}\rangle_2\langle\widetilde{\psi}\rangle_{2\rho}\langle\widetilde{\mathbf{v}}\cdot\mathbf{n}\rangle_{2\rho}S(\mathbf{r},t)$$
$$+\langle\overline{\rho}\rangle_2\langle\widetilde{\psi}^{\star2\rho}(\widetilde{\mathbf{v}}^{\star2\rho}\cdot\mathbf{n})\rangle_{2\rho}S(\mathbf{r},t)$$
$$+\langle\overline{\rho}\widetilde{\psi''(\mathbf{v''}\cdot\mathbf{n})}\rangle_2 S(\mathbf{r},t)+\langle\mathbf{n}\cdot\overline{\mathbf{J}}\rangle_2 S(\mathbf{r},t) \qquad (4.134)$$
$$=\langle\overline{\rho}\rangle_3\langle\widetilde{\phi}\rangle_{3\rho}V(\mathbf{r},t)$$

The composite-averaged conservation equation contains two different values of the conserved property: $\langle\widetilde{\psi}\rangle_{3\rho}$ and $\langle\widetilde{\psi}\rangle_{2\rho}$, and thus is a non-local formulation of the transport equation. The analysis of non-local phenomena requires special attention, as discussed in case of Eq. (4.125). In comparison to Eqs. (4.125) and (4.129), the present equation contains one additional term on the right-hand side: $\langle\overline{\rho}\widetilde{\psi''(\mathbf{v''}\cdot\mathbf{n})}\rangle_2 S(\mathbf{r},t)$. This term represents a flux of the conserved quantity $\psi$ due to its fluctuation and a fluctuation of the normal velocity at surface $S(\mathbf{r},t)$.

### 4.6.3 LOCAL SPACE-AVERAGED CONSERVATION EQUATIONS

A local variability of flow parameters around any location $\mathbf{r}$ can be smoothed out by considering space averaging in an infinitesimally small volume with the center of mass, given by Eq. (4.113), located at that point. Assuming further that the averaging volume is a parallelpiped $\Delta x \Delta y \Delta z$, the first term on the right side of the local instantaneous conservation equation given by Eq. (4.125) becomes,

$$
\frac{S}{V} \left( \langle \rho \rangle_2 \langle \psi \rangle_{2\rho} \langle \mathbf{v} \cdot \mathbf{n} \rangle_{2\rho} \right) =
$$

$$
\frac{\Delta y \Delta z}{\Delta x \Delta y \Delta z} \left( \langle \rho \rangle_2 \langle \psi \rangle_{2\rho} \langle u \rangle_{2\rho} \mid_{x+\Delta x/2} - \langle \rho \rangle_2 \langle \psi \rangle_{2\rho} \langle u \rangle_{2\rho} \mid_{x-\Delta x/2} \right) +
$$

$$
\frac{\Delta x \Delta z}{\Delta x \Delta y \Delta z} \left( \langle \rho \rangle_2 \langle \psi \rangle_{2\rho} \langle v \rangle_{2\rho} \mid_{y+\Delta y/2} - \langle \rho \rangle_2 \langle \psi \rangle_{2\rho} \langle v \rangle_{2\rho} \mid_{y-\Delta y/2} \right) +
$$

$$
\frac{\Delta x \Delta y}{\Delta x \Delta y \Delta z} \left( \langle \rho \rangle_2 \langle \psi \rangle_{2\rho} \langle w \rangle_{2\rho} \mid_{z+\Delta z/2} - \langle \rho \rangle_2 \langle \psi \rangle_{2\rho} \langle w \rangle_{2\rho} \mid_{z-\Delta z/2} \right) = \qquad (4.135)
$$

$$
\frac{\Delta \left( \langle \rho \rangle_2 \langle \psi \rangle_{2\rho} \langle u \rangle_{2\rho} \right)}{\Delta x} + \frac{\Delta \left( \langle \rho \rangle_2 \langle \psi \rangle_{2\rho} \langle v \rangle_{2\rho} \right)}{\Delta y} + \frac{\Delta \left( \langle \rho \rangle_2 \langle \psi \rangle_{2\rho} \langle w \rangle_{2\rho} \right)}{\Delta z} \rightarrow
$$

$$
\nabla \cdot \left( \langle \rho \rangle_3 \langle \psi \rangle_{3\rho} \langle \mathbf{v} \rangle_{3\rho} \right)
$$

An arrow indicates that the limiting value of the term for an infinitesimally small volume is equal to the divergence of the average convective flux of the quantity $\psi$. Here we assumed that the area and volume average quantities are converging to the same values. Performing a similar derivation for the remaining flux terms, Eq. (4.125) becomes,

$$
\frac{\partial \left( \langle \rho \rangle_3 \langle \psi \rangle_{3\rho} \right)}{\partial t} + \nabla \cdot \left( \langle \rho \rangle_3 \langle \psi \rangle_{3\rho} \langle \mathbf{v} \rangle_{3\rho} + \langle \rho \rangle_3 \langle \psi^{\star 3\rho} \mathbf{v}^{\star 3\rho} \rangle_{3\rho} \right) + \nabla \cdot \langle \mathbf{J} \rangle_3 - \langle \rho \rangle_3 \langle \phi \rangle_{3\rho} = 0 \qquad (4.136)
$$

We can notice that the derived space-averaged conservation equation (4.136) is equivalent to the time-averaged differential conservation equation given by Eq. (4.97). The difference is that the time averaging used in Eq. (4.97) is replaced with volume averaging in Eq. (4.136). As a result, in both cases the small-scale flow features are separated from the average-scale flow parameters.

### 4.6.4 LOCAL SPACE-AVERAGED CONSERVATION EQUATIONS IN POROUS MEDIA

In the derivation of Eq. (4.136), we assumed that the averaging volume contained fluid only. We will now extend the equation for flow in a rigid, impermeable porous medium. To distinguish the part of the averaging volume that is available for fluid, we introduce the following indicator function,

$$
F(\mathbf{r}) = \begin{cases} 1 & \text{when} \qquad \mathbf{r} \in \text{fluid} \\ 0 & \text{otherwise} \end{cases} . \qquad (4.137)
$$

Substituting this function to Eq. (4.105) we get:

$$
\langle F \rangle_3 (\mathbf{r}_C) \equiv \frac{1}{V(\mathbf{r}_C)} \iiint_{V(\mathbf{r}_C)} F(\mathbf{r}_C + \mathbf{s}) \, dV(\mathbf{s}) = \frac{V_f(\mathbf{r}_C)}{V(\mathbf{r}_C)} = \varepsilon_3 (\mathbf{r}_C), \qquad (4.138)
$$

where $V_f(\mathbf{r}_C)$ is the volume occupied by a fluid contained in the averaging volume $V(\mathbf{r}_C)$ with a centroid located at $\mathbf{r}_C$, and $\varepsilon_3(\mathbf{r}_C)$ is the *volume porosity*, or just *porosity*, defined around that centroid. We can now introduce the $F$-weighted volume average of a quantity $\psi$ as follows,

$$\langle F\psi\rangle_3(\mathbf{r}_C) \equiv \frac{1}{V(\mathbf{r}_C)} \iiint_{V(\mathbf{r}_C)} \psi F(\mathbf{r}_C + \mathbf{s}) \, dV(\mathbf{s}) = \langle\psi\rangle_{3S}(\mathbf{r}_C). \tag{4.139}$$

Here notation $\langle\psi\rangle_{3S}$ is used to indicate the **superficial average** quantity $\psi$ in a three-dimensional space. This should be distinguished from the **intrinsic average** of quantity $\psi$, which, for $n$-dimensional space, and dropping the notation for the centroid location vector, is defined as,

$$\langle\psi\rangle_{nI} \equiv \frac{\langle F\psi\rangle_n}{\langle F\rangle_n} = \frac{\langle\psi\rangle_{nS}}{\varepsilon_n}, \tag{4.140}$$

where $\varepsilon_n \equiv \langle F\rangle_n$ is a porosity in $n$-dimensional space. Replacing $\psi$ with the density $\rho$ in Eqs. (4.139) and (4.140) yields,

$$\langle F\rho\rangle_n = \langle\rho\rangle_{nS} = \varepsilon_n\langle\rho\rangle_{nI}. \tag{4.141}$$

Here $\langle\rho\rangle_{nS}$ is the superficial space-averaged density and $\langle\rho\rangle_{nI}$ is the intrinsic space-averaged density. The corresponding mass-weighted intrinsic average of variable $\psi$ is defined as,

$$\langle\psi\rangle_{nI\rho} \equiv \frac{\langle F\rho\psi\rangle_n}{\langle F\rho\rangle_n} = \frac{\varepsilon_n\langle\rho\psi\rangle_{nI}}{\varepsilon_n\langle\rho\rangle_{nI}} = \frac{\langle\rho\psi\rangle_{nI}}{\langle\rho\rangle_{nI}}. \tag{4.142}$$

Thus,

$$\langle\rho\psi\rangle_{nI} = \langle\rho\rangle_{nI}\langle\psi\rangle_{nI\rho}, \tag{4.143}$$

which means that an intrinsic average of a product of a density and variable $\psi$ is equal to a product of the intrinsic average density and the density-weighted, intrinsic average of variable $\psi$.

An intrinsic average of a product of two variables $f$ and $g$ can be expressed in terms of a product of averages using the following decomposition of the variables,

$$f = \langle f\rangle_{nI\rho} + f^{\star nI\rho}, \quad g = \langle g\rangle_{nI\rho} + g^{\star nI\rho}. \tag{4.144}$$

The above expressions indicate that instantaneous variables $f$ and $g$ are decomposed into space-averaged values and corresponding spatial deviation terms. Multiplying both sides of equations by a density and applying the intrinsic averaging, in view of Eq. (4.143) we obtain,

$$\langle\rho f^{\star nI\rho}\rangle_{nI} = \langle\rho\rangle_{nI}\langle f^{\star nI\rho}\rangle_{nI\rho} = 0, \quad \langle\rho g^{\star nI\rho}\rangle_{nI} = \langle\rho\rangle_{nI}\langle g^{\star nI\rho}\rangle_{nI\rho} = 0. \tag{4.145}$$

Thus, an intrinsic averaging of a product of a density and variables $f$ and $g$ yields,

$$\langle\rho fg\rangle_{nI\rho} = \langle\rho\rangle_{nI}\langle f\rangle_{nI\rho}\langle g\rangle_{nI\rho} + \langle\rho\rangle_{nI}\langle f^{\star nI\rho}g^{\star nI\rho}\rangle_{nI\rho}, \tag{4.146}$$

where the second term on the right-hand side arises from spatial deviations of variables $f$ and $g$.

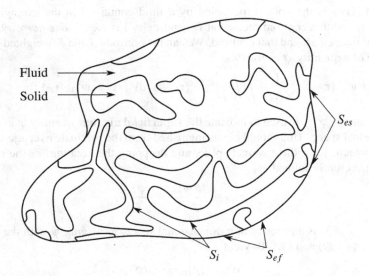

**Figure 4.2**  Porous averaging volume $V = V_f + V_s$ bounded by a surface $S_e = S_{fe} + S_{se}$, and containing internal fluid surface $S_i$.

We turn now our attention to Eq. (4.118) and perform the integration over a volume $V$ bounded by a surface $S$, containing a fluid with volume $V_f$, as shown in Fig. 4.2. The surface $S$ consists of two parts: an external part $S_e$ and an internal part $S_i$. The two surfaces have different properties. Once the external surface is permeable and coincides with the surface $S$ that is bounding the volume $V$, the internal surface is impermeable and separates all the fluid from the solid in the averaging volume $V$. In addition, as shown in Fig. 4.2, the external surface consists of a fluid part $S_{ef}$, and a solid part $S_{es}$.

The convective term of the surface integral in Eq. (4.118) is as follows,

$$
\begin{aligned}
\frac{1}{V} \iint_S \rho\,\psi(\mathbf{v}\cdot\mathbf{n})\mathrm{d}S &= \frac{1}{V} \iint_{S_{ef}} \rho\,\psi(\mathbf{v}\cdot\mathbf{n})\mathrm{d}S + \frac{1}{V} \underbrace{\iint_{S_{es}} \rho\,\psi(\mathbf{v}\cdot\mathbf{n})\mathrm{d}S}_{=0} \\
&\quad + \frac{1}{V} \underbrace{\iint_{S_i} \rho\,\psi(\mathbf{v}\cdot\mathbf{n})\mathrm{d}S}_{=0} \\
&= \frac{S_e}{V}\frac{S_{ef}}{S_e}\frac{1}{S_{ef}} \iint_{S_{ef}} \rho\,\psi(\mathbf{v}\cdot\mathbf{n})\mathrm{d}S = \frac{S_e}{V}\varepsilon_2\langle\rho\,\psi v_n\rangle_{2I} \\
&= \frac{S_e}{V}\varepsilon_2\langle\rho\rangle_{2I}\langle\psi\rangle_{2I\rho}\langle v_n\rangle_{2I\rho} \\
&\quad + \frac{S_e}{V}\varepsilon_2\langle\rho\rangle_{2I}\langle\psi^{\star 2I\rho}v_n^{\star 2I\rho}\rangle_{2I\rho}
\end{aligned}
$$

$$\tag{4.147}$$

where $\varepsilon_2 \equiv S_{ef}/S_e$ is an *area porosity* of the external surface $S_e$ and $v_n = \mathbf{v} \cdot \mathbf{n}$ is a velocity normal to the integration surface $S_{ef}$, positively defined when directed outward of the volume $V_f$.

The diffusive term of the surface integral in Eq. (4.118) contains both internal and external part and is given as,

$$
\begin{aligned}
\frac{1}{V} \iint_S \mathbf{n} \cdot \mathbf{J} dS &= \frac{1}{V} \iint_{S_{ef}} \mathbf{n} \cdot \mathbf{J} dS + \frac{1}{V} \underbrace{\iint_{S_{es}} \mathbf{n} \cdot \mathbf{J} dS}_{=0} + \frac{1}{V} \iint_{S_i} \mathbf{n} \cdot \mathbf{J} dS \\
&= \frac{S_e}{V} \frac{S_{ef}}{S_e} \frac{1}{S_{ef}} \iint_{S_{ef}} \mathbf{n} \cdot \mathbf{J} dS + \frac{S_i}{V} \frac{1}{S_i} \iint_{S_i} \mathbf{n} \cdot \mathbf{J} dS \qquad (4.148) \\
&= \frac{S_e}{V} \varepsilon_2 \langle \mathbf{n} \cdot \mathbf{J} \rangle_{2Ie} + a_i''' \langle \mathbf{n} \cdot \mathbf{J} \rangle_{2Ii}
\end{aligned}
$$

Here $a_i''' \equiv S_i/V$ is an interfacial area density for a porous medium, $\langle \mathbf{n} \cdot \mathbf{J} \rangle_{2Ie}$ is the external area-averaged diffusion flux, and $\langle \mathbf{n} \cdot \mathbf{J} \rangle_{2Ii}$ is the internal area-averaged diffusion flux of quantity $\psi$. The integration over surface $S_{es}$ is not contributing to the diffusive term since it is a solid-solid interface.

An intrinsic area average of the triple product $\rho \psi v_n$ can be decomposed in a similar manner as given in Eq. (4.124):

The term on the left-hand side of Eq. (4.118) can be integrated as follows,

$$
\frac{V_f}{V} \frac{1}{V_f} \iiint_{V_f} \frac{\partial (\rho \psi)}{\partial t} dV = \varepsilon_3 \frac{\partial}{\partial t} \left( \frac{1}{V_f} \iiint_{V_f} \rho \psi dV \right) = \varepsilon_3 \frac{\partial \langle \rho \psi \rangle_{3I}}{\partial t}. \qquad (4.149)
$$

In view of Eq. (4.143), this term can be expressed as,

$$
\varepsilon_3 \frac{\partial \langle \rho \psi \rangle_{3I}}{\partial t} = \varepsilon_3 \frac{\partial \left( \langle \rho \rangle_{3I} \langle \psi \rangle_{3I\rho} \right)}{\partial t}. \qquad (4.150)
$$

A similar derivation can be applied to the last term on the right-hand side of Eq. (4.118) representing a volumetric source of property $\psi$:

$$
\frac{1}{V} \iiint_V \rho \phi dV = \frac{V_f}{V} \frac{1}{V_f} \iiint_{V_f} \rho \phi dV = \varepsilon_3 \langle \rho \phi \rangle_{3I} = \varepsilon_3 \langle \rho \rangle_{3I} \langle \phi \rangle_{3I\rho}. \qquad (4.151)
$$

Here $\langle \phi \rangle_{3I\rho}$ is a density-weighted, intrinsic volume-averaged quantity $\phi$.

Using Eqs. (4.147)–(4.151) in (4.118) yields the following space-smoothed conservation equation in a porous medium,

$$
\begin{aligned}
\varepsilon_3 \frac{\partial \left( \langle \rho \rangle_{3I} \langle \psi \rangle_{3I\rho} \right)}{\partial t} &+ \frac{S_e}{V} \varepsilon_2 \langle \rho \rangle_{2I} \langle \psi \rangle_{2I\rho} \langle v_n \rangle_{2I\rho} + \frac{S_e}{V} \varepsilon_2 \langle \mathbf{n} \cdot \mathbf{J} \rangle_{2Ie} \\
&+ \frac{S_e}{V} \varepsilon_2 \langle \rho \rangle_{2I} \langle \psi^{*2I\rho} v_n^{*2I\rho} \rangle_{2I\rho} + a_{fi}''' \langle \mathbf{n} \cdot \mathbf{J} \rangle_{2Ii} \\
&- \varepsilon_3 \langle \rho \rangle_{3I} \langle \phi \rangle_{3I\rho} = 0.
\end{aligned} \qquad (4.152)
$$

The equation is valid for a finite known volume $V$ containing a rigid non-permeable porous medium, surrounded by a known external surface $S_e$. The equation is non-local, since the intrinsic averaged variables $\langle \rangle_{3I\rho}$ and $\langle \rangle_{2I\rho}$ are defined at different locations of the fluid volume centeroid and the fluid external surface centroid, respectively. In addition, two diffusive terms are present, one representing the external surface, or fluid-fluid, contribution, and the other the internal surface, corresponding to solid-fluid interaction. Physically these terms correspond to the transport of heat or momentum through these bounding surfaces.

We can now consider a special type of a volume $V = \Delta x \Delta y \Delta z$ for which $S_e$ consists of six sides of a parallelpiped with areas $\Delta y \Delta z$, $\Delta x \Delta z$, and $\Delta x \Delta y$ in the directions of $x$, $y$, and $z$ axes. Using a derivation shown in Eq. (4.135), a local form of Eq. (4.152) is obtained,

$$
\begin{aligned}
\varepsilon_3 \frac{\partial \left( \langle \rho \rangle_{3I} \langle \psi \rangle_{3I\rho} \right)}{\partial t} &+ \nabla \cdot \left[ \boldsymbol{\varepsilon}_2 \cdot \left( \langle \rho \rangle_{3I} \langle \psi \rangle_{3I\rho} \langle \mathbf{v} \rangle_{3I\rho} \right) \right] + \nabla \cdot \left[ \boldsymbol{\varepsilon}_2 \cdot \left( \langle \mathbf{J} \rangle_{3I} \right) \right] \\
&+ \nabla \cdot \left[ \boldsymbol{\varepsilon}_2 \cdot \left( \langle \rho \rangle_{3I} \langle \psi^{\star 3I\rho} \mathbf{v}^{\star 3I\rho} \rangle_{3I\rho} \right) \right] + a_i''' \langle \mathbf{n} \cdot \mathbf{J} \rangle_{2Ii} \\
&- \varepsilon_3 \langle \rho \rangle_{3I} \langle \phi \rangle_{3I\rho} = 0
\end{aligned}
\qquad (4.153)
$$

Here we assumed that $\langle \rangle_{2I\rho} \rightarrow \langle \rangle_{3I\rho}$ when all sides of the parallelpiped go to zero, and $\boldsymbol{\varepsilon}_2$ is the *area porosity tensor* given as,

$$
\boldsymbol{\varepsilon}_2 = \begin{bmatrix} \varepsilon_{2x} & 0 & 0 \\ 0 & \varepsilon_{2y} & 0 \\ 0 & 0 & \varepsilon_{2z} \end{bmatrix}.
\qquad (4.154)
$$

The diagonal elements of the tensor represent the area porosity in the direction of each of the Cartesian coordinate.

### 4.6.5   AREA-AVERAGED CONSERVATION EQUATIONS FOR CHANNEL FLOWS

In many applications, a locally smoothed flow is predominantly one-dimensional and rectilinear. The flow simulation can be then significantly simplified by averaging the conservation equations over an area perpendicular to the main flow direction. Without any loss of generality, we can assume that the flow direction and the channel axis coincides with the positive direction of the $z$-axis, where $\mathbf{e}_z$ is a unit vector in the positive $z$-axis direction and $A(z)$ is the cross-section area of the channel at location $z$. We consider a small "slab" of the channel between axial locations $z$ and $z + \Delta z$ with volume $V = \Delta z \left( A(z) + A(z + \Delta z) \right) / 2 \approx A(z) \Delta z$. The outer surface of the averaging volume can be divided into the following three parts: (1)–surface with area $A(z)$ through which fluid is flowing with an average velocity parallel to the $z$-axis and for which the outward normal vector is $\mathbf{n} = -\mathbf{e}_z$; (2)–surface with area $A(z + \Delta z) = A(z) + \Delta A$ through which fluid is flowing with an average velocity parallel to the $z$-axis and for which the outward normal vector is $\mathbf{n} = \mathbf{e}_z$; (3)–surface of a stationary solid wall with area $S_w = P_w \Delta z$ bounding the channel slab, where $P_w$ is the solid wall perimeter. We will use the Favre-averaged generic conservation equation in the integral form, as given by Eq. (4.101). The volume integral on the left-hand side of the equation can be transformed as,

$$\iiint_V \frac{\partial (\overline{\rho}\,\widetilde{\psi})}{\partial t} dV = \int_z^{z+\Delta z} \iint_{A(z)} \frac{\partial (\overline{\rho}\,\widetilde{\psi})}{\partial t} dA dz$$
$$= \frac{\partial}{\partial t} \left( \int_z^{z+\Delta z} \langle \overline{\rho} \rangle_2 \langle \widetilde{\psi} \rangle_{2\rho} A dz \right) \tag{4.155}$$

Here we introduced area-averaged density, $\langle \overline{\rho} \rangle_2$, and area-averaged, density weighted conserved variable, $\langle \widetilde{\psi} \rangle_{2\rho}$. Both these quantities are first time and Favre averaged before the area averaging is applied.

The surface integral on the right-hand side is first considered in area $A(z)$ as follows,

$$-\iint_{A(z)} \overline{\rho}\,\widetilde{\psi}(\widetilde{\mathbf{v}} \cdot \mathbf{n}) dA = -\iint_{A(z)} \overline{\rho}\,\widetilde{\psi}\widetilde{\mathbf{v}} \cdot (-\mathbf{e}_z) dA = \iint_{A(z)} \overline{\rho}\,\widetilde{\psi}\widetilde{v}_n dA$$
$$= \langle \overline{\rho} \rangle_2 \langle \widetilde{\psi}\widetilde{v}_n \rangle_{2\rho} A(z) = C_{\psi v} \langle \overline{\rho} \rangle_2 \langle \widetilde{\psi} \rangle_{2\rho} \langle \widetilde{v}_n \rangle_{2\rho} A \mid_z \tag{4.156}$$

where $v_n = \mathbf{v} \cdot \mathbf{e}_z$ is the velocity normal to the cross-section of the channel and $C_{\psi v}$ is the *covariance coefficient* defined as,

$$C_{\psi v} \equiv \frac{\langle \widetilde{\psi}\widetilde{v}_n \rangle_{2\rho}}{\langle \widetilde{\psi} \rangle_{2\rho} \langle \widetilde{v}_n \rangle_{2\rho}} \tag{4.157}$$

Physically this coefficient represents the effect of velocity and variable $\psi$ (actually the effect of their time-averaged and Favre-averaged values) on the flux of quantity $\psi$ through area $A(z)$. The corresponding term on area $A(z+\Delta z)$ with $\mathbf{n} = \mathbf{e}_z$ is obtained as follows,

$$-\iint_{A(z+\Delta z)} \overline{\rho}\,\widetilde{\psi}(\widetilde{\mathbf{v}} \cdot \mathbf{n}) dA = C_{\psi v} \langle \overline{\rho} \rangle_2 \langle \widetilde{\psi} \rangle_{2\rho} \langle \widetilde{v}_n \rangle_{2\rho} A \mid_{z+\Delta z} \tag{4.158}$$

The turbulent flux resulting from the Favre averaging is as follows,

$$-\iint_{A(z)} \overline{\rho\psi''(\mathbf{v}'' \cdot \mathbf{n})} dA = \iint_{A(z)} \overline{\rho\psi''(\mathbf{v}'' \cdot \mathbf{e}_z)} dA = \iint_{A(z)} \overline{\rho\psi''v_n''} dA$$
$$= \langle \overline{\rho} \rangle_2 \langle \widetilde{\psi''v_n''} \rangle_{2\rho} A \mid_z \tag{4.159}$$

Similarly, for area $A(z+\Delta z)$ w have,

$$-\iint_{A(z+\Delta z)} \overline{\rho\psi''(\mathbf{v}'' \cdot \mathbf{n})} dA = -\langle \overline{\rho} \rangle_2 \langle \widetilde{\psi''v_n''} \rangle_{2\rho} A \mid_{z+\Delta z} \tag{4.160}$$

The last surface integral term over area $A(z)$ in Eq. (4.101) is as follows,

$$-\iint_{A(z)} \mathbf{n} \cdot \overline{\mathbf{J}} dA = -\iint_{A(z)} (-\mathbf{e}_z) \cdot \overline{\mathbf{J}} dA = \iint_{A(z)} \overline{J}_z dA = \langle \overline{J}_z \rangle_2 A \mid_z , \tag{4.161}$$

where $\overline{J}_z = \mathbf{e}_z \cdot \overline{\mathbf{J}}$ is the traction of tensor $\mathbf{J}$ on area $A$. The corresponding term for area $A(z + \Delta z)$ is given as,

$$-\iint_{A(z+\Delta z)} \mathbf{n} \cdot \overline{\mathbf{J}} dA = -\iint_{A(z+\Delta z)} \mathbf{e}_z \cdot \overline{\mathbf{J}} dA = -\iint_{A(z+\Delta z)} \overline{J}_z dA \qquad (4.162)$$
$$= -\langle \overline{J}_z \rangle_2 A \mid_{z+\Delta z}$$

The surface integral over wall surface $S_w$ has only one non-zero term as follows,

$$-\iint_{S_w} \mathbf{n}_w \cdot \overline{\mathbf{J}} dS_w = -\int_z^{z+\Delta z} \int_{P_w} \overline{J}_w dP_w dz = -\int_z^{z+\Delta z} P_w \langle \overline{J}_w \rangle_1 dz . \qquad (4.163)$$

Here we introduced the perimeter-averaged quantity,

$$\langle \overline{J}_w \rangle_1 \equiv \frac{1}{P_w} \int_{P_w} \overline{J}_w dP_w . \qquad (4.164)$$

The source term in Eq. (4.101) is given as,

$$\iiint_V \overline{\rho} \widetilde{\phi} dV = \int_z^{z+\Delta z} \iint_{A(z)} \overline{\rho} \widetilde{\phi} dA dz = \int_z^{z+\Delta z} \langle \overline{\rho} \rangle_2 \langle \widetilde{\phi} \rangle_{2\rho} A dz . \qquad (4.165)$$

Substituting the derived terms given by Eqs. (4.155)–(4.165) into (4.101), dividing both sides of the equation by $\Delta z$, and taking $\Delta z \to 0$ yields the following *area-averaged conservation equation* for flow in a channel:

$$\frac{\partial \left( \langle \overline{\rho} \rangle_2 \langle \widetilde{\psi} \rangle_{2\rho} A \right)}{\partial t} + \frac{\partial}{\partial z} \left( C_{\psi v} \langle \overline{\rho} \rangle_2 \langle \widetilde{\psi} \rangle_{2\rho} \langle \widetilde{v}_n \rangle_{2\rho} A \right) = -\frac{\partial}{\partial z} \left( \langle \overline{J}_z \rangle_2 A \right)$$
$$- P_w \langle \overline{J}_w \rangle_1 - \frac{\partial}{\partial z} \left( \langle \overline{\rho} \rangle_2 \langle \widetilde{\psi'' v_n''} \rangle_{2\rho} A \right) + \langle \overline{\rho} \rangle_2 \langle \widetilde{\phi} \rangle_{2\rho} A \qquad (4.166)$$

In this equation, all field variables are first time or Favre averaged and then averaged over the channel cross-section area to yield a one-dimensional generic conservation equation. It should be noted that the equation is non-local, since $\langle \overline{J}_w \rangle_1$ is a traction of $\mathbf{J}$ averaged along the channel perimeter, $P_w$, whereas all other variables are averaged over the channel flow area $A$. Similar equations for the instantaneous and Reynolds averaged field variables can be obtained by applying the averaging procedure to Eqs. (4.63) and (4.84), respectively. Using definitions of $\psi$, $\mathbf{J}$, and $\phi$ given in Table 4.1, the conservation equations for mass, momentum, and energy can be obtained.

### 4.6.6 AREA-AVERAGED CONSERVATION EQUATIONS FOR FLOW IN CHANNELS WITH PERMEABLE WALLS

In the previous sub-section, we assumed that the channel walls are stationary and impermeable, where no-slip conditions apply. Here we assume that at least a fraction of channel walls is permeable, through which the fluid can flow into the channel or

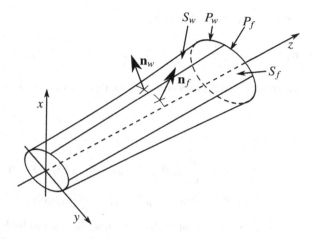

**Figure 4.3** A radially-bounded channel by a solid wall surface $S_w$ with a normal unit vector $\mathbf{n}_w$ and a perimeter $P_w$, and by a fluid surface $S_f$ with a normal unit vector $\mathbf{n}_f$ and a perimeter $P_f$.

out of the channel. Assuming that the channel wall shown in Fig. 4.3 can be divided into two parts, $S_w$ and $S_f$, representing the impermeable and the permeable part, respectively, the convective flux of $\psi$ through $S_f$ is as follows,

$$
\begin{aligned}
-\iint_{S_f} \overline{\rho}\widetilde{\psi}(\widetilde{\mathbf{v}}\cdot\mathbf{n}_f)\mathrm{d}S_f &= -\int_z^{z+\Delta z}\int_{P_f}\overline{\rho}\widetilde{\psi}\widetilde{v}_f\mathrm{d}P_f\mathrm{d}z \\
&= -\int_z^{z+\Delta z}\langle\overline{\rho}\rangle_1\langle\widetilde{\psi}\widetilde{v}_f\rangle_{1\rho}P_f\mathrm{d}z \qquad , \\
&= -\int_z^{z+\Delta z}C_{\psi f}\langle\overline{\rho}\rangle_1\langle\widetilde{\psi}\rangle_{1\rho}\langle\widetilde{v}_f\rangle_{1\rho}P_f\mathrm{d}z
\end{aligned}
\tag{4.167}
$$

where $C_{\psi f}$ is the *covariance coefficient* for the lateral convection of $\psi$ through the permeable surface defined as,

$$
C_{\psi f} \equiv \frac{\langle\widetilde{\psi}\widetilde{v}_f\rangle_{1\rho}}{\langle\widetilde{\psi}\rangle_{1\rho}\langle\widetilde{v}_f\rangle_{1\rho}},
\tag{4.168}
$$

and $\widetilde{v}_f$ is the Favre-averaged velocity normal to the permeable surface. In a similar manner, we obtain an expression for the turbulent flux of $\psi$ through $S_f$ as,

$$
\begin{aligned}
-\iint_{S_f} \overline{\rho}\widetilde{\psi''(\mathbf{v''}\cdot\mathbf{n})}\mathrm{d}S_f &= -\int_z^{z+\Delta z}\int_{P_f}\overline{\rho}\widetilde{\psi''v_f''}\mathrm{d}P_f\mathrm{d}z \\
&= -\int_z^{z+\Delta z}\langle\overline{\rho}\rangle_1\langle\widetilde{\psi''v_f''}\rangle_{1\rho}P_f\mathrm{d}z
\end{aligned}
\tag{4.169}
$$

and an expression for traction of $\mathbf{J}$ over $S_f$ as,

$$-\iint_{S_f} \mathbf{n}_f \cdot \overline{\mathbf{J}} \mathrm{d} S_f = -\int_z^{z+\Delta z} \int_{P_f} \overline{J}_w \mathrm{d} P_f \mathrm{d} z = -\int_z^{z+\Delta z} P_f \langle \overline{J}_f \rangle_1 \mathrm{d} z . \qquad (4.170)$$

The *area-averaged conservation equation* for flow in a channel with a permeable wall becomes,

$$
\begin{aligned}
\frac{\partial \left( \langle \overline{\rho} \rangle_2 \langle \widetilde{\psi} \rangle_{2\rho} A \right)}{\partial t} &+ \frac{\partial}{\partial z} \left( C_{\psi v} \langle \overline{\rho} \rangle_2 \langle \widetilde{\psi} \rangle_{2\rho} \langle \widetilde{v}_n \rangle_{2\rho} A \right) = -\frac{\partial}{\partial z} \left( \langle \overline{J}_z \rangle_2 A \right) \\
&- \langle \overline{J}_w \rangle_1 P_w - \frac{\partial}{\partial z} \left( \langle \overline{\rho} \rangle_2 \langle \widetilde{\psi'' v_n''} \rangle_{2\rho} A \right) + \langle \overline{\rho} \rangle_2 \langle \widetilde{\phi} \rangle_{2\rho} A \\
&- C_{\psi f} \langle \overline{\rho} \rangle_1 \langle \widetilde{\psi} \rangle_{1\rho} \langle \widetilde{v}_f \rangle_{1\rho} P_f - \langle \overline{\rho} \rangle_1 \langle \widetilde{\psi'' v_f''} \rangle_{1\rho} P_f - \langle \overline{J}_f \rangle_1 P_f
\end{aligned} \qquad . \quad (4.171)
$$

Compared to the conservation equation for a channel flow given by Eq. (4.166), this equation incorporates three supplementary terms. These terms originate from the flow traversing the radially bounding fluid surface $S_f$. The first of the two terms represents a net convection of $\psi$ through the surface, whereas the second term represents the turbulent flux of $\psi$. The third term is the traction of $\mathbf{J}$ on surface $S_f$.

## 4.7   CLOSURE RELATIONS

The differential and integral conservation equations derived in §4.4, §4.5, and §4.6, must be supplemented with addition equations in order to be solved. The reason for this is that the conservation equations, which for single-phase flow include one mass conservation equation, three linear momentum conservation equations, and one energy conservation equation, contain more unknowns than five. The unknown variables include the fluid density $\rho$, three components of the velocity vector $\mathbf{v}$, the pressure $p$, nine components of the viscous shear stress tensor $\tau$, the internal energy $e_I$, and three components of the heat flux vector $\mathbf{q}''$. Thus, the system of five conservation equations contains eighteen unknowns.

Such a system of equations in which the number of unknowns is greater than the number of equations is called an underdetermined system. To obtain a determined system of equations for a laminar single-phase flow and heat transfer, additional fifteen equations are needed. Such equations are derived on empirical or semi-empirical grounds and are collectively termed as the *constitutive equations*. A complete formulation of a physically plausible three-dimensional unsteady fluid flow and heat transfer model includes then conservation equations, constitutive equations, and initial and boundary conditions.

### 4.7.1   THE EQUATION OF STATE

The **equation of state** describes a relationship between the pressure $p$, the density $\rho$ or the specific volume $v = 1/\rho$, and the temperature $T$ of a physically homogeneous system in thermodynamic equilibrium. This equation is not derived from thermodynamic relations alone but rather relies on empirical or semi-empirical considerations.

## Ideal Gas

The **ideal gas law** is the equation of state of a hypothetical ideal gas. Even though it has several limitations, the ideal gas law is a good approximation of the behavior of many gases. The ideal gas law is written in terms of a **specific gas constant** as,

$$pv = \rho R_{sp} T, \tag{4.172}$$

where $R_{sp}$ is a specific gas constant obtained from the **universal gas constant** $R_u$ and the gas molar mass $M$ as,

$$R_{sp} = \frac{R_u}{M}, \quad R_u \approx 8.3145 \text{ J mol}^{-1} \text{ K}^{-1}. \tag{4.173}$$

## Real Gases

For real gases virial equations of state are used of a general form given as,

$$pv = \rho R_{sp} T \left( 1 + b(T)/v + c(T)/v^2 + \cdots \right), \tag{4.174}$$

where $b(T)$, $c(T)$,..., are the second, third, etc. virial coefficient. Examples of the equations of state for real gases are given in §3.4.

## Liquids

Liquids, such as pressurized water, show a rather weak dependence of the density on pressure. For small or moderate pressure and temperature variations, the liquid density can be assumed to be constant. With this assumption, the mass conservation equation in the differential and integral form become, respectively,

$$\nabla \cdot \overline{\mathbf{v}} = 0, \tag{4.175}$$

$$\iint_S \overline{\mathbf{v}} \cdot \mathbf{n} dS = 0. \tag{4.176}$$

Thus, the mass conservation equation degenerates to the kinematic condition that the velocity field be *solenoidal* or *divergence-free*. This simplification is used, either implicitly or explicitly, in the formulation of numerous constitutive equations. For example, the general form of the local shear stress given by Eq. (4.19) can be simplified to the form given by Eq. (4.18), if Eq. (4.175) is taken into account.

The integral form of the mass conservation equation suggests that for a closed heated channel with the same cross-section area at the inlet and the outlet, the area-averaged fluid velocity should be the same. Consider, however, a channel with flowing water at pressure 17 MPa, which is heated from 40 K subcooling at the inlet to the saturation temperature at the outlet. A simple calculation shows that the water velocity will increase by about 22% due to the density change. The selected conditions correspond quite well to the coolant flow in a PWR fuel assembly. Thus, a neglect of water density variation in a PWR core would lead to a quite significant error.

For liquids, the density is usually given as a function of temperature and pressure. It is common to express the density in terms of analytic functions, presented for various coolants and other materials in §3, or tables, as provided in Appendix D for water and steam.

## Specific Enthalpy

Some forms of the energy conservation equation are expressed in terms of specific enthalpy. The following generally valid relations can be used to close the system:

$$\bar{i} = \overline{e_I} + \frac{\overline{p}}{\rho}, \tag{4.177}$$

and

$$\frac{D\bar{i}}{Dt} = c_p \frac{D\overline{T}}{Dt} + \frac{1 - \beta \overline{T}}{\rho} \frac{D\overline{p}}{Dt}, \tag{4.178}$$

where $\bar{i}$ is the time average of the specific enthalpy, $c_p$ is the isobaric specific heat capacity, and $\beta$ is the coefficient of thermal expansion,

$$\beta = -\frac{1}{\rho} \left( \frac{\partial \rho}{\partial T} \right)_p. \tag{4.179}$$

## 4.7.2 THE STRESS TENSOR

In the process of averaging the momentum conservation equations, two stress tensors were derived: the time-averaged stress tensor and the Reynolds stress tensor. The former can be expressed using averaged flow variables, while the latter necessitates additional models and relationships for its closed-form expression.

## The Time-Averaged Stress Tensor

For *Newtonian fluids*, the general expression for a shear stress tensor is given by Eq. (4.19). In Cartesian coordinates, the components of the time average shear stress tensor are as follows,

$$\overline{\tau_{xx}} = 2\mu \frac{\partial \overline{u}}{\partial x} - \left( \frac{2}{3}\mu - \mu' \right) \left( \frac{\partial \overline{u}}{\partial x} + \frac{\partial \overline{v}}{\partial y} + \frac{\partial \overline{w}}{\partial z} \right), \tag{4.180}$$

$$\overline{\tau_{yy}} = 2\mu \frac{\partial \overline{v}}{\partial y} - \left( \frac{2}{3}\mu - \mu' \right) \left( \frac{\partial \overline{u}}{\partial x} + \frac{\partial \overline{v}}{\partial y} + \frac{\partial \overline{w}}{\partial z} \right), \tag{4.181}$$

$$\overline{\tau_{zz}} = 2\mu \frac{\partial \overline{w}}{\partial z} - \left( \frac{2}{3}\mu - \mu' \right) \left( \frac{\partial \overline{u}}{\partial x} + \frac{\partial \overline{v}}{\partial y} + \frac{\partial \overline{w}}{\partial z} \right), \tag{4.182}$$

$$\overline{\tau_{xy}} = \overline{\tau_{yx}} = \mu \left( \frac{\partial \overline{v}}{\partial x} + \frac{\partial \overline{u}}{\partial y} \right), \tag{4.183}$$

$$\overline{\tau_{yz}} = \overline{\tau_{zy}} = \mu \left( \frac{\partial \overline{w}}{\partial y} + \frac{\partial \overline{v}}{\partial z} \right), \tag{4.184}$$

$$\overline{\tau_{zx}} = \overline{\tau_{xz}} = \mu \left( \frac{\partial \overline{u}}{\partial z} + \frac{\partial \overline{w}}{\partial x} \right), \tag{4.185}$$

where, using the Stokes hypothesis, we can assume that $\mu' = 0$.

## The Reynolds Stress Tensor

Reynolds averaged momentum conservation equation (4.82) contains a quantity $-\overline{\mathbf{v}'\mathbf{v}'}$ which is known as the *specific Reynolds stress tensor*. Once multiplied by the density, the quantity is termed as **Reynolds stress tensor** and is given as,

$$\tau_t = -\rho \overline{\mathbf{v}'\mathbf{v}'} = -\rho \begin{bmatrix} \overline{u'^2} & \overline{v'u'} & \overline{w'u'} \\ \overline{u'v'} & \overline{v'^2} & \overline{w'v'} \\ \overline{u'w'} & \overline{v'w'} & \overline{w'^2} \end{bmatrix}. \tag{4.186}$$

This expression represents a symmetric second-order tensor with six independent components. The diagonal components $\overline{u'^2}$, $\overline{v'^2}$, and $\overline{w'^2}$ are *turbulent normal stresses*, while the off-diagonal components $\overline{u'v'}$, $\overline{u'w'}$, and $\overline{v'w'}$ are *turbulent shear stresses*.

The Reynolds stresses are often normalized relative to the freestream mean flow velocity $U$ and are referred to as the *relative turbulence intensities*. Turbulent intensities for an incompressible flat-plate boundary layer are shown in Fig. 4.4. As can be seen, the three normal intensities have different values indicating that the turbulence is anisotropic.

The **turbulence kinetic energy** $k(\mathbf{r}, t)$ at location $\mathbf{r}$ and time $t$ is defined to be half the trace of the Reynolds stress tensor,

$$k = \frac{1}{2} \left( \overline{u'^2} + \overline{v'^2} + \overline{w'^2} \right). \tag{4.187}$$

When the turbulence anisotropy is low or cannot be determined, we often specify relative turbulence intensity by assuming the fluctuations are approximately isotropic, that is $\overline{u'^2} \approx \overline{v'^2} \approx \overline{w'^2}$. We then define the relative turbulence intensity in percent as,

$$\text{Tu} \equiv 100 \sqrt{\frac{2}{3} \frac{k}{U^2}}. \tag{4.188}$$

There are many different approaches for the modeling of the Reynolds stress tensor. In the most advanced approach, six transport equations are introduced for the stress tensor components and one additional equation for the turbulence dissipation rate $\varepsilon$ of the turbulent kinetic energy. The advantage of this approach is the ability to capture the flow details with anisotropic turbulence. The equations have to be solved together with the conservation equations, which significantly increases the required computational effort.

The most popular approach for the modeling of the Reynolds stress tensor is to use the *Boussinesq hypothesis*, which claims that there is a general linear constitutive law between the stress and strain tensors, in a direct analogy to the constitutive

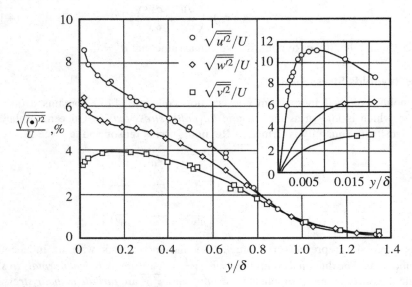

**Figure 4.4** Distribution of turbulence intensities for a flat-plate boundary layer of thickness $\delta$, [128].

relationship for Newtonian fluids. The Reynolds stress tensor is then given as,

$$\tau_t = \mu_t \left[ \nabla \bar{\mathbf{v}} + (\nabla \bar{\mathbf{v}})^T \right]. \tag{4.189}$$

For variable-density fluids this expression is generalized as proposed by Stokes,

$$\tau_t = \mu_t \left[ \nabla \tilde{\mathbf{v}} + (\nabla \tilde{\mathbf{v}})^T \right] + \left( \mu' - \frac{2}{3}\mu_t \right) (\nabla \cdot \tilde{\mathbf{v}}) \mathbf{I}, \tag{4.190}$$

where $\mu_t$ is the **dynamic eddy viscosity** that needs to be determined.

During the past few decades, **two-equation models of turbulence** have gained common acceptance for computation of the turbulence kinetic energy $k$ and the turbulence length scale, represented by either the specific dissipation rate $\omega$, or the dissipation per unit mass $\varepsilon$. Once these quantities are computed, the kinematic eddy viscosity $v_t = \mu_t/\rho$ is found as,

$$v_t = C_\mu \frac{k^2}{\varepsilon}, \tag{4.191}$$

for the $k - \varepsilon$ model, and,

$$v_t = C_\mu \frac{k}{\omega}, \tag{4.192}$$

for the $k - \omega$ model. Here $C_\mu$ is a constant that needs to be established from experimental data.

### 4.7.3  THE HEAT FLUX VECTOR

The heat flux vector is one of the three main components of the energy flux vector introduced in §4.1. Similarly to the shear stress tensor in fluid mechanics, the heat flux vector is obtained from closure laws that express the heat flux in terms of local flow and heat transfer parameters. Usually, these local parameters include a gradient of the time-averaged temperature field and some additional time-averaged turbulent interaction terms. For laminar flows with heat transfer and for heat conduction in solids, the heat flux vector is described in terms of the Fourier law of heat conduction, whereas for turbulent flows with heat transfer various approaches to represent the turbulent heat flux are used.

**The Time-Averaged Heat Flux**

It is commonly accepted that the Fourier law given by Eq. (4.25) is valid for the time average temperature field. Thus, the time average heat flux is given as,

$$\overline{\mathbf{q}''} = -\lambda \nabla \overline{T} = -\lambda \left( \frac{\partial \overline{T}}{\partial x} + \frac{\partial \overline{T}}{\partial y} + \frac{\partial \overline{T}}{\partial z} \right). \tag{4.193}$$

**The Turbulent Heat Flux**

The time-averaged energy conservation equation contains a correlation term of velocity and internal energy fluctuations. This term incorporates the turbulent part of convective heat transfer and for the Reynolds average energy equation (4.83) is given as,

$$\mathbf{q}_t'' = -\rho \overline{e'_{IK} \mathbf{v}'} - \overline{p' \mathbf{v}'} = -\rho \overline{(e'_{IK} + p'/\rho) \mathbf{v}'}. \tag{4.194}$$

## 4.8  USEFUL FORMS OF CONSERVATION EQUATIONS

Certain forms of conservation equations are particularly useful for specific applications. In this section, we collect examples of such formulations.

### 4.8.1  INSTANTANEOUS CONSERVATION EQUATIONS

Instantaneous conservation equations are useful for study the transient behavior of systems in a simplified geometry. These forms of equations are also used for the most detailed solutions when employing the direct numerical simulation of turbulence. A viscous flow is described with a set of equations representing the mass conservation and the conservation of momentum in a three dimensional space $(x, y, z)$. For fluids with temperature-dependent properties, the system of equations has to include the energy equation as well. A constant property flow can be assumed when the variations of the fluid temperature and pressure are insignificant or when the fluid properties have a weak dependence on these two variables. Under such conditions the balance equations are significantly simplified and from the mass conservation equation results that the flow is divergence free. The conservation equations with variable and constant properties are presented below.

## The Mass Conservation Equation

For variable density, the differential mass conservation equation is given as,

$$\frac{\partial \rho}{\partial t} + \frac{\partial}{\partial x}(\rho u) + \frac{\partial}{\partial y}(\rho v) + \frac{\partial}{\partial z}(\rho w) = 0. \tag{4.195}$$

In case of constant physical properties of the fluid, the mass conservation equation becomes,

$$\frac{\partial u}{\partial x} + \frac{\partial v}{\partial y} + \frac{\partial w}{\partial z} = 0. \tag{4.196}$$

## The Momentum Conservation Equation

For fluid flow with variable density and dynamic viscosity, the differential momentum conservation equations are given as,

$$
\begin{aligned}
\rho \frac{Du}{Dt} =& \rho g_x - \frac{\partial p}{\partial x} + \frac{2}{3}\frac{\partial}{\partial x}\left[\mu\left(2\frac{\partial u}{\partial x} - \frac{\partial v}{\partial y} - \frac{\partial w}{\partial z}\right)\right] + \\
& \frac{\partial}{\partial y}\left[\mu\left(\frac{\partial u}{\partial y} + \frac{\partial v}{\partial x}\right)\right] + \frac{\partial}{\partial z}\left[\mu\left(\frac{\partial w}{\partial x} + \frac{\partial u}{\partial z}\right)\right],
\end{aligned} \tag{4.197}
$$

$$
\begin{aligned}
\rho \frac{Dv}{Dt} =& \rho g_y - \frac{\partial p}{\partial y} + \frac{2}{3}\frac{\partial}{\partial y}\left[\mu\left(2\frac{\partial v}{\partial y} - \frac{\partial u}{\partial x} - \frac{\partial w}{\partial z}\right)\right] + \\
& \frac{\partial}{\partial z}\left[\mu\left(\frac{\partial v}{\partial z} + \frac{\partial w}{\partial y}\right)\right] + \frac{\partial}{\partial x}\left[\mu\left(\frac{\partial u}{\partial y} + \frac{\partial v}{\partial x}\right)\right],
\end{aligned} \tag{4.198}
$$

$$
\begin{aligned}
\rho \frac{Dw}{Dt} =& \rho g_z - \frac{\partial p}{\partial z} + \frac{2}{3}\frac{\partial}{\partial z}\left[\mu\left(2\frac{\partial w}{\partial z} - \frac{\partial u}{\partial x} - \frac{\partial v}{\partial y}\right)\right] + \\
& \frac{\partial}{\partial x}\left[\mu\left(\frac{\partial w}{\partial x} + \frac{\partial u}{\partial z}\right)\right] + \frac{\partial}{\partial y}\left[\mu\left(\frac{\partial v}{\partial z} + \frac{\partial w}{\partial y}\right)\right],
\end{aligned} \tag{4.199}
$$

When density and dynamic viscosity are constant, the equations take the following form,

$$\rho \frac{Du}{Dt} = \rho g_x - \frac{\partial p}{\partial x} + \mu\left(\frac{\partial^2 u}{\partial x^2} + \frac{\partial^2 u}{\partial y^2} + \frac{\partial^2 u}{\partial z^2}\right), \tag{4.200}$$

$$\rho \frac{Dv}{Dt} = \rho g_y - \frac{\partial p}{\partial y} + \mu\left(\frac{\partial^2 v}{\partial x^2} + \frac{\partial^2 v}{\partial y^2} + \frac{\partial^2 v}{\partial z^2}\right), \tag{4.201}$$

$$\rho \frac{Dw}{Dt} = \rho g_z - \frac{\partial p}{\partial z} + \mu\left(\frac{\partial^2 w}{\partial x^2} + \frac{\partial^2 w}{\partial y^2} + \frac{\partial^2 w}{\partial z^2}\right), \tag{4.202}$$

## The Energy Conservation Equation

For variable-property fluids, the energy conservation equation is as follows,

$$\rho c_p \frac{DT}{Dt} = \nabla(\lambda\nabla T) - \left(\frac{\partial\ln\rho}{\partial\ln T}\right)_p \frac{Dp}{Dt} + q''' + \mu\Phi. \tag{4.203}$$

Here the viscous dissipation term in the energy equation is given as:

$$\Phi = 2\left[\left(\frac{\partial u}{\partial x}\right)^2 + \left(\frac{\partial v}{\partial y}\right)^2 + \left(\frac{\partial w}{\partial z}\right)^2\right] + \left(\frac{\partial v}{\partial x} + \frac{\partial u}{\partial y}\right)^2 +$$
$$\left(\frac{\partial w}{\partial y} + \frac{\partial v}{\partial z}\right)^2 + \left(\frac{\partial u}{\partial z} + \frac{\partial w}{\partial x}\right)^2 - \frac{2}{3}\left(\frac{\partial u}{\partial x} + \frac{\partial v}{\partial y} + \frac{\partial w}{\partial z}\right)^2. \tag{4.204}$$

In case of constant physical properties of the fluid, the energy balance equation becomes,

$$\rho c_p \frac{DT}{Dt} = \lambda\left(\frac{\partial^2 T}{\partial x^2} + \frac{\partial^2 T}{\partial y^2} + \frac{\partial^2 T}{\partial z^2}\right) + q''' + \mu\Phi_0, \tag{4.205}$$

where,

$$\Phi_0 = 2\left[\left(\frac{\partial u}{\partial x}\right)^2 + \left(\frac{\partial v}{\partial y}\right)^2 + \left(\frac{\partial w}{\partial z}\right)^2\right] +$$
$$\left(\frac{\partial v}{\partial x} + \frac{\partial u}{\partial y}\right)^2 + \left(\frac{\partial w}{\partial y} + \frac{\partial v}{\partial z}\right)^2 + \left(\frac{\partial u}{\partial z} + \frac{\partial w}{\partial x}\right)^2, \tag{4.206}$$

is the viscous dissipation term for incompressible flows. Using Eq. (4.178), the energy balance equation can be written as,

$$\rho c_p \frac{DT}{Dt} = \nabla(\lambda\nabla T) + \beta T\frac{Dp}{Dt} + q''' + \mu\Phi, \tag{4.207}$$

where $\beta$ is the coefficient of thermal expansion given by Eq. (4.179).

### 4.8.2 REYNOLDS-AVERAGED CONSERVATION EQUATIONS

General forms of balance equations for turbulent flows have been derived in §4.4 and §4.5 for the constant property flows and the varying property flows, respectively. In this section we give the balance equations in closed forms, with explicit formulation of the closure relationships for the turbulent momentum and energy fluxes.

## The Mass Conservation Equation

In case of constant physical properties of the fluid, the Reynolds-averaged balance equations of mass becomes as follows:

$$\frac{\partial\bar{u}}{\partial x} + \frac{\partial\bar{v}}{\partial y} + \frac{\partial\bar{w}}{\partial z} = 0. \tag{4.208}$$

This equation is valid for both steady-state and transient conditions.

## The Momentum Conservation Equation

The Reynolds-averaged momentum conservation equations are as follows,

$$\rho \frac{D\overline{u}}{Dt} = \rho g_x - \frac{\partial \overline{p}}{\partial x} + (\mu + \mu_t)\left(\frac{\partial^2 \overline{u}}{\partial x^2} + \frac{\partial^2 \overline{u}}{\partial y^2} + \frac{\partial^2 \overline{u}}{\partial z^2}\right), \qquad (4.209)$$

$$\rho \frac{D\overline{v}}{Dt} = \rho g_y - \frac{\partial \overline{p}}{\partial y} + (\mu + \mu_t)\left(\frac{\partial^2 \overline{v}}{\partial x^2} + \frac{\partial^2 \overline{v}}{\partial y^2} + \frac{\partial^2 \overline{v}}{\partial z^2}\right), \qquad (4.210)$$

$$\rho \frac{D\overline{w}}{Dt} = \rho g_z - \frac{\partial \overline{p}}{\partial z} + (\mu + \mu_t)\left(\frac{\partial^2 \overline{w}}{\partial x^2} + \frac{\partial^2 \overline{w}}{\partial y^2} + \frac{\partial^2 \overline{w}}{\partial z^2}\right). \qquad (4.211)$$

## The Energy Conservation Equation

The Reynolds-averaged energy conservation equation reads,

$$\rho c_p \frac{D\overline{T}}{Dt} = \nabla\left[(\lambda + \lambda_t)\nabla \overline{T}\right] + \beta \overline{T}\frac{D\overline{p}}{Dt} + q''' + \mu \overline{\Phi} + \rho \varepsilon. \qquad (4.212)$$

Here $\rho \varepsilon$ is the turbulent dissipation given as,

$$\rho \varepsilon = \mu \left\{ 2\left[\overline{\left(\frac{\partial u'}{\partial x}\right)^2} + \overline{\left(\frac{\partial v'}{\partial y}\right)^2} + \overline{\left(\frac{\partial w'}{\partial z}\right)^2}\right] + \right.$$
$$\left. \overline{\left(\frac{\partial v'}{\partial x} + \frac{\partial u'}{\partial y}\right)^2} + \overline{\left(\frac{\partial w'}{\partial y} + \frac{\partial v'}{\partial z}\right)^2} + \overline{\left(\frac{\partial u'}{\partial z} + \frac{\partial w'}{\partial x}\right)^2}\right\}. \qquad (4.213)$$

## The Standard $k$-$\varepsilon$ Model

The standard $k$-$\varepsilon$ model applies to nonboundary-layer high-Reynolds-number flows. It consists of a set of two differential equations for the turbulent kinetic energy $k$ and the dissipation rate $\varepsilon$. The equations together with modeling closures and constants are as follows.

Dynamic turbulent viscosity:

$$\mu_t = \rho C_\mu \frac{k^2}{\varepsilon}. \qquad (4.214)$$

Turbulence kinetic energy:

$$\frac{D(\rho k)}{Dt} = \nabla \cdot \left[\left(\mu + \frac{\mu_t}{\sigma_k}\right)\nabla k\right] + P_k + P_b - \rho \varepsilon - Y_M + S_k. \qquad (4.215)$$

Dissipation rate:

$$\frac{D(\rho \varepsilon)}{Dt} = \nabla \cdot \left[\left(\mu + \frac{\mu_t}{\sigma_\varepsilon}\right)\nabla \varepsilon\right] + C_{1\varepsilon}\frac{\varepsilon}{k}(P_k + C_{3\varepsilon}P_b) - C_{2\varepsilon}\rho \frac{\varepsilon^2}{k} + S_\varepsilon. \qquad (4.216)$$

Production of turbulence kinetic energy:

$$P_k = \boldsymbol{\tau} : \nabla \bar{\mathbf{v}}. \tag{4.217}$$

Effect of buoyancy:

$$P_b = \beta \frac{\mu_t}{\mathrm{Pr}_t} \mathbf{g} \cdot \nabla T. \tag{4.218}$$

Model constants:

$$C_{1\varepsilon} = 1.44, \quad C_{2\varepsilon} = 1.92, \quad C_\mu = 0.09, \quad \sigma_k = 1.0, \quad \sigma_\varepsilon = 1.3 \quad . \tag{4.219}$$

The buoyancy effect constant has no standard value but frequently it is taken as $C_{3\varepsilon} = -0.33$. More details on the standard $k$-$\varepsilon$ model can be found in many textbooks, e.g., [179, 233, 235].

### 4.8.3 FAVRE-AVERAGED CONSERVATION EQUATIONS

For compressible or variable-property fluid motion, we must solve the equations governing conservation of mass, momentum, and energy. The instantaneous equations are given in §4.8.1. In this section we provide their Favre-averaged equivalents.

### The Mass Conservation Equation

The Favre-averaged mass conservation equation is as follows:

$$\frac{\partial \bar{\rho}}{\partial t} + \frac{\partial (\bar{\rho}\tilde{u})}{\partial x} + \frac{\partial (\bar{\rho}\tilde{v})}{\partial y} + \frac{\partial (\bar{\rho}\tilde{w})}{\partial z} = 0. \tag{4.220}$$

### The Momentum Conservation Equation

The Favre-averaged momentum conservation equations are as follows,

$$\frac{\mathrm{D}(\bar{\rho}\tilde{u})}{\mathrm{D}t} = \bar{\rho} g_x - \frac{\partial \bar{p}}{\partial x} + (\mu + \mu_t) \left( \frac{\partial^2 \tilde{u}}{\partial x^2} + \frac{\partial^2 \tilde{u}}{\partial y^2} + \frac{\partial^2 \tilde{u}}{\partial z^2} \right), \tag{4.221}$$

$$\frac{\mathrm{D}(\bar{\rho}\tilde{v})}{\mathrm{D}t} = \bar{\rho} g_y - \frac{\partial \bar{p}}{\partial y} + (\mu + \mu_t) \left( \frac{\partial^2 \tilde{v}}{\partial x^2} + \frac{\partial^2 \tilde{v}}{\partial y^2} + \frac{\partial^2 \tilde{v}}{\partial z^2} \right), \tag{4.222}$$

$$\frac{\mathrm{D}(\bar{\rho}\tilde{w})}{\mathrm{D}t} = \bar{\rho} g_z - \frac{\partial \bar{p}}{\partial z} + (\mu + \mu_t) \left( \frac{\partial^2 \tilde{w}}{\partial x^2} + \frac{\partial^2 \tilde{w}}{\partial y^2} + \frac{\partial^2 \tilde{w}}{\partial z^2} \right). \tag{4.223}$$

### The Energy Conservation Equation

The Favre-averaged energy conservation equation reads,

$$\rho c_p \frac{\mathrm{D}\bar{T}}{\mathrm{D}t} = \nabla \left[ (\lambda + \lambda_t) \nabla \bar{T} \right] + \beta \bar{T} \frac{\mathrm{D}\bar{p}}{\mathrm{D}t} + q''' + \mu \bar{\Phi} + \rho \varepsilon. \tag{4.224}$$

Here $\rho\varepsilon$ is the turbulent dissipation given as,

$$\rho\varepsilon = \mu \left\{ 2\left[\left(\overline{\frac{\partial u'}{\partial x}}\right)^2 + \left(\overline{\frac{\partial v'}{\partial y}}\right)^2 + \left(\overline{\frac{\partial w'}{\partial z}}\right)^2\right] + \overline{\left(\frac{\partial v'}{\partial x} + \frac{\partial u'}{\partial y}\right)^2} + \overline{\left(\frac{\partial w'}{\partial y} + \frac{\partial v'}{\partial z}\right)^2} + \overline{\left(\frac{\partial u'}{\partial z} + \frac{\partial w'}{\partial x}\right)^2} \right\}.$$

$$(4.225)$$

## 4.8.4  SELECTED SPECIAL CASES

For gas flows occurring in the presence of radiation, specific forms of the conservation equations can be employed. This is particularly beneficial for examining the coolant flow in gas-cooled nuclear reactors. Another valuable set of conservation equations pertains to two-dimensional flows. These equations can be utilized in the analysis of thermal boundary layers. In this section, we provide the conservation equations for both these cases.

### The Gas Flow

A gas flow can be described with the same set of conservation equations that are valid for a viscous flow. In general, in flows of high-temperature gases heat transfer processes include convection, conduction, and radiation. If the radiation heat transfer is significant, the mass and energy conservation equations should contain additional terms resulting from the radiation processes. The mass conservation equation becomes,

$$\frac{\partial \rho}{\partial t} + \frac{\partial}{\partial x}(\rho u) + \frac{\partial}{\partial y}(\rho v) + \frac{\partial}{\partial z}(\rho w) = \Gamma_r, \qquad (4.226)$$

where $\Gamma_r$ is the mass change resulting from the energy radiation by the gas. In most cases this mass is very small and can be neglected.

The energy equation for a gas with a significant radiation energy transfer becomes,

$$\rho c_p \frac{DT}{Dt} = \nabla(\lambda \nabla T) - \left(\frac{\partial \ln \rho}{\partial \ln T}\right)_p \frac{Dp}{Dt} + q''' + q'''_r + \mu\Phi + \Phi_r, \qquad (4.227)$$

where $q'''_r$ is the energy transport by radiation and $\Phi_r$ is the energy dissipation due to radiation.

The second term on the right-hand side of the equation can be simplified for an ideal gas, since in that case, we have,

$$\left(\frac{\partial \ln \rho}{\partial \ln T}\right)_p = -1. \qquad (4.228)$$

The set of differential equations is closed with algebraic expressions for temperature- and pressure-dependent physical properties,

$$c_p = c_p(T,p), \quad \lambda = \lambda(T,p), \quad \mu = \mu(T,p), \qquad (4.229)$$

and the equation of state,

$$\rho = \rho(T, p). \tag{4.230}$$

## A Two-Dimensional Flow

Two-dimensional flow approximation plays an important role in many applications, such as thermal boundary analysis. For the flow of viscous fluid with variable physical properties in the $xy$ plane, the balance equations are as follows,

$$\frac{\partial \rho}{\partial t} + \frac{\partial(\rho u)}{\partial x} + \frac{\partial(\rho v)}{\partial y} = 0 \tag{4.231}$$

$$\rho\left(\frac{\partial u}{\partial t} + u\frac{\partial u}{\partial x} + v\frac{\partial u}{\partial y}\right) = \rho g_x - \frac{\partial p}{\partial x} + \frac{\partial}{\partial x}\left(2\mu\frac{\partial u}{\partial x}\right) + \frac{\partial}{\partial y}\left[\mu\left(\frac{\partial u}{\partial y} + \frac{\partial v}{\partial x}\right)\right] \tag{4.232}$$

$$\rho\left(\frac{\partial v}{\partial t} + u\frac{\partial v}{\partial x} + v\frac{\partial v}{\partial y}\right) = \rho g_y - \frac{\partial p}{\partial y} + \frac{\partial}{\partial x}\left[\mu\left(\frac{\partial v}{\partial x} + \frac{\partial u}{\partial y}\right)\right] + \frac{\partial}{\partial y}\left(2\mu\frac{\partial v}{\partial y}\right) \tag{4.233}$$

$$\rho c_p\left(\frac{\partial T}{\partial t} + u\frac{\partial T}{\partial x} + v\frac{\partial T}{\partial y}\right) = \frac{\partial}{\partial x}\left(\lambda\frac{\partial T}{\partial x}\right) + \frac{\partial}{\partial y}\left(\lambda\frac{\partial T}{\partial y}\right) + \frac{\partial p}{\partial t} + u\frac{\partial p}{\partial x} + v\frac{\partial p}{\partial y} + q''' + \mu\Phi_1 \tag{4.234}$$

where:

$$\Phi_1 = 2\left[\left(\frac{\partial u}{\partial x}\right)^2 + \left(\frac{\partial v}{\partial y}\right)^2\right] + \left(\frac{\partial u}{\partial y} + \frac{\partial v}{\partial x}\right)^2. \tag{4.235}$$

To close the system of equations, the physical properties have to be specified as a function of temperature and pressure.

For steady-state flow of a fluid with constant physical properties, the balance equations become as follows,

$$\frac{\partial u}{\partial x} + \frac{\partial v}{\partial y} = 0, \tag{4.236}$$

$$u\frac{\partial u}{\partial x} + v\frac{\partial u}{\partial y} = g_x - \frac{1}{\rho}\frac{\partial p}{\partial x} + \nu\left(\frac{\partial^2 u}{\partial x^2} + \frac{\partial^2 u}{\partial y^2}\right), \tag{4.237}$$

$$u\frac{\partial v}{\partial x} + v\frac{\partial v}{\partial y} = g_y - \frac{1}{\rho}\frac{\partial p}{\partial y} + \nu\left(\frac{\partial^2 v}{\partial x^2} + \frac{\partial^2 v}{\partial y^2}\right), \tag{4.238}$$

$$u\frac{\partial T}{\partial x} + v\frac{\partial T}{\partial y} = \frac{\lambda}{\rho c_p}\left(\frac{\partial^2 T}{\partial x^2} + \frac{\partial^2 T}{\partial y^2}\right) + \frac{q'''}{\rho c_p} + \frac{\nu}{c_p}\Phi_1. \tag{4.239}$$

Here $\nu = \mu/\rho$ is the kinematic viscosity of the fluid.

## PROBLEMS

### PROBLEM 4.1

Explain why the volume and composite averaging lead to a non-local formulation of the transport equation.

### PROBLEM 4.2

Using the definition of the covariance coefficient given by Eq. (4.157), derive its value for laminar flow in a circular tube assuming that $\tilde{\psi}$ parameter has the same radial distribution as the velocity.

### PROBLEM 4.3

Using definitions of $\psi$, $\mathbf{J}$, and $\phi$ given in Table 4.1 and substituting in Eq. (4.166), derive the special form of the conservation equations for mass, momentum, and energy for fluid flow in a heated circular tube with a constant cross-section area.

# 5 Single-Phase Flow in Coolant Channels

The distribution of coolant flow within a reactor core is influenced by numerous factors, including the intricate details of the geometry of the core, the coolant flow patterns within the core, and the specific type of coolant used. Generally, the flow distribution should be optimized to ensure efficient and stable cooling of fuel elements while minimizing pressure losses. The cooling efficiency is crucial from both safety and economic perspectives, while pressure loss minimization is primarily driven by economic considerations. The power required to pump coolant through the core increases with pressure losses, which can significantly impact the overall efficiency of the plant. Therefore, single-phase flow analyses are crucial for the thermal-hydraulic design of nuclear reactor cores.

In this chapter we discuss various aspects of single-phase coolant flow in nuclear reactor cores. Particular attention is given to predictions of the pressure losses due to shear stresses.

The importance of pressure and shear stress distributions in a reactor core stems from two reasons. On the one hand, pressure losses in a reactor core make up the greater part of all pressure losses in the primary loop. Thus, their magnitude determines the flow characteristic of the loop during the forced and natural circulation of the coolant. On the other hand, forces acting on the core structure should be well known and low enough to assure stable and rigid geometry of the core.

The distribution of temperature in a reactor core is of great interest, since it is directly related to reactor thermal margins. In particular, the clad surface temperature strongly depends on the heat transfer intensity between the solid wall and the fluid coolant. Traditionally this intensity is expressed in terms of a heat transfer coefficient, which is a complex function of flow characteristics and fluid properties, as it is discussed in §7 and §8 for single-phase flows and two-phase flows, respectively.

In thermal reactors, such fluids as water, heavy water, and various gases are used as coolants. Gases and liquid metals are used for fast reactor cooling, since these fluids do not slow down fission neutrons. Molten salt reactors use fuel and coolant mixed together in a homogeneous fluid that contains various fluoride salts. In all these reactor types the coolant remains in single-phase state during normal operation. Under such conditions, the coolant flow distribution and the heat transfer intensity are determined by the single-phase flow balance equations discussed in §4.

Most water, gas, and liquid metal-cooled reactors have cores that contain parallel fuel rod assemblies, which are organized in either rectangular or hexagonal patterns. These assemblies vary in geometry details, such as the cross-section area and shape, the assembly length, and the fuel rod spacing and their mutual organization. The fuel assemblies can be physically separated by solid walls, as in BWRs, VVERs, and

DOI: 10.1201/9781003255000-5

SFRs, or can be open, allowing coolant to mix between the neighboring assemblies, as in PWRs.

Since fuel elements in assemblies are organized in regular patterns, the assemblies can be partitioned into channels that contain fuel, coolant, and other materials in representative average proportions. Such channels are called **coolant channels** and are used in thermal-hydraulic analyses as various approximations of a fuel assembly or a whole reactor core.

In this chapter we apply the balance equations to determine the velocity, temperature, and pressure distributions in various coolant channels. This task requires a solution of a set of differential equations formulated in terms of independent variables such as velocity, pressure, and temperature. The thermophysical properties usually are functions of the fluid local temperature and pressure, causing a mutual coupling between the equations. In spatially distributed flows the flow variables are functions of spatial coordinates and, in non-steady flows, of time. Depending on the shape of the coolant channel, the Cartesian rectangular coordinate system $(x, y, z)$, the cylindrical coordinate system $(r, \theta, z)$, or the spherical coordinate system $(r, \theta, \phi)$ can be used.

The differential equations represent the conservation principles for mass, momentum, and energy. Their specific forms using various assumptions are presented in §4.8. The equations can be solved analytically for laminar flow in some simple channels as shown in §5.1. In §5.2 we discuss one-dimensional bulk flow equations in heated channels. These equations apply for transient and steady-state flow of coolant and play important role in a simplified thermal-hydraulic analysis of fuel assemblies.

The balance equations for single-phase flow presented in §4 can be written in various useful forms that are applicable for analysis of single-phase laminar flows in coolant channels. To describe the processes of heat transfer in a reactor core it is necessary to solve the following differential equations:

1. Equation of mass balance, which is expressing the mass conservation principle.
2. Equation of motion, which is equivalent to the Newton second law, and represents the conservation of the momentum.
3. Energy equation.

Various formulations and solutions of these equations are presented in this chapter.

## 5.1  CHANNEL FLOW

A fluid flowing through a channel enters the channel at some location and a boundary layer develops along the channel walls. When the boundary layer expands to fill the entire channel, the developing flow becomes a **fully developed flow**, in which flow characteristics no longer change with axial position. The flow can be either laminar or turbulent, depending on a Reynolds number $\mathrm{Re} = UD/v$, where $U$ is the mean flow velocity, $D$ is a suitable "diameter" scale for the channel and $v$ is the fluid kinematic viscosity.

For flows in channels the shear stress exerted by the flowing fluid on the solid walls is usually expressed in terms of the characteristic kinetic energy per unit volume $e_K$ given as,

$$e_k = \frac{1}{2}\rho U^2, \tag{5.1}$$

where $\rho$ is the fluid density. A ratio of the wall shear stress $\tau_w$ to the characteristic kinetic energy $e_k$ is called a **friction factor**. Two definitions of the friction factor are commonly used: the **Fanning friction factor**,

$$C_f = \frac{\tau_w}{\frac{1}{2}\rho U^2}, \tag{5.2}$$

and the **Darcy friction factor**,

$$\lambda = \frac{4\tau_w}{\frac{1}{2}\rho U^2}. \tag{5.3}$$

In a fully-developed flow region, the velocity varies only with the lateral coordinates, whereas the total pressure is a function of the axial coordinate only. It is thus possible to find a unique relationship between the pressure drop and the mean flow velocity in the channel, using the concept of the friction coefficient. We consider three simple cases with flow between parallel plates, in a circular pipe, and in an annulus to demonstrate how such relationships can be derived.

The solution procedure involves the following steps:

1. Choose a coordinate system with one axis pointing in the flow direction. We take as a general rule that the flow velocity is directed in the positive direction of the $z$ axis.
2. Make simplifying assumptions. Almost always the considered cases will involve steady flow of a Newtonian fluid with constant physical properties.
3. Write down the conservation equations striking out the terms that are zero.
4. Write down the boundary conditions.
5. Choose a proper transformation of variables to make the simplified conservation equation and the boundary condition equations non-dimensional.
6. Solve the set of equations and find an unique solution for the lateral velocity distribution.
7. Apply area averaging and find the average flow velocity.
8. Find wall shear stress from the calculated velocity gradient at the wall vicinity.
9. Find the friction factor as a ratio of the wall shear stress and the characteristic kinetic energy per unit volume of the fluid.

In the following sections we apply this procedure to analyze the laminar boundary layer flow and the laminar flow in ducts with various cross sections.

### 5.1.1 THE LAMINAR BOUNDARY LAYER

In a fluid flowing over a stationary flat plate a thin layer is created in which velocity rapidly changes from zero value at the plate, surface to the unperturbed free-stream velocity far from the plate. This layer is called the *boundary layer* and the description of the flow behavior in that layer is called the *boundary-layer theory*. Boundary-layer equations are obtained from a simplification of the Navier-Stokes equations when the ratio of the boundary layer thickness $\delta$ to a reference length $L$, $\delta/L$, is sufficiently small. By using an *order-of-magnitude analysis* it is possible to identify in the balance equations small terms that can be neglected.

For a stationary flow with velocity $U$ in the direction of $x$ and with $y$ perpendicular to the plate, the following order-of-magnitude estimates can be noticed,

$$u \sim U, \quad p \sim \rho U^2, \quad x \sim L, \quad y \sim \delta. \tag{5.4}$$

Using these estimates, we can derive order-of-magnitude estimates for derivatives and other variables that are present in the Navier-Stokes equations as follows,

$$\frac{\partial u}{\partial x} \sim \frac{U}{L}, \quad \frac{\partial u}{\partial y} \sim \frac{U}{\delta}, \quad v \sim \frac{U\delta}{L}, \quad \frac{\partial^2 u}{\partial x^2} \sim \frac{U}{L^2}, \quad \frac{\partial^2 u}{\partial y^2} \sim \frac{U}{\delta^2}. \tag{5.5}$$

In particular, we note that the viscous term $\nu(\partial^2 u/\partial x^2)$ is small compared with $\nu(\partial^2 u/\partial y^2)$ and can be neglected in the $x$-momentum equation (4.237). Performing similar order-of-magnitude estimates and neglecting small terms in the mass and $x$-momentum balance equations, the following *Prandtl boundary layer equations* are obtained,

$$\frac{\partial u}{\partial x} + \frac{\partial v}{\partial y} = 0, \tag{5.6}$$

$$u\frac{\partial u}{\partial x} + v\frac{\partial u}{\partial y} = -\frac{1}{\rho}\frac{dp}{dx} + \nu\frac{\partial^2 u}{\partial y^2}, \tag{5.7}$$

where $-(1/\rho)(dp/dx)$ represents a pressure gradient in the fluid far from the plate surface. For potential steady flow, this term can be replaced with $U(dU/dx)$, where $U(x)$ is the flow velocity far from the plate surface. The boundary conditions for the equations are as follows: $u = v = 0$ at $y = 0$ and the velocity merging at the outer edge of the boundary layer $(u(x,y) \to U)$.

### Example 5.1: Drag Force on a Plate

Using the boundary layer theory, estimate the drag force on a finite plate of width $W = 0.25$ m and length $L = 1.25$ m, wetted from both sides by liquid sodium that is flowing with constant and uniform velocity $U = 0.2$ m/s. Assume the following sodium velocity distribution in the boundary layer:

$$\frac{u}{U} = \begin{cases} \frac{3}{2}\frac{y}{\delta} - \frac{1}{2}\left(\frac{y}{\delta}\right)^3 & \text{for } 0 \leq y \leq \delta(x) \\ 1 & \text{for } y \geq \delta(x) \end{cases} \tag{5.8}$$

Assume a potential flow of sodium for $y > \delta(x)$ and that its temperature is the same everywhere and equal to 450 K.

* * *

*Solution*: The drag force exerted on both sides of the plate can be found as,

$$F_D = 2 \int_0^W \int_0^L \tau_w \mathrm{d}x \mathrm{d}z, \tag{5.9}$$

where $\tau_w$ is the wall shear stress given as,

$$\tau_w = \mu \left. \frac{\partial u}{\partial y} \right|_{y=0}.$$

The $x$-momentum equation can be written as

$$\rho u \frac{\partial u}{\partial x} + \rho v \frac{\partial u}{\partial y} = \rho U \frac{\mathrm{d}U}{\mathrm{d}x} + \frac{\partial \tau}{\partial y}. \tag{5.10}$$

Here we multiplied both sides of Eq. (5.7) with $\rho$ and replaced $-\mathrm{d}p/\mathrm{d}x$ with $\rho U (\mathrm{d}U/\mathrm{d}x)$, and $\rho v (\partial^2 u/\partial y^2)$ with $\partial \tau / \partial y$. Equation (5.10) can be now integrated over the thickness of the boundary layer from 0 to $\delta$ and, after rearranging, we get,

$$\int_0^\delta \frac{\partial \tau}{\partial y} \mathrm{d}y = \rho \int_0^\delta \left( u \frac{\partial u}{\partial x} + v \frac{\partial u}{\partial y} - U \frac{\mathrm{d}U}{\mathrm{d}x} \right) \mathrm{d}y. \tag{5.11}$$

We should note here that the integral on the left hand side is just the wall shear stress with a minus sign,

$$\int_0^\delta \frac{\partial \tau}{\partial y} \mathrm{d}y = \tau(\delta) - \tau(0) = -\tau_w.$$

Multiplying the mass conservation equation by $u$ and adding to the term under integral on the right hand side of the equation gives,

$$\rho \int_0^\delta \left( u \frac{\partial u}{\partial x} + v \frac{\partial u}{\partial y} + u \frac{\partial u}{\partial x} + u \frac{\partial v}{\partial y} - U \frac{\mathrm{d}U}{\mathrm{d}x} \right) \mathrm{d}y =$$
$$\rho \int_0^\delta \left( \frac{\partial u^2}{\partial x} + \frac{\partial (uv)}{\partial y} - U \frac{\mathrm{d}U}{\mathrm{d}x} \right) \mathrm{d}y = \rho \int_0^\delta \left( \frac{\partial u^2}{\partial x} - U \frac{\partial u}{\partial x} - U \frac{\mathrm{d}U}{\mathrm{d}x} \right) \mathrm{d}y \tag{5.12}$$

Here we used the following relationship,

$$\int_0^\delta \frac{\partial (uv)}{\partial y} = U v_e = -U \int_0^\delta \frac{\partial u}{\partial x} \mathrm{d}y,$$

where from an integration of the mass conservation equation (5.6) over the boundary layer thickness we obtained,

$$v_e = - \int_0^\delta \frac{\partial u}{\partial x} \mathrm{d}y.$$

Finally, after rearrangements we obtain the following *von Karman momentum balance* for the boundary layer,

$$\tau_w = \rho \frac{\mathrm{d}}{\mathrm{d}x} \int_0^\delta (U - u) u \mathrm{d}y + \rho \frac{\mathrm{d}U}{\mathrm{d}x} \int_0^\delta (U - u) \mathrm{d}y. \tag{5.13}$$

Since $U = const$, $dU/dx = 0$ and the second term on the right hand side is zero. Substituting the velocity distribution given by Eq. (5.8) into (5.13) yields,

$$\frac{3}{2}\frac{\mu U}{\delta} = \frac{39}{280}\rho U^2 \frac{d\delta}{dx}.$$

This first order differential equation can be now solved for $\delta$ and the following expression for the boundary layer thickness is obtained,

$$\delta(x) = \sqrt{\frac{280}{13}\frac{\mu x}{\rho U}} \approx 4.641\sqrt{\frac{\mu x}{\rho U}}. \tag{5.14}$$

Substituting this expression into Eq. (5.9), the drag force is found as,

$$F_D = \frac{1.293}{\sqrt{Re_L}}\rho U^2 WL, \tag{5.15}$$

where $Re_L$ is the Reynolds number based on the plate length defined as,

$$Re_L = \frac{\rho UL}{\mu}. \tag{5.16}$$

From the correlations given in §3.4 for sodium density and dynamic viscosity, we have $\rho = 909.4$ kg/m$^3$ and $\mu = 4.605 \cdot 10^{-4}$ Pa·s. The Reynolds number is then

$$Re_L = \frac{909.4 \text{ kg/m}^3 \times 0.2 \text{ m/s} \times 1.25 \text{ m}}{4.609 \cdot 10^{-4} \text{ Pa·s}} \approx 4.937 \cdot 10^5, \tag{5.17}$$

and the drag force is found as

$$F_D = \frac{1.293}{\sqrt{4.937 \cdot 10^5}} \times 909.4 \text{ kg/m}^3 \times (0.2 \text{ m/s})^2 \times 0.25 \text{ m} \times 1.25 \text{ m}$$
$$\approx 2.902 \cdot 10^{-2} \text{ N} \tag{5.18}$$

The Prandtl boundary layer equations were first investigated and solved by Blasius, who assumed $dp/dx = 0$ and applied the stream function $\psi$ to represent velocities $u$ and $v$ as follows,

$$u = \frac{\partial \psi}{\partial y}, \quad v = -\frac{\partial \psi}{\partial x}. \tag{5.19}$$

Introducing the following non-dimensional variable

$$\eta = \frac{y}{2}\left(\frac{\rho U}{\mu x}\right)^{1/2}, \tag{5.20}$$

the stream function can be expressed as

$$\psi = \left(\frac{\mu Ux}{\rho}\right)^{1/2} f(\eta), \tag{5.21}$$

where $f(\eta)$ is a new unknown function of $\eta$. Expressing velocities and their derivatives in terms of $f(\eta)$, the following ordinary differential equation is obtained,

$$f''' + ff'' = 0, \tag{5.22}$$

and the boundary conditions become,

$$\begin{aligned} \text{for} \quad \eta = 0 \quad f = 0, \quad f' = 0, \\ \text{for} \quad \eta \to \infty \quad f \to 1, \quad f' \to 2. \end{aligned} \tag{5.23}$$

Equation (5.22) together with boundary conditions (5.23) can be solved numerically and values of $f$, $f'$, and $f''$ can be presented in tables as a function of $\eta$. From these values, the velocities $u$ and $v$ can be obtained as,

$$u = \frac{U}{2} f'(\eta), \quad v = \frac{1}{2} \left( \frac{\mu U}{\rho x} \right)^{1/2} (\eta f'(\eta) - f(\eta)), \tag{5.24}$$

whereas velocity gradients are as follows,

$$\frac{\partial u}{\partial x} = -\frac{U\eta}{4x} f''(\eta), \quad \frac{\partial u}{\partial y} = \frac{U}{4} \left( \frac{U\rho}{\mu x} \right)^{1/2} f''(\eta). \tag{5.25}$$

If we assume that the boundary layer thickness corresponds to the distance from the plate where $u = 0.99U$, the following expression for $\delta(x)$ can be obtained,

$$\delta(x) \approx 4.96 \sqrt{\frac{\mu x}{\rho U}}. \tag{5.26}$$

Using the expression for the velocity gradient, the drag force on a plate of width $W$ and length $L$ can be found as,

$$\begin{aligned} F_D &= 2 \int_0^W \int_0^L \mu \left( \frac{\partial u}{\partial y} \right)_{y=0} \mathrm{d}x\mathrm{d}z \\ &= 2 \int_0^W \int_0^L \left[ \frac{\mu U}{2} \sqrt{\frac{\rho U}{\mu x}} f''(0) \right] \mathrm{d}x\mathrm{d}z \quad , \\ &= \frac{1.328}{\sqrt{Re_L}} \rho U^2 W L \end{aligned} \tag{5.27}$$

where $f''(0) = 1.328$ is used and $Re_L = \rho U L / \mu$ is the length-based Reynolds number.

The *Blasius solution* of the boundary-layer problem presented above agrees very well with experimental data. Because of the neglect of the pressure gradient $\mathrm{d}p/\mathrm{d}x$, the solution is most accurate at large local Reynolds numbers, $Re_x = xU\rho/\mu \gg 1$, where this assumption is valid. However, the initial region for low Reynolds numbers is quite small in most drag calculations. The theory is also limited by an upper value of the local Reynolds number, due to a transition to turbulent flow. The transition is found to begin for the local Reynolds number in a range $3 \cdot 10^5 < Re_x < 3 \cdot 10^6$, depending on the turbulence intensity of the approaching stream.

In addition to the exact boundary layer solutions, in which all parameters of interest are predicted rather than assumed, there exist approximate solutions, in which the velocity profile in the boundary layer is assumed. Such an approximate solution is presented in Example 5.1. It is interesting to note that the approximate solution yields quite similar results in terms of the boundary layer thickness and the drag force on the plate, and gives results approximately only 3% below the exact solution.

### 5.1.2  LAMINAR CHANNEL FLOW

Laminar flow occurs in cases where the channel is relatively small, the fluid is moving slowly, and its viscosity is relatively high. At a sufficient length of the channel, the hydrodynamic characteristics of the flow change from arbitrary inlet to stabilized, or fully developed flow values, which are determined by the geometrical shape of the channel. For several simple shapes the Navier-Stokes equations can be solved analytically, providing valuable information on velocity distribution and wall shear stress values. In this section we discuss some important results obtained for fully developed flows and developing flows. A fully developed flow is established in a channel after a certain entrance length, which for laminar flows is fairly large. This length depends both on the channel diameter and the Reynolds number.

#### Fully Developed Flow

We will first investigate fully developed laminar flow in circular tubes since the obtained expressions are particularly simple. These results will be next extended to laminar flows in non-circular channels.

#### Circular Channels

We will find the radial velocity distribution and the wall shear stress for a fully-developed laminar upward flow through a circular channel. For the specified conditions, the momentum equation can be written in the cylindrical coordinate system as follows:

$$0 = -g\rho - \frac{dp}{dz} + \frac{\mu}{r}\frac{d}{dr}\left(r\frac{dw}{dr}\right),$$   (5.28)

where $r$–the tube radius, $w$–the $r$-dependent axial velocity, $p$–pressure, $\mu$–dynamic viscosity, $\rho$–density, and $g$–acceleration of gravity. The boundary conditions are as follows:

$$\frac{dw}{dr} = 0 \quad \text{at} \quad r = 0,$$   (5.29)

$$w = 0 \quad \text{at} \quad r = R,$$   (5.30)

where $R$ is the pipe radius.

The differential equation and the boundary conditions can be expressed in a dimensionless form by introducing the following transformation of variables,

$$r = \xi R, \quad w = \vartheta U.$$   (5.31)

Here $\xi$ and $\vartheta$ are new dimensionless variables, and $U$ is a velocity scale taken to be equal to the mean velocity over the pipe cross section area,

$$U = \langle w \rangle_2 \equiv \frac{1}{\pi R^2} \int_0^R 2\pi r w(r) \mathrm{d}r. \tag{5.32}$$

Substituting expressions (5.31) into Eqs. (5.28)–(5.30) yields,

$$\frac{1}{\xi} \frac{\mathrm{d}}{\mathrm{d}\xi} \left( \xi \frac{\mathrm{d}\vartheta}{\mathrm{d}\xi} \right) = \frac{R^2 \left( \frac{\mathrm{d}p}{\mathrm{d}z} + \rho g \right)}{\mu U} = -P, \tag{5.33}$$

$$\frac{\mathrm{d}\vartheta}{\mathrm{d}\xi} = 0 \quad \text{at} \quad \xi = 0, \tag{5.34}$$

$$\vartheta = 0 \quad \text{at} \quad \xi = 1. \tag{5.35}$$

The general solution of Eq. (5.33) is as follows,

$$\vartheta = -\frac{P}{4} \xi^2 + C_1 \ln \xi + C_2, \tag{5.36}$$

where $P$ is a non-dimensional parameter defined by Eq. (5.33) and $C_1$, $C_2$ are integration constants. These constants can be determined from the boundary conditions given by Eqs. (5.34) and (5.35) and the following dimensionless solution is obtained,

$$\vartheta = \frac{P}{4} \left( 1 - \xi^2 \right). \tag{5.37}$$

We can now find the cross-section mean value of $\vartheta(\xi)$ as,

$$\vartheta_m \equiv 1 = \frac{1}{\pi} \int_0^1 2\pi \xi \vartheta(\xi) \mathrm{d}\xi = \frac{P}{8}, \tag{5.38}$$

where we noted that the mean value $\vartheta_m$ must be equal to one. Thus, the distribution of non-dimensional velocity in the pipe is,

$$\vartheta = 2 \left( 1 - \xi^2 \right). \tag{5.39}$$

Using the dimensional variables, the solution is as follows,

$$w(r) = 2U \left[ 1 - \left( \frac{r}{R} \right)^2 \right]. \tag{5.40}$$

As can be seen, the local fluid velocity at the centerline is equal to twice the average velocity: $w_0 = w(0) = 2U$. With known radial velocity distribution, the wall shear stress can be found as,

$$\tau_w = -\mu \frac{\mathrm{d}w}{\mathrm{d}r}\bigg|_{r=R} = -\frac{R^2}{2} \left( \frac{\mathrm{d}p}{\mathrm{d}z} + \rho g \right) = \frac{P U \mu}{2 R} = \frac{4 U \mu}{R}. \tag{5.41}$$

We note here that for the vertical upward flow $\frac{dp}{dz} + \rho g < 0$ and $\tau_w > 0$. Opposite signs should be taken for the vertical downward flow. For stagnant liquid we have $\frac{dp}{dz} + \rho g = 0$ and $\tau_w = 0$, as expected. Substituting the obtained expression for the wall shear stress into the definitions of friction factors, Eqs. (5.2) and (5.3), we get the Fanning and Darcy friction factors for a circular pipe as follows,

$$C_f = \frac{16}{\text{Re}}, \quad \lambda = \frac{64}{\text{Re}}, \tag{5.42}$$

where Re is the Reynolds number based on the pipe diameter.

### Annuli

A cylindrical annulus consists of two coaxial circular cylinders of different radii. We investigate a fully-developed axisymmetric laminar upward flow through the annulus with outer radius $R$ and inner radius $R_i = \kappa R$, where $0 < \kappa < 1$. We use the cylindrical coordinate system $(r, \theta, z)$, with the corresponding velocity components $(u, v, w)$. Since the flow is assumed to be fully developed and axisymmetric, $u = v = 0$ and $w = w(r)$, the only non-trivial conservation equation is the following radial component of the momentum equation,

$$0 = -g\rho - \frac{dp}{dz} + \frac{\mu}{r} \frac{d}{dr}\left(r \frac{dw}{dr}\right), \tag{5.43}$$

with the no-slip boundary conditions at the walls,

$$w(r)|_{r=\kappa R} = 0 \quad \text{and} \quad w(r)|_{r=R} = 0. \tag{5.44}$$

Using a variable transformation given by Eq. (5.31), the ordinary differential equation to be solved is as given by Eq. (5.33), and the boundary conditions become,

$$\vartheta = 0 \quad \text{at} \quad \xi = \kappa \quad \text{and} \quad \vartheta = 0 \quad \text{at} \quad \xi = 1. \tag{5.45}$$

Substituting the boundary conditions into the solution given by Eq. (5.36), the non-dimensional velocity distribution is as follows,

$$\vartheta(\xi) = \frac{P}{4}\left[1 - \xi^2 + \left(\kappa^2 - 1\right)\frac{\ln \xi}{\ln \kappa}\right]. \tag{5.46}$$

The mean non-dimensional velocity in the cross section of the annulus is found as,

$$\vartheta_m \equiv 1 = \frac{1}{\pi(1 - \kappa^2)} \int_\kappa^1 2\pi \xi \, \vartheta(\xi) d\xi = \frac{P}{8}\left[\frac{1 - \kappa^4}{1 - \kappa^2} - \frac{1 - \kappa^2}{\ln(1/\kappa)}\right]. \tag{5.47}$$

Thus,

$$P = 8\left[\frac{1 - \kappa^4}{1 - \kappa^2} - \frac{1 - \kappa^2}{\ln(1/\kappa)}\right]^{-1}, \tag{5.48}$$

and the non-dimensional velocity distribution can be written as,

$$\vartheta(\xi) = 2\left[1 - \xi^2 + (\kappa^2 - 1)\frac{\ln\xi}{\ln\kappa}\right]\left[\frac{1 - \kappa^4}{1 - \kappa^2} - \frac{1 - \kappa^2}{\ln(1/\kappa)}\right]^{-1}. \tag{5.49}$$

The radial velocity is thus obtained as,

$$w(r) = 2U\left[1 - \left(\frac{r}{R}\right)^2 + (\kappa^2 - 1)\frac{\ln(r/R)}{\ln\kappa}\right]\left[\frac{1 - \kappa^4}{1 - \kappa^2} - \frac{1 - \kappa^2}{\ln(1/\kappa)}\right]^{-1}. \tag{5.50}$$

The shear stress on the inner and outer wall can be found as,

$$\tau_{wi} = \mu\frac{dw}{dr}\bigg|_{r=\kappa R} = -\frac{PU\mu}{4R}\left[2\kappa - \frac{\kappa^2 - 1}{\ln\kappa}\right], \tag{5.51}$$

and

$$\tau_{wo} = -\mu\frac{dw}{dr}\bigg|_{r=R} = \frac{PU\mu}{4R}\left[2 - \frac{\kappa^2 - 1}{\ln\kappa}\right], \tag{5.52}$$

respectively. The "effective" wall shear stress in a cross section is found as a weighted mean value as follows,

$$\tau_{we} = \frac{\kappa\tau_{wi} + \tau_{wo}}{1 + \kappa} = \frac{PU\mu(1 - \kappa)}{2R}. \tag{5.53}$$

The corresponding "effective" Fanning and Darcy friction factors are found as,

$$C_{fe} = \frac{2P}{\text{Re}}(1 - \kappa) = \frac{16}{\text{Re}}\frac{1 - \kappa}{\frac{1-\kappa^4}{1-\kappa^2} - \frac{1-\kappa^2}{\ln(1/\kappa)}}, \quad \lambda = \frac{64}{\text{Re}}\frac{1 - \kappa}{\frac{1-\kappa^4}{1-\kappa^2} - \frac{1-\kappa^2}{\ln(1/\kappa)}}. \tag{5.54}$$

For $\kappa = 0$ the expressions become identical with the ones derived for laminar flow in a pipe.

## Non-Circular Channels

As a first example of fully-developed laminar flow in a non-circular channel, we consider a vertical upward flow between two parallel plates, whose length and width in $z$ and $x$ directions, respectively, is very large when compared with their separation in the $y$ direction. Assuming the steady laminar flow of a Newtonian fluid with constant properties in the positive direction of $z$ axis, far from the inlet the flow will be fully developed and the flow variables will have the following values:

$$u = v = 0, \quad w = w(y), \quad p = p(z), \tag{5.55}$$

where $u, v, w$ are velocity components in the positive directions of axes $x, y$ and $z$, respectively, and $p$ is the pressure.

Taking into account the assumptions, the only non-trivial conservation equation, from mass and momentum conservation equations specified in §4.8, is the following momentum conservation equation in $z$-direction,

$$0 = -g\rho - \frac{dp}{dz} + \mu\frac{d^2w}{dy^2}, \tag{5.56}$$

where $y$–the lateral coordinate normal to the plates, $w$–the axial velocity, $p$–pressure, $\mu$–dynamic viscosity, $\rho$–density, and $g$–acceleration of gravity. The solution has to satisfy the following boundary conditions,

$$\frac{dw}{dy} = 0 \quad \text{at} \quad y = 0, \tag{5.57}$$

$$w = 0 \quad \text{at} \quad y = H. \tag{5.58}$$

Here the separation between the plates is assumed to be equal to $2H$.

A more compact solution of Eqs. (5.56)–(5.58) will be obtained if we introduce the following non-dimensional variables,

$$y = \eta H, \quad w = \omega U. \tag{5.59}$$

Here $\eta$ and $\omega$ are new dimensionless variables, and $U$ is a velocity scale taken to be equal to the mean velocity over the channel cross-section area,

$$U = \langle w \rangle_2 \equiv \frac{1}{H} \int_0^H w(y) dy. \tag{5.60}$$

Substituting expressions (5.59) into Eqs. (5.56)–(5.58) yields,

$$\frac{d^2\omega}{d\eta^2} = \frac{H^2 \left( \frac{dp}{dz} + \rho g \right)}{\mu U} = -P, \tag{5.61}$$

$$\frac{d\omega}{d\eta} = 0 \quad \text{at} \quad \eta = 0, \tag{5.62}$$

$$\omega = 0 \quad \text{at} \quad \eta = 1. \tag{5.63}$$

The general solution of Eq. (5.61) is as follows,

$$\omega = -\frac{P}{2}\eta^2 + C_2\eta + C_1, \tag{5.64}$$

where $P$ is a non-dimensional parameter defined by Eq. (5.61) and $C_1, C_2$ are integration constants. These constants can be determined from the boundary conditions given by Eqs. (5.62) and (5.63) and the following dimensionless solution is obtained,

$$\omega = \frac{P}{2} \left( 1 - \eta^2 \right). \tag{5.65}$$

We can now find the cross-section mean value of $\omega(\eta)$ as,

$$\omega_m \equiv 1 = \int_0^1 \omega(\eta) d\eta = \frac{P}{3}, \tag{5.66}$$

where we noted that the mean value $\omega_m$ must be equal to one. Thus, the distribution of non-dimensional velocity between two parallel plates is,

$$\omega(\eta) = \frac{3}{2} \left( 1 - \eta^2 \right). \tag{5.67}$$

Using the dimensional variables, the solution is as follows,

$$w(y) = \frac{3U}{2}\left[1 - \left(\frac{y}{H}\right)^2\right]. \tag{5.68}$$

As can be seen, a parabolic velocity distribution is obtained with the maximum value equal to $w(0) = 3U/2$.

The wall shear stress is obtained as,

$$\tau_w = -\mu\frac{dw}{dy}\bigg|_{y=H} = \frac{3U\mu}{H}, \tag{5.69}$$

and the corresponding Fanning and Darcy friction factors are

$$C_f \equiv \frac{\tau_w}{\frac{1}{2}\rho U} = \frac{6\mu}{HU\rho}, \quad \lambda \equiv \frac{4\tau_w}{\frac{1}{2}\rho U} = \frac{24\mu}{HU\rho}. \tag{5.70}$$

The friction factors are usually expressed in terms of a Reynolds number $\mathrm{Re} = \rho UD/\mu$, where $D$ is a hydraulic diameter of the channel. Since for flow between parallel channels $D_h = 4H$, the friction factors become,

$$C_f = \frac{24}{\mathrm{Re}}, \quad \lambda = \frac{96}{\mathrm{Re}}. \tag{5.71}$$

For flow in rod bundles with rod diameter $D_R$ and triangular lattice with pitch $P$, the friction factor can be determined similarly to the circular channel, using an effective diameter that is provided as follows [129],

$$D_{eff} = \frac{2\varepsilon}{(1-\varepsilon)^2}\left(\frac{\varepsilon}{2} - \frac{3}{2} - \frac{\ln\varepsilon}{1-\varepsilon}\right)D_h, \tag{5.72}$$

where $\varepsilon$ is a fraction of the channel cross-section area occupied by the rods, which for the triangular lattice is found as,

$$\varepsilon = \frac{\pi}{2\sqrt{3}}\left(\frac{D_R}{P}\right)^2. \tag{5.73}$$

The method is applicable to rod bundles with $P/D_R > 1.3$.

## Developing Flow

Just downstream of the channel entrance the velocity changes from uniform to fully developed distribution. The channel region where this velocity distribution takes place is called the entrance or developing section and the length of this section is called the **hydrodynamic entrance length**. Velocity distribution and pressure drop in the hydrodynamic entrance region in a round tube with uniform velocity distribution at the tube entrance can be approximated with the following expressions [247],

$$\frac{w}{U} = 2(1-\eta^2) - 4\sum_{n=0}^{\infty}\frac{1}{\beta_n^2}\left[1 - \frac{J_0(\beta_n\eta)}{J_0(\beta_n)}\right]\exp\left(-4\beta_n^2\zeta\right), \tag{5.74}$$

$$\frac{\Delta p}{\rho U^2/2} = 64\zeta + \frac{2}{3} - 8\sum_{n=0}^{\infty} \frac{1}{\beta_n^2} \exp\left(-4\beta_n^2 \zeta\right), \tag{5.75}$$

where $J_0$ is the Bessel function of first kind and zero order, $\beta_n$ ($n = 1, 2, ...$) are roots of the Bessel function of first kind and second order $J_2$, $\zeta = z/D\mathrm{Re}$, $\eta = r/R$, $w$ is the axial local velocity, and $U$ is the mean velocity. The Fanning friction factor is given as,

$$C_f = \frac{16}{\mathrm{Re}} + \frac{8}{\mathrm{Re}} \sum_{n=0}^{\infty} \exp\left(-4\beta_n^2 \zeta\right). \tag{5.76}$$

For $\zeta < 0.001$ a simple approximation for the Fanning friction factor is given as,

$$C_{f,z} = \frac{1.72}{\mathrm{Re}\sqrt{\zeta}}. \tag{5.77}$$

For a channel with an arbitrary cross-section, pressure drop in the whole length of the hydraulically developing section can be estimated from the following expression,

$$\frac{\Delta p}{\rho U^2/2} = \frac{C_1}{\mathrm{Re}} \frac{z}{D_h} + C_2, \tag{5.78}$$

where $C_1$ is the coefficient in the Darcy friction factor $\lambda = C_1/\mathrm{Re}$ for fully-developed flow in the channel and $C_2$ is obtained as,

$$C_2 = \frac{2}{A} \int_A \left[\left(\frac{w}{U}\right)^3 - \left(\frac{w}{U}\right)^2\right] dA, \tag{5.79}$$

where $w$ is the fully-developed velocity distribution in the channel cross-section area and $U$ is the mean velocity. The total length of the hydrodynamic entrance section $L_h$ in a channel can be estimated from the following expression,

$$L_h/D_h = 0.055 \, \mathrm{Re}, \tag{5.80}$$

where $D_h$ is the equivalent (hydraulic) diameter of the channel.

The entrance region of a channel with laminar flow is characterized by a changing velocity profile due to developing shear layers along the channel walls. The excess pressure drop in this region is due to both increased shear in the boundary layers and the acceleration of the core flow. Assuming that the point where the developing centerline velocity equals 99 percent of the maximum value of the fully developed flow corresponds the end of the developing region, the so called *hydrodynamic entrance length $L_h$* can be found as [199],

$$\frac{L_h}{D} = \frac{0.6}{1 + 0.035\mathrm{Re}} + 0.056\mathrm{Re}. \tag{5.81}$$

The following formula for friction coefficient, valid for many duct shapes, has been proposed [197],

$$C_{f,z}\mathrm{Re} = \frac{3.44}{\sqrt{\zeta}} + \frac{C_f\mathrm{Re} + K_\infty/4\zeta - 3.44/\sqrt{\zeta}}{1 + c/\zeta^2}, \tag{5.82}$$

where $\zeta = (z/D)/\text{Re}$. The constants $C_f\text{Re}$, $K_\infty$ and $c$ depend on the duct shape and are equal to 16, 1.25 and $2.12\times10^{-4}$ for a circular pipe and 1.23, 1.43, and $2.9\times10^{-4}$ for a square, respectively.

### 5.1.3   THE TURBULENT BOUNDARY LAYER

For fully-developed turbulent flows in channels, the axial velocity significantly changes in the region close to the walls. This region, called a turbulent boundary layer, is very important for the mass, momentum, and heat transfer processes. Experiments and theoretical investigations show that the mean velocity and temperature profiles in the turbulent boundary layer follow some general principles when expressed in terms of length and velocity *wall units*.

The distance from the wall measured in wall units is denoted by,

$$y^+ \equiv \frac{y}{\delta_v} = \frac{u_\tau y}{\nu},\tag{5.83}$$

where the **viscous lengthscale** is defined as,

$$\delta_v \equiv \nu\sqrt{\frac{\rho}{\tau_w}} = \frac{\nu}{u_\tau},\tag{5.84}$$

and the **friction velocity** is given as,

$$u_\tau = \sqrt{\frac{\tau_w}{\rho}}.\tag{5.85}$$

Here $\tau_w$ is the **wall shear stress**, which is entirely due to viscous contribution and can be defined in terms of the mean axial velocity gradient at the wall surface as,

$$\tau_w \equiv \mu\left(\frac{d\bar{u}}{dy}\right)_{y=0}.\tag{5.86}$$

### The Mean Velocity Profile

At a high Reynolds number there is an **inner layer** close to the wall in which the mean velocity profile is determined by the viscous scales, independent of the channel radius and the bulk velocity. As a result, a dimensionless velocity in the inner layer,

$$u^+ \equiv \frac{\bar{u}}{u_\tau},\tag{5.87}$$

depends solely on $y^+$. The relationship,

$$u^+ = f_w\left(y^+\right),\tag{5.88}$$

is called the **law of the wall**. Here $f_w$ is a universal function for boundary layers, channel flows, and pipe flows. In the **viscous sublayer**, when $y^+ < 5$, the following linear relation is valid,

$$u^+ = y^+.\tag{5.89}$$

For larger $y^+$ the velocity profile is such that the mean velocity gradient is,

$$\frac{du^+}{dy^+} = \frac{1}{\kappa y^+},\tag{5.90}$$

where $\kappa$ is the von Kármán constant. Equation (5.90) integrates to,

$$u^+ = \frac{1}{\kappa}\ln y^+ + B,\tag{5.91}$$

where $B$ is a constant. There is some variation in the values ascribed to constants $\kappa$ and $B$, but generally they are taken as,

$$\kappa = 0.41, \quad B = 5.2.\tag{5.92}$$

The logarithmic equation (5.91) together with constants (5.92) is referred to as the **log law**. The mean velocity profile and the various wall regions and layers are shown in Fig. 5.1.

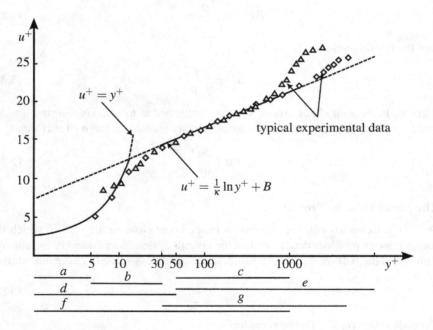

**Figure 5.1**   The mean velocity profile and various wall regions and layers defined in terms of $y^+$ and $y/\delta$ for turbulent channel flow at high Reynolds number: $a$–viscous sublayer ($y^+ < 5$), $b$–buffer layer ($5 < y^+ < 30$), $c$–overlap region ($y^+ > 50$), $d$–viscous wall region ($y^+ < 50$), $e$–outer layer ($y^+ > 50$, $f$–inner layer ($y/\delta < 0.1$), $g$–log-law region ($y^+ > 30$, $y/\delta < 0.3$).

There are many attempts to develop velocity profile correlations that are valid through several wall regions and layers shown in Fig. 5.1. One example is the Reichardt correlation [185],

$$u^+ = \frac{1}{\kappa} \ln(1 + \kappa y^+) + 7.8 \left[ 1 - e^{-y^+/11} - \frac{y^+}{11} e^{-y^+/3} \right]. \tag{5.93}$$

For entire wall-related region, the following correlation was proposed by Spalding [210],

$$y^+ = u^+ + e^{-\kappa B} \left[ e^{\kappa u^+} - 1 - \kappa u^+ - \frac{(\kappa u^+)^2}{2} - \frac{(\kappa u^+)^3}{6} \right]. \tag{5.94}$$

This expression is an excellent fit to inner-law data from the wall to after $y^+ > 300$.

### The Mean Temperature Profile

In analogy to the mean velocity profile in the turbulent boundary layer, a universal mean temperature profile can be formulated for turbulent flows with heat transfer. With constant properties $\lambda$, $\rho$, $c_p$, and $\nu$, we can introduce the **wall conduction temperature** as follows,

$$T^* = \frac{\overline{q_w''}}{\rho c_p u_\tau}, \tag{5.95}$$

where $\overline{q_w''}$ is the Reynolds averaged wall heat flux. This heat flux is assumed to be constant throughout the boundary layer and to consist of two components,

$$\overline{q_w''} = \overline{q_\lambda''} + q_t'', \tag{5.96}$$

where the molecular heat flux is defined as

$$\overline{q_\lambda''} \equiv -\lambda \frac{\partial \overline{T}}{\partial y}, \tag{5.97}$$

and the turbulent heat flux is given as

$$q_t'' \equiv \rho c_p \overline{T'v'}. \tag{5.98}$$

Defining the dimensionless temperature and the dimensionless turbulent heat flux as, respectively,

$$T^+ \equiv \frac{T_w - \overline{T}}{T^*}, \quad (q_t'')^+ \equiv \frac{q_t''}{q_w''}, \tag{5.99}$$

the following equation for the wall layer is valid,

$$\frac{1}{\mathrm{Pr}} \frac{dT^+}{dy^+} + (q_t'')^+ = 1, \tag{5.100}$$

with the boundary conditions,

$$y^+ = 0, \quad (q_t'')^+ = 0. \tag{5.101}$$

Several formulations of the temperature log-law have been proposed in the literature, assuming the following general relationship,

$$T^+(y^+, \mathrm{Pr}) = \begin{cases} \mathrm{Pr}\, y^+ & \text{in viscous sublayer} & y^+ < 5 \\ \frac{1}{\kappa_T}\ln y^+ + \beta(\mathrm{Pr}) & \text{in logarithmic region} & 30 < y^+ \end{cases}, \qquad (5.102)$$

where $\mathrm{Pr} = \mu c_p/\lambda$ and $\kappa_T$ and $\beta(\mathrm{Pr})$ are given as [193],

$$\kappa_T = 0.47, \quad \beta(\mathrm{Pr}) = 13.7\,\mathrm{Pr}^{2/3} - 7.5, \qquad (5.103)$$

or [117],

$$\kappa_T = 0.47, \quad \beta(\mathrm{Pr}) = \left(3.85\,\mathrm{Pr}^{1/3} - 1.3\right)^2 + 2.12\ln\mathrm{Pr}. \qquad (5.104)$$

A correlation that is valid in the near wall region, in the logarithmic region, and across the overlap boundary layer is given as [118],

$$T^+ = \mathrm{Pr}y^+ e^{-L} + \left[2.12\ln(1+y^+) + \beta\right] e^{-1/L}, \qquad (5.105)$$

where

$$\beta = \left(3.85\mathrm{Pr}^{1/3} - 1.3\right)^2 + 2.12\ln(\mathrm{Pr}), \qquad (5.106)$$

and

$$L = \frac{0.01\,(\mathrm{Pr}y^+)^4}{1 + 5\mathrm{Pr}^3 y^+}. \qquad (5.107)$$

## 5.1.4 TURBULENT CHANNEL FLOW

In laminar flow the fluid velocity is low enough to prevent unstable, random, and unpredictable movements of fluid particles. Thanks to this, no time smoothing is necessary, and the balance equations can be expressed in terms of instantaneous values of flow variables. For general engineering purpose, it can be assumed that in a round tube, laminar flow prevails when $\mathrm{Re} < 2100$ and turbulent flow exists when $\mathrm{Re} > 4000$. For $2100 < \mathrm{Re} < 4000$ transitional flow takes place. When flow becomes turbulent the flow becomes inherently unstable. For such flows steady distributions of flow parameters can be obtained by proper time averaging of their instantaneous values. Experiments show that steady distributions of flow parameters for turbulent flow are significantly different from their laminar counterparts. For example, the radial distribution of the mean velocity in a circular pipe no longer follows the parabolic function and the presence of a thin turbulent boundary layer can be observed.

The transition from laminar to turbulent flow drastically changes the mixing, transport, and drag properties of fluids. Unlike laminar flows, turbulent flows cannot be represented in terms of analytical solutions, even when considered in the simplest geometries, such as flows in round tubes or flows between infinite parallel plates. Thus, turbulent flow theory relies almost entirely on experimental investigations, from which proper computational models are derived.

## Fully Developed Flow

Just like in laminar flows, the hydrodynamic characteristics of turbulent flow in a channel transition from arbitrary inlet conditions to stabilized, fully-developed values once a sufficient length of the channel is reached. In this sub-section we discuss the main characteristics of fully-developed turbulent flow in circular and non-circular channels.

### Circular Channel

In the early 1930s it was established experimentally by Nikuradse that irrespective of the Reynolds number of mean flow, velocity distribution for smooth pipes is always given by the formula,

$$u^+ = 5.75 \log y^+ + 5.5, \tag{5.108}$$

where $u^+$ and $y^+$ are the dimensionless velocity and the distance from the wall, respectively. More recent measurements of the turbulent flow of various fluids in circular pipes provide a somewhat more involved expression as follows,

$$u^+ = \begin{cases} y^+ \left[1 - \frac{1}{4}(y^+/14.5)^3\right] & \text{in viscous sublayer} & y^+ < 5 \\ 5\ln(y^+ + 0.205) - 3.27 & \text{in buffer layer} & 5 < y^+ < 30 \\ 2.5\ln y^+ + 5.5 & \text{in log-law region} & 30 < y^+ \end{cases} . \tag{5.109}$$

As an alternative to the logarithmic velocity distribution, the **power law** can be used. For turbulent flow in a pipe with radius $R$, the velocity distribution is roughly described by the following expression,

$$\frac{w(r)}{w_0} \approx \left(1 - \frac{r}{R}\right)^{1/n}, \tag{5.110}$$

where $w_0$ is the velocity at the center of the pipe and the exponent $n$ varies with the Reynolds number. This power law agrees reasonably well with experimental data for $r/R > 0.04$, but is not applicable close to the wall, where the velocity profile gives an infinite velocity gradient.

The ratio of the average velocity to the centerline velocity may be calculated for the power-law velocity profile and the result is,

$$\frac{U}{w_0} = \frac{n^2}{(n+1)(2n+1)}. \tag{5.111}$$

This expression shows that with increasing $n$ the ratio of the average velocity to the centerline velocity increases. As a representative value for fully developed turbulent flow, 7 is often used for the exponent, giving rise to the term "a one-seventh power profile".

The friction factor for turbulent flow is obtained from experimental data. One of the first correlations for the friction factor was obtained by Blasius who used data obtained in smooth tubes and employed dimensional analysis, and proposed the following relationship,

$$C_f = \frac{0.0791}{\mathrm{Re}^{1/4}}. \tag{5.112}$$

This formula has a rather narrow range of applicability limited to smooth circular tubes for turbulent flow with Raynolds number in a range from $4 \times 10^3$ to $10^5$. For any Reynolds number greater than $4 \times 10^3$, the following formula can be used [233],

$$\frac{1}{\sqrt{C_f}} = 4.0 \log \left( \text{Re} \sqrt{C_f} \right) - 0.396. \tag{5.113}$$

This formula was derived by Prandtl and can be used for smooth-wall turbulent pipe flow instead of the Blasius formula.

The Prandtl formula is implicit, since for given Reynolds number, one has to iterate to compute $C_f$. This annoyance is avoided by instead using the following formula [65],

$$C_f = \frac{1}{4 \left( 1.82 \log \text{Re} - 1.64 \right)^2}, \tag{5.114}$$

which gives practically the same results as the Prandtl formula for Re in the range from $10^4$ to $5 \times 10^6$.

## Non-Circular Channels

For turbulent flow between parallel plates with smooth surfaces, the Fanning friction factor can be calculated as [233],

$$\frac{1}{\sqrt{C_f}} = 4.0 \log \left( \text{Re} \sqrt{C_f} \right) - 1.18, \tag{5.115}$$

where $\text{Re} = U D_h / \nu$ and $D_h = 4H$. Here $H$ is the half-width between plates and $D_h$ is the hydraulic diameter of the channel.

A general rule for estimating turbulent friction in non-circular channels is to use pipe-friction law given by Eq. (5.113) based on an *effective Reynolds number* [116],

$$\text{Re}_{eff} = \frac{U D_{eff}}{\nu} \quad \text{where} \quad D_{eff} = D_h \frac{16}{\left( C_f \text{Re} \right)_{lam}}, \tag{5.116}$$

where $\left( C_f \text{Re} \right)_{lam}$ is found for laminar flow condition in the same non-circular channel and $D_h$ is the hydraulic diameter of the channel.

For turbulent flow in a rod bundle with triangular lattice and $1.0 \leq P/D_R \leq 1.5$, the friction factor can be found from Eq. (5.114) multiplied by the correction factor $(0.96 P/D_R + 0.63)$. For both the triangle and the square lattice, the following formula is applicable [2],

$$C_f = A \left( \frac{P_{wC}}{P_{wR}} \right) \left( \frac{A_C}{A_R} \right)^m \text{Re}^{-0.25}, \tag{5.117}$$

where $P_{wC}$ and $P_{wR}$–wetted perimeter of the channel and the rod, respectively, $A_C$ and $A_R$–cross-section area of the channel and the rod, respectively, and $A$ and $m$ are constants given as:

For triangular lattice and $4 \times 10^3 \leq \text{Re} \leq 10^5$: $A = 0.1175$, $m = 0.35$,
For square lattice and $10^3 \leq \text{Re} \leq 5 \times 10^4$: $A = 0.095$, $m = 0.45$ .

*Effect of Roughness*

Wall roughness has little influence upon laminar flow, however, in turbulent flow, even a small roughness will break up the thin viscous sublayer and significantly increase the wall friction. As a result, both the friction pressure drop and the heat transfer intensity in a channel are affected.

For turbulent flow in pipes with rough walls the following formula has been derived by C.F. Colebrook and later plotted by L.F. Moody [165],

$$\frac{1}{\sqrt{C_f}} = -4.0 \log \left( \frac{k/D}{3.7} + \frac{1.255}{\text{Re} \sqrt{C_f}} \right), \tag{5.118}$$

where $k$ is the average roughness height. This formula is implicit and iterations are required. An explicit formula of the Colebrook-Moody correlation was proposed by Haaland [82],

$$\frac{1}{\sqrt{C_f}} = -3.6 \log \left[ \left( \frac{(k/D)}{3.7} \right)^{1.11} + \frac{6.9}{\text{Re}} \right]. \tag{5.119}$$

More correlations for friction factors are provided in Appendix C.

## 5.2 ONE-DIMENSIONAL BULK FLOW

One-dimensional *coolant channels* are used for an approximate calculation of the axial distribution of pressure, temperature, and enthalpy in a core. In such channels it is assumed that the lateral distribution of flow parameters can be neglected and that their axial distribution can be expressed in terms of the values averaged over the cross-section area. In this section, we discuss the main futures of one-dimensional single-phase flows in channels.

### 5.2.1 CONSERVATION EQUATIONS

Using area-averaging of instantaneous conservation equations and making some simplifications, the following set of one-dimensional conservation equations is obtained,

Continuity equation

$$\frac{\partial \rho}{\partial t} + \frac{1}{A} \frac{\partial (\rho U A)}{\partial z} = 0, \tag{5.120}$$

Linear momentum equation

$$\frac{\partial (\rho U)}{\partial t} + \frac{1}{A} \frac{\partial}{\partial z} (\rho U^2 A) + \frac{\partial p}{\partial z} + \frac{\partial \tau_{zz,e}}{\partial z} + \frac{P_w \tau_w}{A} + \rho g_z = 0, \tag{5.121}$$

Energy equation

$$\frac{\partial}{\partial t} \left[ \left( i + \frac{U^2}{2} \right) \rho \right] + \frac{1}{A} \frac{\partial}{\partial z} \left[ \left( i + \frac{U^2}{2} \right) \rho U A \right] - \frac{q_w'' P_H}{A} - q''' +$$
$$\frac{1}{A} \frac{\partial}{\partial z} (\tau_{zz} U A) + \frac{1}{A} \frac{\partial}{\partial z} (q_{zz,e}'' A) + g_z \rho U + \frac{Dp}{Dt} = 0 \tag{5.122}$$

In these equation $\rho$ is the area-averaged density, $U$ is the area-averaged and density-weighted velocity, and i is the area-averaged and density-weighted specific enthalpy of the fluid. We will now consider a special case of these conservation equations that is particularly useful in thermal-hydraulics of nuclear systems to predict total pressure drop in a channel.

## Pressure Drop in Steady-State Adiabatic Flow

For steady-state flows, the mass conservation equation becomes,

$$\frac{\partial(\rho U A)}{\partial z} = 0. \tag{5.123}$$

It is convenient to introduce the *mass flow rate* as,

$$W \equiv \rho U A, \tag{5.124}$$

and the steady-state mass conservation equation becomes,

$$\frac{dW}{dz} = 0, \quad W = \text{constant} . \tag{5.125}$$

Thus, the steady-state mass conservation equation requires that the mass flow rate is constant along the channel, irrespective of density changes, velocity changes, and cross-section area changes.

We can note here that a similar result is obtained when we assume that the fluid density is constant and thus its time derivative is zero. In such conditions, the mass conservation equation will be satisfied when the *volumetric flow rate*, defined as,

$$Q \equiv UA, \tag{5.126}$$

is constant along the channel. Finally, when the channel cross-section area is constant, the mass conservation equation will be satisfied when the *mass flux*, defined as,

$$G \equiv \rho U, \tag{5.127}$$

is constant along the channel.

Neglecting the axial equivalent shear stress, $\tau_{zz,e}$, the steady-state linear momentum equation becomes,

$$\frac{W^2}{A} \frac{d}{dz}\left(\frac{1}{\rho A}\right) + \frac{dp}{dz} + \frac{P_w \tau_w}{A} + \rho g_z = 0, \tag{5.128}$$

where we used the fact that $W$ is constant along the channel. Introducing the friction factor given by Eq. (5.2) and the hydraulic diameter defined in Eq. (5.134), we can express the wall shear stress term as

$$\frac{P_w \tau_w}{A} = \frac{4}{D_h} \cdot \frac{C_f}{2} \rho U^2 = \frac{4C_f}{D_h} \frac{W^2}{2\rho A^2} . \tag{5.129}$$

The pressure gradient along the channel is now give as,

$$-\frac{\mathrm{d}p}{\mathrm{d}z} = \frac{W^2}{A}\frac{\mathrm{d}}{\mathrm{d}z}\left(\frac{1}{\rho A}\right) + \frac{4C_f}{D_h}\frac{W|W|}{2\rho A^2} + \rho g_z. \tag{5.130}$$

In this equation, we replaced $W^2$ with $W|W|$ to have a correct sign of this term both for flow in the channel positive direction ($W > 0$), when due to friction the pressure decreases with increasing $z$, and for flow in the channel negative direction ($W < 0$), when due to friction the pressure increases with increasing $z$.

Integration of Eq. (5.130) along the channel length yields the total pressure drop in the channel. Assuming a constant cross-section area of the channel we get,

$$-(p_2 - p_1) = \left(\frac{W}{A}\right)^2\left(\frac{1}{\rho_2} - \frac{1}{\rho_1}\right) + \frac{4C_f}{D_h}\frac{W|W|}{2A^2}\int_{z_1}^{z_2}\frac{\mathrm{d}z}{\rho} + g_z\int_{z_1}^{z_2}\rho\,\mathrm{d}z. \tag{5.131}$$

The first term on the right-hand side of the equation represents a pressure drop due to fluid acceleration, the second term represents an irreversible pressure loss due to friction, and the last term is a pressure drop due to gravity. We should note that the acceleration term in a channel with a constant cross-section area and a constant fluid density will be equal to zero. Under such conditions, the total pressure drop in the channel is,

$$-(p_2 - p_1) = \frac{4C_f L}{D_h}\frac{W|W|}{2\rho A^2} + \rho g_z L, \tag{5.132}$$

where $L = z_2 - z_1$ is the channel length.

With a constant density and a variable channel cross-section area the acceleration term will not be zero and the corresponding pressure drop can be found as,

$$\begin{aligned}-(p_2 - p_1)_{acc} &= \frac{W^2}{\rho}\int_{z_1}^{z_2}\frac{1}{A}\frac{\mathrm{d}}{\mathrm{d}z}\left(\frac{1}{A}\right)\mathrm{d}z = \frac{W^2}{2\rho}\int_{z_1}^{z_2}\mathrm{d}\left(\frac{1}{A^2}\right) \\ &= \frac{W^2}{2\rho}\left(\frac{1}{A_2^2} - \frac{1}{A_1^2}\right)\end{aligned} \tag{5.133}$$

As can be seen, the pressure drop due to acceleration relies solely on the cross-sectional areas of the channel's inlet and outlet. Interestingly, the nature of the area variation throughout the channel does not influence it.

## 5.2.2 ONE-DIMENSIONAL GEOMETRY APPROXIMATIONS

The channel geometry is described by equivalent parameters such as the flow cross-section area $A$, wetted perimeter $P_w$, heated perimeter $P_H$, hydraulic diameter $D_h$, and heated perimeter $P_H$. These equivalent quantities are derived from the primary geometry parameters characterizing a fuel rod assembly. Examples of possible coolant channels in square and triangular lattices are shown in Fig. 5.2. Assuming uniform flow and power distributions in rod bundles, all subchannels shown in Fig. 5.2 are

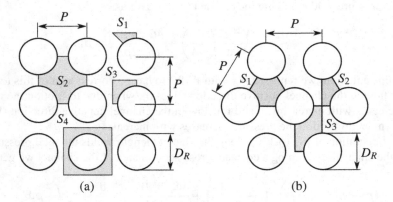

**Figure 5.2**   Various subchannel shapes in a square (a) and hexagonal (b) lattice: $D_R$ - rod diameter, $P$ - rod pitch.

equivalent from the point of view of thermal-hydraulics. Thus, an infinite and uniform rod bundle can be accurately represented by a single subchannel, since all velocity and temperature gradients on subchannel boundaries are equal to zero. Such an ideal approximation is called the *isolated subchannel model*.

In reality, however, the power and coolant flow are distributed non-uniformly within a fuel assembly, giving rise to non-uniform distribution of other parameters such as coolant velocity, density, and temperature. In such situations a better one-dimensional approximation of the fuel assembly should include all fuel rods in the assembly leading to the *isolated assembly model*.

### Isolated Subchannel Model

An equivalent *coolant channel* to represent non-circular subchannels can be derived as shown in Fig. 5.3. Using a circularization, subchannels can be represented by an equivalent tube with the diameter $D_h$, as shown in Fig. 5.3 (a) and (c), or as an annulus with the inner diameter $D_i$ and outer diameter $D_o$, as shown in Fig. 5.3 (b). The equivalent tube diameter to represent a subchannel with flow area $A$ and wetted perimeter $P_w$ can be found from the following formula,

$$D_h = \frac{4A}{P_w}. \tag{5.134}$$

Using the subchannel shapes shown in Fig. 5.3, the flow area can be found as,

$$A = \begin{cases} P^2 - \frac{\pi D_R^2}{4} = \frac{\pi}{4}\left(D_o^2 - D_i^2\right) & \text{for a square lattice} \\[2mm] \frac{\sqrt{3}}{4}P^2 - \frac{\pi D_R^2}{8} & \text{for a triangular lattice} \end{cases}, \tag{5.135}$$

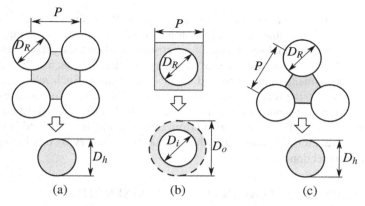

**Figure 5.3** Circularization of non-circular subchannels: (a) coolant-centered square subchannel, (b) rod-centered square subchannel, (c) coolant-centered triangular subchannel.

and the wetted perimeters are,

$$P_w = \begin{cases} \pi D_R = \pi D_i & \text{for a square lattice} \\ \frac{\pi D_R}{2} & \text{for a triangular lattice} \end{cases}. \qquad (5.136)$$

From Eqs. (5.135) and (5.136) we find the outer diameter of the annular subchannel as,

$$D_o = P\sqrt{\frac{4}{\pi}}, \qquad (5.137)$$

and the hydraulic diameter of the subchannels as,

$$D_h = \begin{cases} \left[ \frac{4}{\pi} \left( \frac{P}{D_R} \right)^2 - 1 \right] D_R & \text{for a square lattice} \\ \left[ \frac{2\sqrt{3}}{\pi} \left( \frac{P}{D_R} \right)^2 - 1 \right] D_R & \text{for a triangular lattice} \end{cases}. \qquad (5.138)$$

Subchannels shown in Fig. 5.3 are examples of systems that cannot be analyzed by solving the balance equations analytically. We need to take experimental data, or rely on numerical predictions, if the distributions of flow parameters, such as velocity, pressure, and wall shear stress, are needed. However, one can use the dimensional analysis of the balance equations and by proper scaling up or down of the system, the solution can be derived from another one based on the similarity principle. It can be shown that two systems have exactly the same solution if they have geometric similarity and dynamic similarity. Two systems are **geometrically similar** if they have the same shape and the ratios of all dimensions are the same. The systems are **dynamically similar** if the dimensionless groups (such as the Reynolds number) in the differential equations and in the boundary conditions are the same.

The circularization approach violates the principle of the geometric similarity, thus, the solution of balance equations in the circular geometry is not equivalent to the solution in the actual subchannel geometry, even when the dynamic similarity is preserved. Nevertheless, the circular geometry is often used as a reference system to obtain experimental data or to find an analytic solution. Such reference solution can be applied to the real subchannel geometry as a base solution, whose accuracy is improved by using proper correction coefficients. For example, most correlations for friction loss coefficient are derived from experimental data obtained in tubes, and then the correlations are adapted to the subchannel geometry by employing modifications or correction factors.

## 5.3 TURBULENT FLOW IN FUEL ROD ASSEMBLIES

Experimental data indicate that turbulent flow in fuel rod assemblies exhibits distinct characteristics compared to turbulent flow in circular tubes. The main parameter that influences the flow pattern is the pitch-to-diameter ratio, $P/D_R$. With decreasing $P/D_R$ ratio, both the turbulence intensity and the dominant frequency of turbulent fluctuations increase. The distribution of wall shear stress on the rod surface is non-uniform in the azimuthal direction, with a minimum value at a point facing the gap and a maximum value at a point facing the channel. The distribution of turbulence kinetic energy in the normal-to-wall direction is characterized by a maximum value near the wall and minimum values at the gap and subchannel centers. However, the turbulent kinetic energy at the gap center is significantly higher than that at the subchannel center. In addition, for $P/D_R < 1.2$, significant periodic flow pulsations are observed through the gaps between neighboring rods.

Due to the complex pattern of the turbulence in fuel rod assemblies, the linear eddy viscosity Reynolds averaged Navier-Stokes (RANS) model predictions are inaccurate. To properly capture the turbulence pattern, it is necessary to use a higher-order turbulence model, such as unsteady RANS with the Reynolds stress model, Large Eddy Simulations (LES), or Direct Numerical Simulations (DNS).

## PROBLEMS

### PROBLEM 5.1

Simplifying and solving the energy conservation for flow in a channel given by Eq. (5.122), find the axial enthalpy distribution for uniform and cosine axial distributions of the heat flux.

### PROBLEM 5.2

Using the isolated subchannel model, find the pressure drop in a triangular subchannel with a rod diameter equal to 10 mm and a pitch-to-diameter ratio equal to 1.2. Assume water at 17.5 MPa pressure and 550 K temperature flowing with mass flux equal to 2500 kg m$^{-2}$ s$^{-1}$.

## PROBLEM 5.3

Assume that the velocity distribution in a laminar boundary layer can be represented as $u/U = a(y/\delta) + b(y/\delta)^2 + c(y/\delta)^3$, where $u$ is the velocity in the boundary layer, $U$ is the velocity at the outer edge of the boundary layer, $y$ is the distance from the surface, $\delta$ is the boundary layer thickness, and $a$, $b$, and $c$ are constants. Apply boundary conditions at the surface: $y = 0$, $u = 0$, $d^2u/dy^2 = 0$ and at the outer edge of the boundary layer: $y = \delta$, $u = U$, $du/dy = 0$ and show that the velocity profile is given by Eq. (5.8).

# 6 Multiphase Flows in Channels

Multiphase flow is a term used to describe the concurrent flow of more than one phase of matter, which interacts via shared interfaces. A common example of this is the co-flow of liquid and gas, such as water and vapor moving together within a conduit. This particular instance of multiphase flow is frequently observed both in natural environments and industrial systems. Generally, a multiphase flow can involve the coexistence of solid, liquid, and gas phases. This results in three types of interfaces: between liquid and gas, between solid and gas, and between liquid and solid.

In multiphase flow, an *interface* is defined as a thin region separating two domains, each occupied by a distinct phase. Within this interface, physical properties transition from one phase to the other. For an interface to form, the adjacent phases must be immiscible. If one of the phases is solid, the interface's shape remains constant despite changes in flow conditions, and is determined by the solid domain's surface shape. Conversely, when both phases are non-solid, the interface's shape is dictated by local flow conditions. In such instances, the interface's shape must be determined in conjunction with the velocity, pressure, and temperature fields of the adjacent fluids.

The numerical approaches for solving multiphase flow problems can roughly be divided into two categories: an interface resolving method using a single set of conservation equations for both phases and a phase-averaging method using a set of conservation equations for each phase separately. While the former allows one to follow the evolution in time and space of the interface and provides instantaneous values of flow parameters as a single continuous field, the latter provides the phase-averaged flow parameters as a continuous field for each phase separately.

The method of resolving the interface applies the same approach to each single-phase domain of a multiphase flow as discussed for single-phase flows in chapter 4. Since only one flow field is considered throughout the domain, the method is referred to as a single-field method. To resolve the interface, either interface tracking or interface capturing method is used.

In *interface tracking methods*, the position of the interface is explicitly tracked, which requires meshes that are updated as the flow evolves. By aligning the mesh cell boundary with the interface, each cell belongs to a single phase only in the flow. The boundary integral method, front-tracking method, and immersed boundary method are examples of the interface tracking methods.

In *interface capturing methods* the interface is not tracked explicitly, but instead is implicitly defined through an interface function, such as the level-set, color, or phase-field function. The advantage of these methods is that there is no need of updating the meshes with the evolving flows.

DOI: 10.1201/9781003255000-6

## 6.1   INSTANTANEOUS CONSERVATION EQUATIONS

The instantaneous conservation equations derived in §4.2 for single-phase flows are also valid for multiphase flows for all locations in the flow domain that do not contain the interface. The interface has to be excluded from the solution domain since various flow variables suffer discontinuities when it is crossed. Clearly, a special treatment of the interface in multiphase flows is required. The following section discusses the main methods that are applied for the inclusion of the interfacial effects into the instantaneous conservation equations.

### 6.1.1   INTERFACE TREATMENT

A primary function of the *phase interface* is to separate phases with different properties or different behavior of the material. There are two main groups of continuum models for the phase interface. One group of models represents it as a three-dimensional region, while the other depicts it as a two-dimensional surface transported in a three-dimensional space. The models are shortly considered below.

**Three-Dimensional Interface**

There is currently considerable evidence that fluid properties continuously vary across the interface and that it can be regarded as a three-dimensional region with several molecule diameters in thickness. In the three-dimensional model of an interface, a region of finite thickness is assumed in which the properties or behavior of the material differ from those of the adjoining phases. The stress tensor is a function of the rate of deformation tensor and includes additional variables, such as the gradient of density. The stress and velocity have continuous distributions across the interface region. The three-dimensional model of the interface is more realistic and has a greater potential to predict the properties and behavior of the material within the interface region.

**Two-Dimensional Interface**

In an approach proposed by Gibbs, there is a hypothetical two-dimensional *dividing surface* that lies within or near the interfacial region and separates two homogeneous phases. Any excess mass, momentum, and energy not accounted for by the adjoining homogeneous phases can be assigned to the dividing surface. For any variable such as density or viscosity, there are two distributions to be considered on either side of the dividing surface. The advantage of the two-dimensional model is simplicity in solving problems and ease in analysis of experimental data [205].

It is commonly accepted that the dividing surface coincides with the phase interface. This is our primary assumption in this text and in continuation we will use the terms "interface" and "phase interface" as synonymous with the dividing surface.

## Unit Normal Vector

Assuming that the time-dependent position of the interface in three-dimensional space is given as $f(x,y,z,t) = 0$, a *unit normal vector* to the interface is given by,

$$\mathbf{n} = \frac{\nabla f}{|\nabla f|}. \tag{6.1}$$

Here it is assumed that $\nabla f$ is defined and $|\nabla f| \neq 0$ at each point of the surface. The outward unit normal vector from bulk fluid $k$ is denoted as $\mathbf{n}_k$. Thus for an interface with adjoining phases $k$ and $l$ we have,

$$\mathbf{n}_k = -\mathbf{n}_l. \tag{6.2}$$

## Speed of Displacement

Upon differentiating function $f$ with respect to time following a given point on the interface we have,

$$\frac{\partial f}{\partial t} + \nabla f \cdot \mathbf{v}_i = 0, \tag{6.3}$$

where $\mathbf{v}_i$ is a velocity vector of the point on the interface, which in general has both normal and tangential components. Combining Eqs. (6.1) and (6.3) we obtain the *speed of displacement* of the surface as,

$$v_{in} \equiv \mathbf{v}_i \cdot \mathbf{n} = -\frac{\partial f/\partial t}{|\nabla f|}. \tag{6.4}$$

### 6.1.2  PHASIC CONSERVATION EQUATION

Following a similar approach as used in §4.3 for single-phase flow, the generic integral and differential conservation equations are obtained as follows,

$$\frac{\mathrm{d}}{\mathrm{d}t}\left(\iiint_{V(t)} \rho_k \psi_k \mathrm{d}V\right) = -\iint_{S(t)} [\rho_k \psi_k (\mathbf{v}_k \cdot \mathbf{n}_k) + \mathbf{n}_k \cdot \mathbf{J}_k]\,\mathrm{d}S + \iiint_{V(t)} \rho_k \phi_k \mathrm{d}V, \tag{6.5}$$

$$\frac{\partial (\rho_k \psi_k)}{\partial t} + \nabla \cdot (\rho_k \psi_k \mathbf{v}_k + \mathbf{J}_k) - \rho_k \phi_k = 0, \tag{6.6}$$

where the definitions of the terms in the equations are given in Table 6.1.

### 6.1.3  JUMP CONDITIONS AT THE INTERFACE

The instantaneous differential conservation equations derived in §4.2 are not valid at the interface due to discontinuities of flow variables. However, the conservation equations can be expressed in terms of so called *jump conditions* which relate values of conserved quantities on both sides of the interface. By considering an integral

**TABLE 6.1**

**Definitions of Terms in the Generic Conservation Equations (6.5) and (6.6)**

| Conserved Quantity | $\psi_k$ | $\mathbf{J}_k$ | $\phi_k$ |
|---|---|---|---|
| Mass | 1 | 0 | 0 |
| Linear momentum | $\mathbf{v}_k$ | $-\mathbf{T}_k \equiv p_k\mathbf{I} - \boldsymbol{\tau}_k$ | $\mathbf{b}_k$ |
| Angular momentum | $\mathbf{r} \times \mathbf{v}_k$ | $-(\mathbf{r} \times \mathbf{T}_k)^T$ | $\mathbf{r} \times \mathbf{b}_k$ |
| Total energy | $e_{IK,k} \equiv e_{Ik} + \frac{1}{2}v_k^2$ | $\mathbf{q}_k'' - \mathbf{T}_k \cdot \mathbf{v}_k$ | $\mathbf{b}_k \cdot \mathbf{v}_k + \frac{1}{\rho_k}q_k'''$ |
| Entropy | $s_k$ | $\frac{1}{T}\mathbf{q}_k''$ | $\frac{1}{\rho_k}\Delta_{sk}$ |

balance equation for quantity $\psi$ over a material volume divided by a non-material interface into two parts, the following jump conditions can be derived [52, 112, 113],

$$\frac{\mathrm{d}_s \psi_i}{\mathrm{d}t} + \psi_i \nabla_s \cdot \mathbf{v}_i = \sum_{k=1}^{2} [\rho_k \psi_k \mathbf{n}_k \cdot (\mathbf{v}_k - \mathbf{v}_i) + \mathbf{n}_k \cdot \mathbf{J}_k] - A^{\alpha\beta} g_{ln} \left( t_\alpha^n \mathbf{J}_i^{l*} \right)_{,\beta} + \phi_i. \quad (6.7)$$

Here $\psi_k$ is the conserved quantity, $\rho_k$ is the fluid density, $\mathbf{v}_k$ is the velocity vector, and $\mathbf{J}_k$ is the flux of quantity $\psi$ in phase $k$. The symbols $\mathrm{d}_s/\mathrm{d}t$, $\nabla_s$, $A^{\alpha\beta}$, $g_{ln}$, $t_\alpha^n$, and $()_{,\beta}$ denote the convective derivative with the surface velocity vector $\mathbf{v}_i$, the surface divergence operator, the surface metric tensor, the space metric tensor, the hybrid tensor, and the surface covariant derivative, respectively. The left-hand side of Eq. (6.7) represents the time rate of change of $\psi_i$ on the interface including the effect of surface dilatation. The terms on the right-hand side of the equation are the convective and molecular fluxes from the bulk phases, the surface flux, and the surface source, respectively.

## 6.2  SINGLE-FIELD METHODS

Single-field methods are based on a solution of single-phase conservation equations supplemented with an interface propagation model. As already mentioned, the interface propagation models can roughly be divided into two categories: *interface tracking* and *interface capturing methods*. In the interface tracking methods, the position of the interface is explicitly tracked with mesh being updated as the flow evolves. Boundary integral methods, front-tracking methods, and immersed boundary methods are examples of these types of approaches. In the interface capturing methods, the interface is not tracked explicitly but instead is implicitly defined through an interface function that is propagated in the flow domain based on fluid particle motion. Choices for the interface capturing method include level set methods, volume of fluid methods, and phase-field methods. We shortly present selected methods in the following subsections.

### 6.2.1 LEVEL SET METHODS

*Level set methods* are computational techniques that rely on an implicit representation of the interface, which is characterized by a dedicated equation of motion. The speed of the evolving interface may sensitively depend on local properties such as curvature and normal direction, as well as complex physics near the front, including internal jump and boundary conditions determined by the interface location. Level set methods are particularly designed for multidimensional problems with changing topology of the evolving interface. The numerical development of these techniques and their application to problems in fluid mechanics and bubble dynamics have been addressed in several review articles and books [174, 175, 195, 196].

In the level set method, a motion of two immiscible fluids that are governed by the incompressible Navier-Stokes equations can be described as follows,

$$\rho \frac{D\mathbf{v}}{Dt} = -\nabla p + \nabla \cdot (2\mu \mathbf{D}) - \sigma\kappa\delta(d)\mathbf{n} + \rho\mathbf{g}, \quad \nabla \cdot \mathbf{v} = 0, \quad \mathbf{r} \in V, \quad (6.8)$$

where $V$ is the domain containing both fluids, $\mathbf{v}$ is the velocity vector, $p$ is the pressure, $\rho$ is the density, $D/Dt$ is the material derivative, $\mathbf{D}$ is the deformation tensor, and $\mathbf{g}$ is the acceleration vector due to gravity. The third term on the right-hand side of Eq. (6.8) incorporates surface tension, $\sigma$, as a force concentrated on the interface, $S_i$, where $\mathbf{n}$ is the unit normal of the interface drawn outward from the gas to the liquid, $\kappa = \nabla \cdot \mathbf{n}$ is the curvature of the interface, $\delta$ is the Dirac delta function, and $d$ is the signed-distance function from the interface, which is defined as follows: at a point $\mathbf{r}$ in liquid, $d$ is the distance to the closest point on the interface. In the gas, $d$ is the negative of this quantity. The boundary conditions at the interface $S_i$, between the phases are:

$$(2\mu_l\mathbf{D}_l - 2\mu_g\mathbf{D}_g) \cdot \mathbf{n} = (p_l - p_g + \sigma\kappa)\mathbf{n}, \quad \mathbf{v}_l = \mathbf{v}_g, \quad \mathbf{r} \in S_i, \quad (6.9)$$

where subscripts $l$ and $g$ are used to denote variables in the liquid and gas phases, respectively. Thus, the variables present in Eq. (6.8) should be interpreted as,

$$\mathbf{v}, \mathbf{D}, \rho, \mu, p = \begin{cases} \mathbf{v}_l, \mathbf{D}_l, \rho_l, \mu_l, p_l & \mathbf{r} \in \text{the liquid} \\ \mathbf{v}_g, \mathbf{D}_g, \rho_g, \mu_g, p_g & \mathbf{r} \in \text{the gas} \end{cases}. \quad (6.10)$$

The *level set function*, $\phi(\mathbf{r},t)$, is used to define the gas-liquid interface as follows:

$$S_i = \{\mathbf{r} \mid \phi(\mathbf{r},t) = 0\}, \quad (6.11)$$

that is, the gas-liquid interface is a collection of points in space that corresponds to the zero-level set of $\phi$. We also take $\phi < 0$ in the gas region and $\phi > 0$ in the liquid region. The density and viscosity can be written as,

$$\rho(\phi) = \rho_g + (\rho_l - \rho_g)\theta(\phi), \quad \mu(\phi) = \mu_g + (\mu_l - \mu_g)\theta(\phi), \quad (6.12)$$

where $\theta(\phi)$ is the Heaviside function given by,

$$\theta(\phi) = \begin{cases} 0 & \text{if } \phi < 0 \\ \frac{1}{2} & \text{if } \phi = 0 \\ 1 & \text{if } \phi > 0 \end{cases}. \quad (6.13)$$

Since the sharp changes of fluid properties across the interface can cause numerical difficulties, the Heaviside function can be replaced by

$$\theta_\varepsilon(\phi) = \begin{cases} 0 & \text{if } \phi < -\varepsilon \\ \frac{1}{2}\left[1 + \frac{\phi}{\varepsilon} + \frac{1}{\pi}\sin\left(\frac{\pi\phi}{\varepsilon}\right)\right] & \text{if } |\phi| \le \varepsilon \\ 1 & \text{if } \phi > \varepsilon \end{cases} \tag{6.14}$$

With this function the interface has a thickness of approximately $\frac{2\varepsilon}{|\nabla\phi|}$. The corresponding "smoothed" delta function that should be used in Eq. (6.8) is given by,

$$\delta_\varepsilon(\phi) = \frac{\mathrm{d}\theta_\varepsilon}{\mathrm{d}\phi}. \tag{6.15}$$

The time evolution of the level set function is described by the following differential equation,

$$\frac{\partial\phi}{\partial t} + \mathbf{v}\cdot\nabla\phi = 0, \tag{6.16}$$

since the interface moves with the fluid particles.

## 6.2.2 VOLUME-OF-FLUID METHODS

*Volume-of-fluid methods* are based on the concept of the so-called fraction function (or color function) $C$. It is a scalar function defined as an integral of a fluid's characteristic function in the control volume. When a cell is empty of the tracked phase, the value of $C$ is zero. Correspondingly, when $C$ is equal to unity, the cell is entirely occupied by the tracked phase. For cells containing the interface, $0 < C < 1$.

The evolution of the fraction function $C$ is governed by a transport equation which is the same equation that has to be fulfilled by the level set function:

$$\frac{\partial C}{\partial t} + \mathbf{v}\cdot\nabla C = 0. \tag{6.17}$$

For an incompressible flow of two immiscible fluids Eq. (6.8) has to be solved with fluid properties given as,

$$\rho = \rho_g + (\rho_l - \rho_g)C, \quad \mu = \mu_g + (\mu_l - \mu_g)C, \tag{6.18}$$

where $C = 0$ and $C = 1$ correspond to the gas phase and the liquid phase, respectively. A more detailed description of the volume of fluid methods and their extensions to compressible fluids can be found in several review articles and books [90, 159, 223].

## 6.2.3 PHASE-FIELD METHODS

Phase-field methods also referred to as the diffuse-interface methods, are powerful methods to simulate many types of multiphase flows that involve drop coalescence, drop break-up, contact line dynamics, and dynamics of interfaces with surfactant adsorption and thermocapillary effects. Phase-field methods are based on models of free energy of fluid where the multiphase flow is treated as a flow of one fluid with variable material properties. An order parameter is employed to characterize the different phases, which varies continuously over thin interface layers and is mostly uniform in the bulk phases.

Phase-field methods are similar to level set methods with one important difference. In level set methods, the choice of level set function is arbitrary. In phase-field methods the exact profile of the phase function is important in obtaining the correct interface motion. Since the phase function changes quickly in the neighborhood of the interface, a large number of grid points is needed in that region.

## 6.3 MULTI-FIELD METHODS

Multi-field methods are using a set of averaged conservation equations for each fluid component. The methods require closure relationships for interfacial transfer of mass, momentum, and energy. The theory of multi-field methods is described in many review articles and books [53, 59, 113]. In this section, a short introduction to the topic is provided.

Using the definition of a phasic time average given by Eq. (B.16) in Appendix B, we can introduce additional averages useful in the theory of multiphase flows. The *phasic weighted average* of quantity $f$ for phase $k$ is defined as follows,

$$\overline{f}_k^X \equiv \frac{\overline{fX_k}^T}{\overline{X_k}^T} = \frac{\overline{fX_k}^T}{\alpha_k}, \tag{6.19}$$

where $\alpha_k = \overline{X_k}^T$ is called the volume fraction of phase $k$. Similarly, the density weighted average of quantity $f$ for phase $k$ is defined as,

$$\overline{f}_k^{X\rho} \equiv \frac{\overline{fX_k\rho}^T}{\overline{X_k\rho}^T} = \frac{\overline{fX_k\rho}^T}{\alpha_k\overline{\rho_k}^X}. \tag{6.20}$$

Multiplying the generic conservation equation in the differential form by the phasic characteristic function, $X_k$, and time averaging the product terms yields the following time average equation,

$$\frac{\partial \left( \alpha_k \overline{\rho_k}^X \overline{\psi_k}^{X\rho} \right)}{\partial t} + \nabla \cdot \left( \alpha_k \overline{\rho_k}^X \overline{\psi_k}^{X\rho} \overline{v_k}^{X\rho} \right) = -\nabla \cdot \left( \alpha_k \overline{J}_k^X \right) + \\ \alpha_k \overline{\rho_k}^X \overline{\phi_k}^{X\rho} - \nabla \cdot \left( \overline{\rho_k \psi_k'' v_k''}^T \right) - \frac{1}{T} \sum_j \frac{1}{|n_{ki} \cdot v_i|} \left[ \rho_k \psi_k (v_k - v_i) + J_k \right] \cdot n_{ki} \tag{6.21}$$

The right-hand side of the equation contains the turbulent fluctuations of the conserved quantities defined as follows,

$$\psi_k'' = \psi_k - \overline{\psi_k}^{X\rho}, \quad v_k'' = v_k - \overline{v_k}^{X\rho}. \tag{6.22}$$

The time average turbulent flux $\overline{\rho_k \psi_k'' \mathbf{v}_k''}^T$ can be transformed as follows,

$$\overline{\rho_k \psi_k'' \mathbf{v}_k''}^T = \frac{\overline{\rho_k \psi_k'' \mathbf{v}_k''}^T}{\overline{\rho_k}} \frac{\overline{\rho_k}}{\alpha_k} \alpha_k = \alpha_k \overline{\rho_k}^X \overline{\psi_k'' \mathbf{v}_k''}^{X\rho}. \tag{6.23}$$

Introducing the turbulent flux of quantity $\psi_k$ as,

$$\mathbf{J}_k^t \equiv \overline{\rho_k}^X \overline{\psi_k'' \mathbf{v}_k''}^{X\rho}, \tag{6.24}$$

the time average turbulent flux of quantity $\psi_k$ can be written as,

$$\overline{\rho_k \psi_k'' \mathbf{v}_k''}^T = \alpha_k \mathbf{J}_k^t. \tag{6.25}$$

The last term on the right-hand side of Eq. (6.21) represents a transfer of quantity $\psi_k$ through the interface. The term can be partitioned into two parts describing the convective and diffusive transfer as follows,

$$
\begin{aligned}
\Pi_{ki} &\equiv -\frac{1}{T} \sum_j \frac{1}{|\mathbf{n}_{ki} \cdot \mathbf{v}_i|} \left[ \rho_k \psi_k (\mathbf{v}_k - \mathbf{v}_i) + \mathbf{J}_k \right] \cdot \mathbf{n}_{ki} \\
&= -\frac{1}{T} \sum_j \frac{\rho_k \psi_k (\mathbf{v}_k - \mathbf{v}_i) \cdot \mathbf{n}_{ki}}{|\mathbf{n}_{ki} \cdot \mathbf{v}_i|} - \frac{1}{T} \sum_j \frac{\mathbf{J}_k \cdot \mathbf{n}_{ki}}{|\mathbf{n}_{ki} \cdot \mathbf{v}_i|} \equiv \Pi_{ki}^C + \Pi_{ki}^D
\end{aligned} \tag{6.26}
$$

where $\Pi_{ki}^C$ and $\Pi_{ki}^D$ denote the convective transfer and the diffusive transfer through the interface, respectively.

The convective flux of quantity $\psi_k$ through the interface depends on the values of this quantity at the interface at all time instants $t_j$ when the interface is crossing a given point in space. We will find the average value of quantity $\psi_k$ at the interface. To this end, we first note that for the mass conservation equation we have $\psi_k = 1$ and $\mathbf{J}_k = 0$ so that the convective flux corresponds to the mass flux, $\Gamma_{ki}$, through the interface:

$$\Gamma_{ki} \equiv -\frac{1}{T} \sum_j \frac{\rho_k (\mathbf{v}_k - \mathbf{v}_i) \cdot \mathbf{n}_{ki}}{|\mathbf{n}_{ki} \cdot \mathbf{v}_i|}. \tag{6.27}$$

Introducing the mass-flux weighted average value of quantity $\psi_k$ at the interface as,

$$\overline{\psi_k}^{i\rho} \equiv \frac{\sum_j \frac{\rho_k \psi_k (\mathbf{v}_k - \mathbf{v}_i) \cdot \mathbf{n}_{ki}}{|\mathbf{n}_{ki} \cdot \mathbf{v}_i|}}{\sum_j \frac{\rho_k (\mathbf{v}_k - \mathbf{v}_i) \cdot \mathbf{n}_{ki}}{|\mathbf{n}_{ki} \cdot \mathbf{v}_i|}} = \frac{\Pi_{ki}^C}{\Gamma_{ki}}, \tag{6.28}$$

the convective transfer of quantity $\psi_k$ through the interface can be written as,

$$\Pi_{ki}^C = \overline{\psi_k}^{i\rho} \Gamma_{ki}. \tag{6.29}$$

The diffusive part of the interfacial transfer term, $\Pi_{ki}^D$, result from the transfer of quantity $\psi_k$ on the molecular level with flux $\mathbf{J}_k$. Since this flux contributes to the interfacial transfer at time instants $t_j$ when the interface is crossing a given point in space, we can find a time average of the flux at the interface as follows,

$$\overline{\mathbf{J}_k}^i \equiv \frac{1}{N_j} \sum_j \mathbf{J}_k, \tag{6.30}$$

where $N_j$ is the number of interface crossings during the averaging period. Thus, the instantaneous value of the flux at the interface, $\mathbf{J}_k$, can be expressed in terms of the average value of the flux at the interface, $\overline{\mathbf{J}_k}^i$, and the fluctuating part of the flux, $\mathbf{J}_k'$, as follows,

$$\mathbf{J}_k = \overline{\mathbf{J}_k}^i + \mathbf{J}_k'. \tag{6.31}$$

Thus, the diffusive transfer of quantity $\psi_k$ through the interface can be written as,

$$\Pi_{ki}^D \equiv -\frac{1}{T} \sum_j \frac{\mathbf{J}_k \cdot \mathbf{n}_{ki}}{|\mathbf{n}_{ki} \cdot \mathbf{v}_i|} = \overline{\mathbf{J}_k}^i \cdot \left( -\frac{1}{T} \sum_j \frac{\mathbf{n}_{ki}}{|\mathbf{n}_{ki} \cdot \mathbf{v}_i|} \right) - \frac{1}{T} \sum_j \frac{\mathbf{J}_k' \cdot \mathbf{n}_{ki}}{|\mathbf{n}_{ki} \cdot \mathbf{v}_i|} = \tag{6.32}$$
$$\overline{\mathbf{J}_k}^i \cdot \nabla \alpha_k + \Pi_{ki}^{Df}$$

Here,

$$\Pi_{ki}^{Df} \equiv -\frac{1}{T} \sum_j \frac{\mathbf{J}_k' \cdot \mathbf{n}_{ki}}{|\mathbf{n}_{ki} \cdot \mathbf{v}_i|}, \tag{6.33}$$

is the flux of quantity $\psi_k$ through the interface due to fluctuations of flux $\mathbf{J}_k$ at the interface and we used the following relationship,

$$\nabla \alpha_k = -\frac{1}{T} \sum_j \frac{\mathbf{n}_{ki}}{|\mathbf{n}_{ki} \cdot \mathbf{v}_i|}. \tag{6.34}$$

We can now introduce the derived expressions for time average turbulence term and the interfacial transfer terms into Eq. (6.21) to obtain,

$$\frac{\partial \left( \alpha_k \overline{\rho_k}^X \overline{\psi_k}^{X\rho} \right)}{\partial t} + \nabla \cdot \left( \alpha_k \overline{\rho_k}^X \overline{\psi_k}^{X\rho} \overline{\mathbf{v}_k}^{X\rho} \right) = -\nabla \cdot \left[ \alpha_k \left( \overline{\mathbf{J}_k}^X + \mathbf{J}_k' \right) \right]$$
$$+ \alpha_k \overline{\rho_k}^X \overline{\phi_k}^{X\rho} + \overline{\psi_k}^{i\rho} \Gamma_{ki} + \nabla \alpha_k \cdot \overline{\mathbf{J}_k}^i + \Pi_{ki}^{Df} \tag{6.35}$$

The derived generic equation can be written in a specific form for conservation of mass, linear momentum, and total energy by replacing $\psi_k$, $\mathbf{J}_k$, and $\phi_k$ with quantities provided in Table 6.1.

**BOX 6.1   MULTI-FIELD AVERAGED CONSERVATION EQUATIONS**

For the mass conservation equation, the quantities in Eq. (6.35) are as follows:

$$\psi_k = \overline{\psi_k}^{i\rho} = 1, \quad \mathbf{J}_k = \overline{\mathbf{J}_k}^i = \mathbf{J}_k' = 0, \quad \phi_k = 0, \quad \Pi_{ki}^{Df} = 0, \tag{6.36}$$

thus, the mass conservation equation becomes,

$$\frac{\partial \left( \alpha_k \overline{\rho_k}^X \right)}{\partial t} + \nabla \cdot \left( \alpha_k \overline{\rho_k}^X \overline{\mathbf{v}_k}^{X\rho} \right) = \Gamma_{ki} \, . \tag{6.37}$$

For the linear momentum equation, we have:

$$\psi_k = \mathbf{v}_k, \quad \overline{\psi_k}^{X\rho} = \overline{\mathbf{v}_k}^{X\rho}, \quad \overline{\psi_k}^{i\rho} = \overline{\mathbf{v}_k}^{i\rho}, \tag{6.38}$$

$$\mathbf{J}_k = p_k \mathbf{I} - \boldsymbol{\tau}_k, \quad \overline{\mathbf{J}_k}^X = \overline{p_k}^X \mathbf{I} - \overline{\boldsymbol{\tau}_k}^X, \quad \overline{\mathbf{J}_k}^i = \overline{p_k}^i \mathbf{I} - \overline{\boldsymbol{\tau}_k}^i, \quad \mathbf{J}_k^t = \boldsymbol{\tau}_k^t, \tag{6.39}$$

$$\phi_k = \mathbf{g}, \quad \Pi_{ki}^{Df} \equiv \mathbf{M}_{ki}, \tag{6.40}$$

and the linear momentum conservation equation is as follows,

$$\frac{\partial \left( \alpha_k \overline{\rho_k}^X \overline{\mathbf{v}_k}^{X\rho} \right)}{\partial t} + \nabla \cdot \left( \alpha_k \overline{\rho_k}^X \overline{\mathbf{v}_k}^{X\rho} \overline{\mathbf{v}_k}^{X\rho} \right) = -\nabla \cdot \left[ \alpha_k \left( \overline{p_k}^X \mathbf{I} - \overline{\boldsymbol{\tau}_k}^X - \boldsymbol{\tau}_k^t \right) \right] \\ + \alpha_k \overline{\rho_k}^X \mathbf{g} + \overline{\mathbf{v}_k}^{i\rho} \Gamma_{ki} + \nabla \alpha_k \cdot \overline{\boldsymbol{\tau}_k}^i + \mathbf{M}_{ki} \tag{6.41}$$

For the total energy conservation equation, the quantities in Eq. (6.35) are as follow,

$$\psi_k = e_{IK,k} = e_{I,k} + \frac{1}{2} v_k^2, \quad \phi_k = \mathbf{g} \cdot \mathbf{v}_k + \frac{q_k'''}{\rho_k}, \tag{6.42}$$

$$\mathbf{J}_k = \mathbf{q}_k'' + (p_k \mathbf{I} - \boldsymbol{\tau}_k) \cdot \mathbf{v}_k, \quad \overline{\mathbf{J}_k}^X = \overline{\mathbf{q}_k''}^X + \overline{(p_k \mathbf{I} - \boldsymbol{\tau}_k) \cdot \mathbf{v}_k}^X, \tag{6.43}$$

$$\overline{\mathbf{J}_k}^i = \overline{\mathbf{q}_k''}^i + \overline{(p_k \mathbf{I} - \boldsymbol{\tau}_k) \cdot \mathbf{v}_k}^i, \quad \mathbf{J}_k^t = \overline{\rho_k}^X \overline{e_{IK,k}'' \mathbf{v}_k''}^{X\rho}, \tag{6.44}$$

which yields the following total energy conservation equation,

$$\frac{\partial \left( \alpha_k \overline{\rho_k}^X \overline{e_{IK,k}}^{X\rho} \right)}{\partial t} + \nabla \cdot \left( \alpha_k \overline{\rho_k}^X \overline{e_{IK,k}}^{X\rho} \overline{\mathbf{v}_k}^{X\rho} \right) = \\ - \nabla \cdot \left[ \alpha_k \left( \overline{\mathbf{q}_k''}^X + \overline{\rho_k}^X \overline{e_{IK,k}'' \mathbf{v}_k''}^{X\rho} \right) \right] + \alpha_k \overline{\rho_k}^X \mathbf{g} \cdot \overline{\mathbf{v}_k}^{X\rho} + q_k''' \cdot \\ + \overline{e_{IK,k}}^{i\rho} \Gamma_{ki} + \nabla \alpha_k \cdot \left[ \overline{\mathbf{q}_k''}^i + \overline{(p_k \mathbf{I} - \boldsymbol{\tau}_k) \cdot \mathbf{v}_k}^i \right] + E_{ki} \tag{6.45}$$

Here $E_{ki}$ is the energy flux through the interface due to fluctuations of interfacial heat flux and work done by fluctuating pressure and shear stress at the interface.

## 6.4 ONE-DIMENSIONAL MODELS

The analysis of multiphase flow in channels can be significantly simplified when the flow parameters are averaged over the channel cross-section area. Such

one-dimensional two-phase flow models play a very important role in nuclear reactor safety analyses, since, on the one hand, they are simple enough to be solved
over large systems of interconnected channels, and on the other hand, they can be
used to predict a wide range of steady-state and transient phenomena by employing
proper closure relationships. A particular model is characterized by the number of
conservation equations that are simultaneously solved and the formulation of closure
relationships for interfacial transfers and wall interactions.

The typical unknowns for one-dimensional models are cross-section averaged
velocities, temperatures or specific enthalpies, pressures, and volume fractions. All
these flow variables can have a specific value for each phase or can take one value
that is representative for all phases. In the simplest formulation of a multiphase flow
referred to as the *homogeneous equilibrium model* a perfect multiphase mixture is
assumed and only one value of velocity, pressure, and specific enthalpy is considered at each time instant and point in the flow domain. When these flow variables
are considered separately for each phase, the *two-fluid model* is obtained. Clearly, a
whole spectrum of formulations is possible assuming either a separate field or mixture model for specific conservation principles. For example, a whole class of *drift
flux models* have been developed in which mass and energy conservation are considered separately for each phase whereas a mixture model is assumed in the momentum
equation formulation.

In this section we discuss several useful formulations of one-dimensional multiphase flows, starting with definitions of important one-dimensional flow parameters.

### 6.4.1   DEFINITIONS OF AREA-AVERAGED FLOW PARAMETERS

Many quantities used in the nuclear reactor thermal-hydraulics, such as quality and
superficial velocity, are originating from consideration of two-phase flow in channels. Thus, we introduce here some important definitions of average flow parameters.

**Volume Fraction and Void Fraction**

Assuming that a certain volume $V$ of a channel with cross-section area $A$ and length
$\Delta z$ contains phase $k$ with volume $V_k$, the volume fraction of phase $k$ in that channel
segment is,

$$\alpha_k = \frac{V_k}{V}. \tag{6.46}$$

Assuming that $\Delta z \to 0$, we obtain an expression for the volume fraction of phase $k$
in terms of the area ratio,

$$\alpha_k = \frac{A_k}{A}, \tag{6.47}$$

where $A_k$ is the pipe cross-section area occupied by phase $k$. Clearly, $\alpha_k$ is a non-
dimensional parameter and $0 \leq \alpha_k \leq 1$. In particular, in gas-liquid flows, $\alpha$ represents
the *void fraction*, which is equivalent to the volume fraction of the gas phase.

## Phasic Velocity

The cross-section averaged velocity of phase $k$ in the pipe can be found as,

$$U_k = \frac{1}{A_k} \iint_{A_k} w_k dA_k, \tag{6.48}$$

where $w_k$ is the local velocity of phase $k$ in the channel axial direction and $A_k$ is the total channel cross-section area occupied by phase $k$.

## Volumetric Flow Rate

The volumetric flow rate of phase $k$ can be found as,

$$Q_k = \iint_{A_k} w_k dA_k. \tag{6.49}$$

Here $Q_k$ is the volumetric flow rate of phase $k$ in $m^3\ s^{-1}$, $w_k$ is velocity of phase $k$ in the channel axial direction and $A_k$ is the total channel cross-section area occupied by phase $k$.

## Volumetric Flux - Superficial Velocity

The volumetric flux of phase $k$, frequently referred to as the **superficial velocity**, is defined as follows,

$$J_k = \frac{Q_k}{A} = \frac{1}{A} \iint_{A_k} u_k dA_k = \frac{A_k}{A} \frac{1}{A_k} \iint_{A_k} u_k dA_k = \alpha_k U_k. \tag{6.50}$$

As can be seen, the superficial velocity has units $m\ s^{-1}$ and can be interpreted for each phase as that velocity, with which the phase would move if only that phase was present in the channel.

## Mass Flow Rate and Mass Flux

The total *mass flow rate* of phase $k$ can be found as,

$$W_k = \iint_{A_k} \rho_k w_k dA_k, \tag{6.51}$$

where $W_k$ is the mass flow rate of phase $k$ in $kg\ s^{-1}$, $\rho_k$ is the mass density of phase $k$, $w_k$ is velocity of phase $k$ in the channel axial direction, and $A_k$ is the total channel cross-section area occupied by phase $k$. In general both $\rho_k$ and $w_k$ vary in the cross section area of the channel and the integral of their product is not equal to the product of their individual integrals. To separate the two averaged variables, we can introduce the following average, density-weighted phasic velocity,

$$U_{k\rho} = \frac{\iint_{A_k} \rho_k w_k dA_k}{\iint_{A_k} \rho_k dA_k} = \frac{\iint_{A_k} \rho_k w_k dA_k}{A_k \langle \rho_k \rangle_2}, \tag{6.52}$$

where $\langle \rho_k \rangle_2$ is the area-averaged mass density of phase $k$ defined as follows,

$$\langle \rho_k \rangle_2 \equiv \frac{1}{A_k} \iint_{A_k} \rho_k dA_k. \tag{6.53}$$

Now the mass flow rate of phase $k$ can be expressed in terms of $U_{k\rho}$ as follows,

$$W_k = \langle \rho_k \rangle_2 U_{k\rho} A_k, \tag{6.54}$$

or, using definition of the volume fraction of phase $k$, the mass flow rate is expressed as,

$$W_k = \alpha_k \langle \rho_k \rangle_2 U_{k\rho} A. \tag{6.55}$$

In analogy to the volumetric flux, the *mass flux* for phase $k$ is defined as,

$$G_k = \frac{W_k}{A} = \alpha_k \langle \rho_k \rangle_2 U_{k\rho}. \tag{6.56}$$

Similarly, the superficial velocity of phase $k$ can be defined in terms of $U_{k\rho}$ as,

$$J_{k\rho} = \frac{G_k}{\langle \rho_k \rangle_2} = \alpha_k U_{k\rho}. \tag{6.57}$$

For constant density, $\rho_k$, we have $U_{k\rho} = U_k$ and $J_{k\rho} = J_k$.

### Actual or Flow Quality

The *actual* or *flow quality* for gas-liquid two phase mixture flow is defined as,

$$x_a = \frac{W_v}{W_v + W_l} = \frac{\alpha_v \langle \rho_v \rangle_2 U_{v\rho}}{\alpha_v \langle \rho_v \rangle_2 U_{v\rho} + \alpha_l \langle \rho_l \rangle_2 U_{l\rho}}, \tag{6.58}$$

where subscripts $v$ and $l$ are used to indicate the gas (vapor) and liquid phase, respectively. Assuming further that $\alpha_v = \alpha$ and $\alpha_l = 1 - \alpha$, we get,

$$x_a = \frac{W_v}{W_v + W_l} = \frac{\alpha \langle \rho_v \rangle_2 U_{v\rho}}{\alpha \langle \rho_v \rangle_2 U_{v\rho} + (1 - \alpha) \langle \rho_l \rangle_2 U_{l\rho}}. \tag{6.59}$$

This equation can be also written as,

$$x_a = \frac{1}{1 + \frac{(1-\alpha)}{(\alpha)} \frac{\langle \rho_l \rangle_2}{\langle \rho_v \rangle_2} \frac{U_{l\rho}}{U_{v\rho}}}. \tag{6.60}$$

We can notice that the actual quality is related to the void fraction of the gas phase and the ratios of densities and mean velocities of both phases.

## Thermodynamic Equilibrium Quality

The quality defined in the previous section is based on the mass ratio of the gas phase to the total mass of the mixture. This definition requires that the mass conservation equation for each phase is solved. However, if we imagine a two-phase mixture containing phases at saturation, that is, at thermodynamic equilibrium, there is a unique relationship between the quality of the mixture and its specific enthalpy. The specific enthalpy of the mixture can be found as,

$$i_m = \frac{W_v i_g + W_l i_f}{W_v + W_l} = x_a i_g + (1 - x_a) i_f = x_a (i_g - i_f) + i_f. \tag{6.61}$$

Thus, the actual quality can be expressed in terms of specific enthalpies as,

$$x_a = x_e \equiv \frac{i_m - i_f}{i_g - i_f} = \frac{i_m - i_f}{i_{fg}}. \tag{6.62}$$

Here subscripts $f$ and $g$ are used to indicate the saturation conditions for the phases and $i_{fg}$ is the latent heat. The right-hand-side of the equation represents the *thermodynamic equilibrium quality* and the equation states that for two-phase flow, where the phases are in the thermodynamic equilibrium, the mass-based quality is equal to the thermodynamic equilibrium quality. To distinguish between the two definitions of the two-phase mixture quality, $x_e$ is used to denote the thermodynamic equilibrium quality.

The relationship between the two types of mixture quality for non-equilibrium conditions can be obtained as follows. Let us first assume that the liquid phase is slightly subcooled, whereas the vapor phase is slightly superheated. Such situation can exist in a boiling channel, where initially subcooled liquid enters the heated section and gradually approaches the saturation condition. The mixture specific enthalpy can be found as,

$$i_m = \frac{W_v(i_g + \delta i_v) + W_l(i_f - \delta i_l)}{W_v + W_l} = x_a \left[ i_g - i_f + \delta i_v + \delta i_l \right] + i_f - \delta i_l. \tag{6.63}$$

Here $\delta i_v$ and $\delta i_l$ represent specific enthalpies by which the phases depart from the saturation conditions and $x_a$ is the actual quality. From this equation, the following relationship between the thermodynamic equilibrium and the actual qualities is obtained,

$$x_a = \phi(x_e + \delta x_l), \tag{6.64}$$

where,

$$\phi = \frac{i_{fg}}{i_{fg} + \delta i_v + \delta i_l}, \tag{6.65}$$

$$\delta x_l = \frac{\delta i_l}{i_{fg}}, \quad \delta x_v = \frac{\delta i_v}{i_{fg}} \tag{6.66}$$

$$x_e = \frac{i_m - i_f}{i_{fg}}. \tag{6.67}$$

The above analysis indicates that when the mixture is in the thermodynamic non-equilibrium, the actual quality is different from the equilibrium quality. In particular, when $\delta i_l > 0$ and $\delta i_v = 0$ the actual quality can be positive even for slightly negative or zero equilibrium quality:

$$x_a = \frac{i_{fg}}{i_{fg} + \delta i_v + \delta i_l}(x_e + \delta x_l) = \frac{x_e + \delta x_l}{1 + \delta x_l}. \tag{6.68}$$

For superheated vapor flowing with saturated water, we have $\delta i_v > 0$ and $\delta i_l = 0$, and the relationship between the actual and the thermodynamic equilibrium quality becomes,

$$x_a = \frac{i_{fg}}{i_{fg} + \delta i_v + \delta i_l}(x_e + \delta x_l) = \frac{x_e}{1 + \delta x_v}. \tag{6.69}$$

Thus, the actual quality is less than the thermodynamic equilibrium quality when the vapor is superheated in the presence of the saturated liquid.

## Void-Quality Relationship

The expression for the actual quality given by Eq. (6.60) can be solved for the void fraction as follows,

$$\alpha = \frac{1}{1 + \frac{(1-x_a)}{(x_a)}\frac{\langle\rho_v\rangle_2}{\langle\rho_l\rangle_2}\frac{U_{v\rho}}{U_{l\rho}}}. \tag{6.70}$$

This equation gives an exact expression for the cross-section averaged void fraction in terms of the actual quality. The velocity ratio $S = U_{v\rho}/U_{l\rho}$ is referred to as the *slip ratio* and it turns out to be flow-pattern dependent. For a homogeneous two-phase mixture, the slip ratio is equal to unity, $S = 1$, the void fraction is a function of the quality and the density ratio only:

$$\alpha = \frac{1}{1 + \frac{(1-x_a)}{(x_a)}\frac{\langle\rho_v\rangle_2}{\langle\rho_l\rangle_2}}. \tag{6.71}$$

The relationship given by Eq. (6.70) can be expressed in terms of the superficial velocities $J_{v\rho}$ and $J_{l\rho}$ rather than the actual quality $x_a$. Since,

$$x_a = \frac{G_v}{G_v + G_l} = \frac{\langle\rho_v\rangle_2 J_{v\rho}}{\langle\rho_v\rangle_2 J_{v\rho} + \langle\rho_l\rangle_2 J_{l\rho}} \tag{6.72}$$

thus,

$$\frac{1 - x_a}{x_a} = \frac{\langle\rho_v\rangle_2 J_{v\rho} + \langle\rho_l\rangle_2 J_{l\rho}}{\langle\rho_v\rangle_2 J_{v\rho}} - 1 = \frac{\langle\rho_l\rangle_2 J_{l\rho}}{\langle\rho_v\rangle_2 J_{v\rho}}. \tag{6.73}$$

Now, Eq. (6.70) can be written as,

$$\alpha = \frac{1}{1 + \frac{\langle\rho_l\rangle_2 J_{l\rho}}{\langle\rho_v\rangle_2 J_{v\rho}}\frac{\langle\rho_v\rangle_2}{\langle\rho_l\rangle_2}\frac{U_{v\rho}}{U_{l\rho}}} = \frac{J_{v\rho}}{J_{v\rho} + J_{l\rho}\frac{U_{v\rho}}{U_{l\rho}}}. \tag{6.74}$$

Another useful form of the equation can be obtained by replacing $J_{l\rho}$ with $J_\rho - J_{v\rho}$ which yields,

$$\alpha = \frac{J_{v\rho}}{\frac{U_{v\rho}}{U_{l\rho}}J_\rho + J_{v\rho}\left(1 - \frac{U_{v\rho}}{U_{l\rho}}\right)}. \tag{6.75}$$

The equation can be used to calculate the mean void fraction for the known superficial gas velocity, the total superficial velocity and the slip ratio. This equation and Eq. (6.70) are equivalent to each other and both require the slip ratio, $S$, to be known. For a homogeneous two-phase mixture, $S = 1$ and the void fraction is equal to a ratio of the gas superficial velocity to the total superficial velocity:

$$\alpha = \frac{J_{v\rho}}{J_\rho}. \tag{6.76}$$

So far we have discussed a one-parameter void-quality relationship, where the only parameter is the slip ratio. Initially, various correlations were developed to determine the slip ratio in terms of other flow parameters. This approach was not very successful, however, and it became clear that with only one parameter it is difficult to obtain good agreement of predictions with experimental data for a wide range of operational conditions.

A more advanced void-quality relationship can be obtained if we allow for two independent parameters. Using the form of the relationship as shown in Eq. (6.75), the void fraction can be expressed as follows,

$$\alpha = \frac{J_{v\rho}}{C_0 J_\rho + U_{vj}}, \tag{6.77}$$

where $C_0$ and $U_{vj}$ are two parameters referred to as the *distribution parameter* and the *drift velocity*, respectively. The coefficients $C_0$ and $U_{vj}$ have been subjects of intensive theoretical and empirical investigations and are usually provided separately for various flow patterns and channel configurations. Examples of expressions for $C_0$ and $U_{vj}$ are provided in §6.4.3.

### 6.4.2 HOMOGENEOUS EQUILIBRIUM MODEL

The *homogeneous equilibrium model* is one of the simplest models for two-phase flow. The model assumes that both thermal and mechanical equilibrium prevail and that the two phases travel at equal velocities. Due to its simplicity, the model is particularly useful for calculating the liquid holdup and pressure drop in channels. The simplest formulation of the model is based on the conservation of mass, linear momentum, and total energy for the mixture. To take into account the thermal nonequilibrium effects, the continuity equation for the gas phase can be also included. Here we consider the model with three conservation equations only.

### Conservation Equations

The basic conservation equations of mass, momentum and energy for transient homogeneous two-phase flow in a channel are as follows:

The *mixture continuity equation*

$$\frac{\partial \rho_m}{\partial t} + \frac{\partial G}{\partial z} = 0, \tag{6.78}$$

The *mixture momentum equation*

$$\frac{\partial G}{\partial t} + \frac{\partial}{\partial z}\left(\frac{G^2}{\rho_m}\right) + \frac{\partial p_m}{\partial z} + \left[\frac{4C_f}{D_h} + \sum_{j=1}^{N} \xi_j \delta(z - z_j)\right] \frac{|G|G}{2\rho_m} + \rho_m g_z = 0 \tag{6.79}$$

The *mixture energy equation*

$$\frac{\partial}{\partial t}\left(\rho_m i_m - p_m\right) + \frac{\partial}{\partial z}\left(G i_m\right) = \frac{q'}{A}, \tag{6.80}$$

where the mixture density, $\rho_m$, is given as,

$$\rho_m = \rho_f(1 - \alpha) + \rho_g \alpha = \frac{1}{v_f + x_e v_{fg}}, \tag{6.81}$$

where $v_f = 1/\rho_f$, $v_g = 1/\rho_g$, $v_{fg} = v_g - v_f$, the void fraction, $\alpha$, is found from Eq. (6.71) and the mixture quality, $x_e$, is obtained from Eq. (6.67). All remaining variables in Eqs. (6.78)–(6.80) describe the two-phase mixture properties and in particular: $G$ is the mass flux, $i_m$ is the specific enthalpy, $p_m$ is the pressure, $t$ is the time, $z$ is the distance along the channel, $A$ is the channel cross-section area, and $q' = q_w'' P_H$ is the linear power, where $q_w''$ is the wall heat flux and $P_H$ is the channel heated perimeter. The friction and the local pressure loss, as well as the gravity pressure drop terms contain $C_f$ - friction coefficient, $\xi_j$ - local pressure loss coefficient at obstacle $j$, $z_j$ - location of obstacle $j$, $D_h$ - hydraulic diameter, $g_z$ - gravity acceleration projected on the channel axis.

## Pressure Drop in a Boiling Channel

We will apply the homogeneous equilibrium model to predict pressure drop in a boiling channel with steady-state two-phase flow. For such conditions, Eqs. (6.78)–(6.80) are as follows,

$$\frac{dG}{dz} = 0, \tag{6.82}$$

$$\frac{d}{dz}\left(\frac{G^2}{\rho_m}\right) + \frac{dp_m}{dz} + \left[\frac{4C_f}{D_h} + \sum_{j=1}^{N} \xi_j \delta(z - z_j)\right] \frac{|G|G}{2\rho_m} + \rho_m g_z = 0, \tag{6.83}$$

$$\frac{d(G i_m)}{dz} = \frac{q'}{A}. \tag{6.84}$$

From Eq. (6.82) we obtain,

$$G(z) = const = G_{in}, \qquad (6.85)$$

where $G_{in}$ is the inlet mass flux to the channel. Assuming a uniform axial power distribution, the energy equation, Eq. (6.84), yields,

$$i_{m,ex} = i_{m,in} + \frac{q'L}{AG_{in}}, \qquad (6.86)$$

where $L$ is the channel total length and $i_{m,in}$, $i_{m,ex}$ are specific enthalpies at the inlet and outlet, respectively. In a similar way, we can obtain specific enthalpies $i_{m,j}$ at all local obstacles in the channel,

$$i_{m,j} = i_{m,in} + \frac{q'z_j}{AG_{in}}. \qquad (6.87)$$

Once integrating the momentum conservation equation, it is convenient to divide the assembly into two parts: a single-phase-flow part and a two-phase-flow part of the channel. According to the homogeneous equilibrium model, the length of the single-phase-flow part, $L_{1\phi} = \lambda$, can be found from the energy balance as,

$$L_{1\phi} = \lambda = \frac{AG_{in}(i_f - i_{m,in})}{q'}. \qquad (6.88)$$

Now we can integrate the momentum equation, Eq. (6.83), along the single-phase-flow part of the channel as follows,

$$\frac{G_{in}^2}{\rho_f} - \frac{G_{in}^2}{\rho_{in}} + (p_{m,\lambda} - p_{m,in}) + \int_0^\lambda \frac{4C_{f,1\phi}}{D_h} \frac{|G_{in}|G_{in}}{2\rho_m} dz +$$
$$\sum_{j=1}^{N_\lambda} \xi_{1\phi,j} \frac{|G_{in}|G_{in}}{2\rho_{m,j}} + \int_0^\lambda \rho_m g_z dz = 0, \qquad (6.89)$$

where $N_\lambda$ is the number of local obstacles in the single-phase-flow part of the channel, $C_{f,1\phi}$ is the friction factor for single-phase flow, and $\xi_{1\phi}$ is the local pressure loss coefficient for single-phase flow. In a boiling channel such as a fuel assembly, the inlet local loss coefficient, $\xi_{1\phi,in}$, and the local loss coefficients, $\xi_{1\phi,j}$, caused by spacers have to be taken into account.

The acceleration pressure drop term results from a slightly varying liquid density in the single-phase-flow part of the channel. This effect is, however, very small and can be neglected, thus,

$$\Delta p_{1\phi,acc} = \frac{G_{in}^2}{\rho_f} - \frac{G_{in}^2}{\rho_{m,in}} \approx 0. \qquad (6.90)$$

In the friction pressure drop term both the friction loss coefficient $C_f$ and the liquid density change along the channel because of the liquid temperature change.

However, this effect is also small and we assume that the density in the friction pressure drop term is constant and equal to $\rho_f$, thus,

$$\Delta p_{1\phi,fric} = \int_0^\lambda \frac{4C_{f,1\phi}}{D_h} \frac{|G_{in}|G_{in}}{2\rho_m} \mathrm{d}z \approx \frac{4C_{f,1\phi}\lambda}{D_h} \frac{|G_{in}|G_{in}}{2\rho_f}. \tag{6.91}$$

The local pressure loss term is approximated as,

$$\Delta p_{1\phi,loc} = \sum_{j=1}^{N_\lambda} \xi_{1\phi,j} \frac{|G_{in}|G_{in}}{2\rho_{m,j}} \approx \left( \xi_{1\phi,in} + \sum_{j=1}^{N_\lambda} \xi_{1\phi,j} \right) \frac{|G_{in}|G_{in}}{2\rho_f}. \tag{6.92}$$

Finally, the gravity pressure drop term is as follows,

$$\Delta p_{1\phi,grav} = \int_0^\lambda \rho_m g_z \mathrm{d}z \approx \lambda \rho_f g_z. \tag{6.93}$$

Combining all pressure drop terms and substituting to Eq. (6.89) yields,

$$-\Delta p_{1\phi} = -(p_{m,\lambda} - p_{m,in}) =$$
$$\left( \xi_{1\phi,in} + \sum_{j=1}^{N_\lambda} \xi_{1\phi,j} + \frac{4C_{f,1\phi}\lambda}{D_h} \right) \frac{|G_{in}|G_{in}}{2\rho_f} + \lambda \rho_f g_z. \tag{6.94}$$

Taking into account Eq. (6.88), we obtain the following expression for the total pressure drop in the single-phase-flow part,

$$-\Delta p_{1\phi} = \left( \xi_{1\phi,in} + \sum_{j=1}^{N_\lambda} \xi_{1\phi,j} \right) \frac{|G_{in}|G_{in}}{2\rho_f} + \frac{4C_{f,1\phi}A|G_{in}|G_{in}^2(i_f - i_{m,in})}{2D_h \rho_f q'} +$$
$$\frac{AG_{in}(i_f - i_{m,in})\rho_f g_z}{q'}. \tag{6.95}$$

We can note that with increasing linear power, the friction pressure loss term and the gravity pressure drop term are decreasing. These terms will be equal to zero when $i_{m,in} = i_f$, since then no single-phase part of the channel will exist.

A similar derivation can be performed for the boiling part of the channel, however, since the mixture density is now significantly changing with the changing mixture specific enthalpy, this effect has to be taken into account. We integrate the momentum equation, Eq. (6.83), along the boiling part of the assembly to obtain,

$$\frac{G_{in}^2}{\rho_{m,ex}} - \frac{G_{in}^2}{\rho_f} + (p_{m,ex} - p_{m,\lambda}) + \int_\lambda^L \frac{4C_{f,2\phi}}{D_h} \frac{|G_{in}|G_{in}}{2\rho_m} \mathrm{d}z +$$
$$\sum_{j=N_\lambda+1}^{N} \xi_{2\phi,j} \frac{|G_{in}|G_{in}}{2\rho_{m,j}} + \int_\lambda^L \rho_m g_z \mathrm{d}z = 0. \tag{6.96}$$

Here $C_{f,2\phi}$ and $\xi_{2\phi,j}$ are the friction coefficient and local loss coefficient for two-phase flow, respectively. The acceleration pressure drop in the boiling region is,

$$\Delta p_{2\phi,acc} = \frac{G_{in}^2}{\rho_{m,ex}} - \frac{G_{in}^2}{\rho_f} = \frac{G_{in}^2}{\rho_f} \left( \frac{\rho_f}{\rho_{m,ex}} - 1 \right) = r_2 \frac{G_{in}^2}{\rho_f}, \tag{6.97}$$

where we introduce here the *acceleration pressure drop multiplier,*

$$r_2 \equiv \frac{\rho_f}{\rho_{m,ex}} - 1 = \rho_f(\upsilon_f + x_{m,ex}\upsilon_{fg}) - 1 = x_{m,ex}\left(\frac{\rho_f}{\rho_g} - 1\right). \qquad (6.98)$$

The friction pressure loss is as follows,

$$\Delta p_{2\phi,fric} = \int_\lambda^L \frac{4C_{f,2\phi}}{D_h} \frac{|G_{in}|G_{in}}{2\rho_m} dz =$$
$$\frac{4C_{f,1\phi}(L-\lambda)}{D_h} \frac{|G_{in}|G_{in}}{2\rho_f} \frac{1}{L-\lambda} \int_\lambda^L \frac{C_{f,2\phi}}{C_{f,1\phi}} \frac{\rho_f}{\rho_m} dz. \qquad (6.99)$$

By introducing the *friction pressure drop multiplier,*

$$r_3 \equiv \frac{1}{L-\lambda} \int_\lambda^L \frac{C_{f,2\phi}}{C_{f,1\phi}} \frac{\rho_f}{\rho_m} dz, \qquad (6.100)$$

the two-phase friction pressure loss becomes,

$$\Delta p_{2\phi,fric} = r_3 \frac{4C_{f,1\phi}(L-\lambda)}{D_h} \frac{|G_{in}|G_{in}}{2\rho_f}. \qquad (6.101)$$

The local pressure loss term in the boiling part of the channel is as follows,

$$\Delta p_{2\phi,loc} = \sum_{j=N_\lambda+1}^N \xi_{2\phi,j} \frac{|G_{in}|G_{in}}{2\rho_j} \approx$$
$$\left(\xi_{2\phi,ex} \frac{\rho_f}{\rho_{m,ex}} + \sum_{j=N_\lambda+1}^N \xi_{2\phi,j} \frac{\rho_f}{\rho_{m,j}}\right) \frac{|G_{in}|G_{in}}{2\rho_f}, \qquad (6.102)$$

where the summation goes over all flow obstacles that are present in the boiling part of the channel. Here, as we can see, the density varies at the local loss positions and the local pressure loss terms cannot be lumped to a single term as was the case in the non-boiling part of the channel. Introducing a local two-phase pressure drop multiplier,

$$\phi_{lo,loc}^2 \equiv \frac{\rho_f}{\rho_m}, \qquad (6.103)$$

we have,

$$\Delta p_{2\phi,loc} = \left(\phi_{lo,loc,ex}^2 \xi_{2\phi,ex} + \sum_{j=N_\lambda+1}^N \phi_{lo,loc,j}^2 \xi_{2\phi,j}\right) \frac{|G_{in}|G_{in}}{2\rho_f}, \qquad (6.104)$$

where $\xi_{2\phi,ex}$ is the local loss coefficient at the channel outlet. The last term, representing the gravity pressure drop in the boiling part, is as follows,

$$\Delta p_{2\phi,grav} \equiv \int_\lambda^L \rho g_z dz = (L-\lambda)\rho_f g_z \frac{1}{(L-\lambda)\rho_f} \int_\lambda^L \rho dz. \qquad (6.105)$$

Introducing the two-phase *gravity pressure drop multiplier*,

$$r_4 \equiv \frac{1}{(L-\lambda)\rho_f} \int_\lambda^L \rho \, dz, \tag{6.106}$$

we have,

$$\Delta p_{2\phi,grav} = r_4 (L-\lambda)\rho_f g_z. \tag{6.107}$$

Substituting all terms into Eq. (6.96) yields,

$$-\Delta p_{2\phi} = -(p_{m,ex} - p_{m,\lambda}) = r_2 \frac{G^2}{\rho_f} + \left[ r_3 \frac{4C_{f,1\phi}(L-\lambda)}{D_h} + \right.$$
$$\left. \sum_{j=N_\lambda+1}^{N} \phi_{lo,loc,j}^2 \xi_{2\phi,j} + \phi_{lo,loc,ex}^2 \xi_{2\phi,ex} \right] \frac{|G_{in}|G_{in}}{2\rho_f} + \tag{6.108}$$
$$r_4 (L-\lambda)\rho_f g_z$$

Combining this equation with Eq. (6.88) yields,

$$-\Delta p_{2\phi} = -r_3 \frac{4C_{f,1\phi}A(i_f - i_{m,in})}{D_h q'} \frac{|G_{in}|G_{in}^2}{2\rho_f} + r_2 \frac{G^2}{\rho_f} + \left( r_3 \frac{4C_{f,lo}L}{D_h} + \right.$$
$$\left. \sum_{j=N_\lambda+1}^{N} \phi_{lo,loc,j}^2 \xi_{2\phi,j} + \phi_{lo,loc,ex}^2 \xi_{2\phi,ex} \right) \frac{|G_{in}|G_{in}}{2\rho_f} - \tag{6.109}$$
$$r_4 \frac{\rho_f g_z A(i_f - i_{m,in})}{q'} G_{in} + r_4 L \rho_f g_z,$$

The total pressure drop in the boiling channel is obtained by combining Eqs. (6.95) and (6.109). As a result, assuming $G_{in} = G$, we obtain the following expression for the total pressure drop in a boiling channel as a function of the mass flux,

$$-\Delta p_{tot} = aG^3 + bG^2 + cG + d, \tag{6.110}$$

where,

$$a = \pm \frac{2(1-r_3)C_{f,1\phi}A(i_f - i_{m,in})}{\rho_f q' D_h} \tag{6.111}$$

$$b = \frac{1}{2\rho_f} \left[ 2r_2 \pm \left( \xi_{1\phi,in} + r_3 \frac{4C_{f,1\phi}L}{D_h} + \sum_{j=1}^{N_\lambda} \xi_{1\phi,j} + \right. \right.$$
$$\left. \left. \sum_{j=N_\lambda+1}^{N} \phi_{lo,loc,j}^2 \xi_{2\phi,j} + \phi_{lo,loc,ex}^2 \xi_{2\phi,ex} \right) \right] \tag{6.112}$$

$$c = (1-r_4) \frac{\rho_f g_z A(i_f - i_{in})}{q'}, \tag{6.113}$$

$$d = r_4 L \rho_f g_z, \tag{6.114}$$

where a plus sign in coefficients $a$ and $b$ should be used for $G > 0$ and a minus sign for $G < 0$.

Equation (6.110) describes the pressure drop characteristic of a boiling channel. The characteristic is not exactly a third-degree polynomial since coefficients $a$ and $b$ are mass-flux dependent.

## Two-Phase Flow in Parallel Boiling Channels

If we consider flow in parallel channels, the following equations have to be satisfied,

$$\Delta p_1 = \Delta p_2 = \cdots = \Delta p_N, \tag{6.115}$$

and

$$A_1 G_1 + A_2 G_2 + \cdots + A_N G_N = W_{tot}. \tag{6.116}$$

Here $G_i$ is the mass flux in channel $i$, $W_{tot}$ is the total mass flow rate in all parallel channels, and $A_i$ is the cross-section area in the $i$th channel. The above system of nonlinear algebraic equations with $N$ unknowns can be solved only numerically due to the complexity of involved expressions for pressure drop characteristic of boiling channels. Guessing the initial flow distribution for known (and possibly skewed) power distribution between channels, the Newton method leads to the following linear system of equations,

$$\mathbf{A} \cdot \delta \mathbf{G} = \mathbf{B}, \tag{6.117}$$

where $\mathbf{A}$ is a Jacobian matrix. For example, assuming four parallel channels, the Jacobian matrix is given as,

$$\mathbf{A} = \begin{bmatrix} \dfrac{\partial \Delta p_1}{\partial G_1} & -\dfrac{\partial \Delta p_2}{\partial G_2} & 0 & \cdots & 0 \\ 0 & \dfrac{\partial \Delta p_2}{\partial G_2} & -\dfrac{\partial \Delta p_3}{\partial G_3} & \cdots & 0 \\ 0 & 0 & \cdots & \dfrac{\partial \Delta p_3}{\partial G_3} & -\dfrac{\partial \Delta p_4}{\partial G_4} \\ 1 & 1 & \cdots & 1 \end{bmatrix}, \tag{6.118}$$

and the right-hand-side vector $\mathbf{B}$ is as follows

$$\mathbf{B} = \begin{bmatrix} \Delta p_2 - \Delta p_1 & \Delta p_3 - \Delta p_2 & \Delta p_4 - \Delta p_3 & 0 \end{bmatrix}^T, \tag{6.119}$$

with a vector of unknown mass flux corrections $\delta \mathbf{G}$ to satisfy the total flow rate and pressure drop conditions,

$$\delta \mathbf{G} = \begin{bmatrix} \delta G_1 & \delta G_2 & \delta G_3 & \delta G_4 \end{bmatrix}^T. \tag{6.120}$$

Since the Jacobian matrix $\mathbf{A}$ depends on the flow distribution between parallel channels, an iterative process is required to obtain convergence when $\delta \mathbf{G} \approx \mathbf{0}$.

### 6.4.3  DRIFT FLUX MODEL

The *drift flux model* is a widely used model for two-phase flows, particularly in the context of vapor-liquid interaction in channels. The model provides a semi-empirical methodology for modeling the gas-liquid slip in one-dimensional flows. It requires two adjustable parameters, which can be found analytically for some idealized cases and are more often obtained empirically. One of the important features of the drift flux model is the usage of a single linear momentum equation, resulting in significant savings in computational cost, as compared to the two-fluid model. Using the drift flux model, some major difficulties associated with the two-fluid model can be avoided. In particular, difficulties with the flow-regime-dependent constitutive relations for the interfacial transport of momentum are eliminated.

The standard drift flux model is typically expressed in terms of four field equations:

1. Mixture continuity equation that represents the conservation of mass for the mixture.
2. Momentum equation that represents the conservation of linear momentum for the mixture.
3. Energy equation that represents the conservation of total energy for the mixture.
4. Gas continuity equation that represents the conservation of mass for the gas phase.

This model formulation is particularly useful to describe two-phase flow dynamics when the dynamics of both components are closely coupled and mainly concern the low-velocity wave propagation flow [113]. In this subsection we present the drift flux conservation equations for one-dimensional two-phase.

### Conservation Equations

The basic concept of the drift flux model is to use the two-phase mixture field equations that can be derived from instantaneous conservation equations properly averaged over time and space. For practical applications, several important simplifications are needed, as discussed in more detail in [113]. A system of equations particularly useful for studies of two-phase flow dynamics in fuel rod assemblies is as follows.

The *mixture continuity equation*

$$\frac{\partial \rho_m}{\partial t} + \frac{\partial}{\partial z} (\rho_m U_m) = 0, \tag{6.121}$$

The *continuity equation for the gas phase*

$$\frac{\partial}{\partial t} (\alpha \rho_v) + \frac{\partial}{\partial z} (\alpha \rho_v U_m) + \frac{\partial}{\partial z} \left( \frac{\alpha \rho_v \rho_l}{\rho_m} \bar{U}_{vj} \right) = \Gamma_v, \tag{6.122}$$

The *mixture momentum equation*

$$\frac{\partial}{\partial t}(\rho_m U_m) + \frac{\partial}{\partial z}(\rho_m U_m^2) + \frac{\partial p_m}{\partial z} + \frac{\partial}{\partial z}\left[\frac{\alpha \rho_v \rho_l}{(1-\alpha)\rho_m}\bar{U}_{vj}^2\right] +$$

$$\left[\frac{4C_f}{D_h} + \sum_{j=1}^{N}\xi_j\delta(z-z_j)\right]\frac{\rho_m|U_m|U_m}{2} + \rho_m g_z = 0 \quad , \tag{6.123}$$

The *mixture energy equation*

$$\frac{\partial}{\partial t}(\rho_m i_m - p_m) + \frac{\partial}{\partial z}(\rho_m i_m U_m) + \frac{\partial}{\partial z}\left(\frac{\alpha \rho_v \rho_l}{\rho_m}\Delta i_{vl}\bar{U}_{vj}\right) = \frac{q_w'' P_H}{A} +$$

$$\left[U_m + \frac{\alpha(\rho_l - \rho_g)}{\rho_m}\bar{U}_{vj}\right]\frac{\partial p_m}{\partial z} \quad . \tag{6.124}$$

In these equations $U_m$ is the area-averaged and density-weighted mixture velocity, $\bar{U}_{vj}$ is the *mean drift velocity*, $\Delta i_{vl} = i_v - i_l$ is the specific enthalpy difference between the phases, $\Gamma_v$ is the mass transfer across the interface, and the remaining variables have the same meaning as in the homogeneous equilibrium model described in §6.4.2.

The mean drift velocity is expressed in terms of the *weighted mean drift velocity* of the gas phase, $U_{vj}$, the *distribution parameter*, $C_0$, and the volumetric flux of the mixture, $J$, as follows,

$$\bar{U}_{vj} = U_{vj} + (C_0 - 1)J. \tag{6.125}$$

The volumetric flux of the mixture can be related to the mixture velocity, $U_m$, as,

$$J = U_m + \frac{\alpha(\rho_l - \rho_v)}{\rho_m}\bar{U}_{vj}. \tag{6.126}$$

where the mixture density, $\rho_m$, is given as,

$$\rho_m = \rho_l(1-\alpha) + \rho_v \alpha \,. \tag{6.127}$$

## Constitutive Relationships

To close the drift flux model, expressions for $C_0$ and $U_{vj}$ are needed. These parameters depend on the channel geometry and two-phase flow conditions. An extensive summary of correlations for $C_0$ and $U_{vj}$ applicable for a wide range of conditions is provided by Ishii and Hibiki [113]. For example, for boiling two-phase flow in a rod bundle with a square lattice, they provide the following relationships,

$$C_0 = \begin{cases} \left(1.03 - 0.03\sqrt{\rho_v/\rho_l}\right)\left(1 - e^{-26.3\alpha^{0.780}}\right) & \text{for } D_R/P = 0.3 \\ \left(1.04 - 0.04\sqrt{\rho_v/\rho_l}\right)\left(1 - e^{-21.2\alpha^{0.762}}\right) & \text{for } D_R/P = 0.5 \\ \left(1.05 - 0.05\sqrt{\rho_v/\rho_l}\right)\left(1 - e^{-34.1\alpha^{0.925}}\right) & \text{for } D_R/P = 0.7 \end{cases} \tag{6.128}$$

where $\alpha$ is the void fraction and $D_R/P$ is the ratio of the rod diameter to the lattice pitch. The weighted mean drift velocity of the gas phase can be calculated as,

$$U_{vj} = B_{sf} \left( \frac{4 \Delta \rho g \sigma}{\rho_l^2} \right)^{1/4} (1 - \alpha)^{1.75}, \tag{6.129}$$

where $B_{fs}$ is the bubble size factor taking into account the rod wall effect on the bubble rise velocity, given as,

$$B_{sf} = \begin{cases} 1 - \frac{d_B}{0.9 L_m} & \text{for } \frac{d_B}{L_m} < 0.6 \\ 0.12 \left( \frac{d_B}{L_m} \right)^{-2} & \text{for } \frac{d_B}{L_m} \geq 0.6 \end{cases}, \tag{6.130}$$

where $L_m = \sqrt{2} P - D_R$ and $d_B$ is the bubble equivalent diameter.

### 6.4.4 PHENOMENOLOGICAL MODEL OF ANNULAR FLOW

In vertical boiling channels, as the void fraction increases from zero to unity, various *two-phase flow regimes*, also referred to as the two-phase flow patterns, can appear. Initially single-phase flow with only liquid present will transit into bubbly flow as vapor bubbles are generated on a heated wall. As the void fraction increases, the bubbles coalesce to form larger bubbles or "slugs", leading to *slug flow*. This type of two-phase flow is characterized by altering liquid "plugs" and large gas slugs. Further increase in void fraction leads to churn-turbulent flow, where the gas slugs are so large that they span the channel diameter and induce strong liquid circulations. With still increased void fraction, liquid plugs break up into droplets and the flow transits into wispy-annular and later into annular two-phase flow. These various two-phase flow regimes are illustrated in Fig. 6.1. The exact conditions for transition between these regimes can depend on many factors including fluid properties, channel diameter, and flow rates.

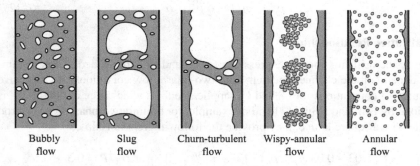

<table>
<tr><td>Bubbly<br>flow</td><td>Slug<br>flow</td><td>Churn-turbulent<br>flow</td><td>Wispy-annular<br>flow</td><td>Annular<br>flow</td></tr>
</table>

**Figure 6.1**   Two-phase flow patterns for upward co-current flow in a vertical pipe.

Two-phase annular flow is a common flow regime in the upper parts of fuel assemblies of boiling water reactors. The flow is characterized by the presence of a liquid film flowing on the channel wall while gas flows as a continuous phase up in the center of the channel. The liquid film is wavy and can be broken up into droplets, which are entrained by the gas core. The droplet size distribution and entrainment rate are influenced by various factors such as liquid viscosity, surface tension, and gas velocity.

The thickness of the liquid film in annular two-phase flow is influenced by many factors, including the liquid and gas flow rates, the fluid properties, and the channel geometry. In addition, the gas-liquid interface is highly dynamic with various types of waves traveling on it. Due to these reasons, predicting the liquid film thickness is a challenging tasks that requires advanced computational models and experimental data for model validation. In this section we discuss a phenomenological model of the liquid film in annular flow that has the capability to predict the film dryout phenomenon.

## Mass Conservation Equations

In its simplest form, the model employs mass conservation equations for the liquid film and the gas phase as follows,

$$\frac{dG_{LF}}{dz} = \frac{P_F}{A}(D - E - \Gamma), \tag{6.131}$$

$$\frac{dG_G}{dz} = \frac{P_F}{A}\Gamma. \tag{6.132}$$

Here $G_{LF}$ is the mass flux of the liquid film, $G_G$ is the mass flux of the gas phase, $P_F$ is the film perimeter, $A$ is the channel cross-section area, $D$ is the droplet deposition rate on the film surface, $E$ is the droplet entrainment rate from the film surface, and $\Gamma$ is the film evaporation rate that can be determined from the energy balance as

$$\Gamma = \frac{q_w''}{i_{fg}}. \tag{6.133}$$

In this equation $q_w''$ is the heat flux, $i_{fg}$ is the latent heat and we assumed that the film contains a saturated liquid.

## Droplet Entrainment and Deposition

The entrainment and deposition rates are obtained from experimental data and Hewitt and Govan proposed the following set of correlations [89]:

The *entrainment rate*

$$E = 5.75 \cdot 10^{-5} G_G \left[ (G_{LF} - G_{LF,crit}) \frac{D_h \rho_l}{\sigma \rho_v^2} \right]^{0.316} \quad \text{for} \quad G_{LF} > G_{LF,cit}, \tag{6.134}$$

where the *critical film mass flux* for the onset of entrainment is given by,

$$G_{LF,crit} = \exp\left( 5.8504 + 0.4249 \frac{\mu_v}{\mu_l} \sqrt{\frac{\rho_l}{\rho_v}} \right) \frac{\mu_l}{D_h}. \tag{6.135}$$

The *deposition rate*

$$D = kC,$$ (6.136)

where the *deposition coefficient* is calculated as,

$$k = \begin{cases} \dfrac{0.18}{\sqrt{\frac{\rho_v D_h}{\sigma}}} & \text{for } \frac{C}{\rho_v} \le 0.3 \\[3ex] \dfrac{0.083}{\sqrt{\frac{\rho_v D_h}{\sigma}}} \left(\dfrac{C}{\rho_v}\right)^{-0.65} & \text{for } \frac{C}{\rho_v} > 0.3 \end{cases},$$ (6.137)

and the droplet concentration in the gas core is given as,

$$C = \frac{G_{LE}}{\frac{G_{LE}}{\rho_l} + \frac{G_G}{\rho_v}}.$$ (6.138)

Here $G_{LE}$ is the mass flux of the entrained liquid. To solve the annular flow model given by Eqs. (6.131)–(6.138), it is necessary to provide the mass flux distribution at the point of the annular flow onset.

### Onset of Annular Flow

Wallis used entrainement data which were taken in a 25.4 mm tube for air and water flow at atmospheric pressure and derived the following correlation for the onset of annular flow [229],

$$J_v^* = 0.4 + 0.6 J_l^*,$$ (6.139)

where

$$J_v^* \equiv \frac{J_v \rho_v^{1/2}}{[g D_h (\rho_l - \rho_v)]^{1/2}}, \quad J_l^* \equiv \frac{J_l \rho_l^{1/2}}{[g D_h (\rho_l - \rho_v)]^{1/2}}.$$ (6.140)

Here the superficial velocities of the liquid phase, $J_l$, and the gas phase phase, $J_v$, are calculated as,

$$J_l = \frac{(1-x)G}{\rho_l}, \quad J_v = \frac{xG}{\rho_v},$$ (6.141)

$x$ is the two-phase mixture quality, $G$ is the total mass flux, and $D_h$ is the channel hydraulic diameter. Combining Eqs. (6.139)–(6.141), the following correlation for the mixture quality at the onset of annular flow is obtained,

$$x_{OAF} = \frac{0.6 + 0.4\sqrt{g D_h \rho_l (\rho_l - \rho_v)}/G}{0.6 + \sqrt{\rho_l/\rho_v}}.$$ (6.142)

Thus, the liquid and gas mass fluxes at the onset of annular flow can be found as,

$$G_{L,OAF} = (1 - x_{OAF})G, \quad G_{G,OAF} = x_{OAF} G.$$ (6.143)

**Initial Entrained Fraction**

The initial entrained fraction at the onset of annular flow is defined as,

$$\Phi_{IEF} \equiv \frac{G_{LE,OAF}}{G_{L,OAF}}, \tag{6.144}$$

where $G_{LE,OAF}$ is the mass flux of the entrained liquid at the onset of annular flow. Based on the air-water data obtained at atmospheric pressure, the following correlation for the initial entrained rate is proposed [5],

$$\Phi_{IEF} = \begin{cases} 0 & \text{for} \quad J_l^* \leq 0.025 \\ \min\left[\frac{8.4 \cdot 10^6 (J_l^* - 0.025)}{Re_l^{3/2}}, 1\right] & \text{for} \quad J_l^* > 0.025 \end{cases}, \tag{6.145}$$

where

$$Re_l = \frac{GD_h}{\mu_l}. \tag{6.146}$$

The correlation given by Eq. (6.145) should be used together with the following criterion for the onset of annular flow,

$$J_v^* = \begin{cases} 1.0 & \text{for} \quad J_l^* \leq 0.5 \\ 1.0 + 0.56(J_l^* - 0.5) & \text{for} \quad J_l^* > 0.5 \end{cases}. \tag{6.147}$$

## 6.5 THREE-DIMENSIONAL MULTIPHASE FLOW IN ROD BUNDLES

The importance of a thorough understanding of multiphase flow within fuel rod assemblies cannot be overstated for several reasons. Firstly, the distribution of phases between sub-channels significantly influences the local margin to the onset of critical heat flux. Secondly, the internal structure of multiphase flow has profound effects on the overall instability of multiphase flow. With a comprehensive understanding of these governing phenomena, it is possible to influence these characteristics of multiphase flow through appropriate design measures.

Conducting an in-depth study of multiphase flow within a fuel rod assembly presents significant challenges for two primary reasons. On the one hand, the intricate geometry of a fuel rod assembly, which includes design details with dimensions spanning from fractions of a millimeter to several meters, adds to the complexity. On the other hand, the phenomena associated with multiphase flow are quite involved, encompassing complex physical processes such as the formation of bubbles, their migration, and their coalescence into larger structures. These phenomena, coupled with the inherent complexity of turbulence within fuel rod bundles, can result in a wide variety of flow patterns.

Undoubtedly, due to the reasons previously mentioned, there is a necessity for experimental investigations of multiphase flow in fuel rod bundles. However, for certain specific cases, such as bubbly two-phase flow or dispersed annular two-phase flow, numerical approaches can be employed. These applications will be elaborated upon in greater detail in the subsequent subsections.

### 6.5.1 LAGRANGE PARTICLE TRACKING IN GAS CORE FLOW

Within the context of fuel bundle applications, the use of Lagrange particle tracking proves to be especially beneficial in predicting the paths of droplet migration and their rates of deposition on walls. Specifically, this approach offers a mechanistic tool for examining the impact of spacers and their design details on droplet behavior [147].

In the Eulerian-Lagrangian framework, the gas phase is considered a continuum, with the Reynolds averaged Navier-Stokes equations being solved, while the droplets are addressed through Lagrangian particle tracking. These droplets have the ability to exchange mass, momentum, and energy with the gas phase, for which a two-way coupling method is utilized. This method also takes into account the influence of droplets on gas flow through mutual interactions. However, interactions between droplets, such as collision and coalescence, are usually not considered due to the lack of well-established models to handle their complexity. Consequently, the conservation equations for the continuous gas phase are as follows [145]:

Mass conservation equation

$$\frac{\partial \rho}{\partial t} + \nabla \cdot (\rho \mathbf{v}) = \Gamma_G \,, \tag{6.148}$$

Momentum conservation equation

$$\frac{\partial (\rho \mathbf{v})}{\partial t} + \nabla \cdot (\rho \mathbf{v} \mathbf{v}) = -\nabla p + \nabla \cdot \boldsymbol{\tau}_e + \rho \mathbf{g} + \Gamma_G \mathbf{v}_F + \mathbf{M}_D \,, \tag{6.149}$$

Energy conservation equation

$$\frac{\partial (\rho i)}{\partial t} + \nabla \cdot (\rho i \mathbf{v}) = -\nabla \cdot \mathbf{q}_e'' + \Gamma_G i_F + q_D''' \,. \tag{6.150}$$

In these equations $\Gamma_G$ is the gas mass source due to the evaporation of a liquid film, $\mathbf{v}_F$ is the liquid film velocity, $\boldsymbol{\tau}_e$ is the effective stress tensor that includes turbulent effects, $\mathbf{M}_D$ is the momentum transfer term due to interactions with droplets, $\mathbf{q}_e''$ is the effective heat flux vector that includes turbulent effects, $i_F$ is the specific enthalpy of the liquid film, and $q_D'''$ is the energy transfer term due to gas-droplet heat transfer.

The droplet motion is tracked individually according to the following set of equations,

$$\frac{d\mathbf{r}_D}{dt} = \mathbf{v}_D \,, \tag{6.151}$$

$$m_D \frac{d\mathbf{v}_D}{dt} = \mathbf{F}_D + \mathbf{F}_L + m_D \mathbf{g} \,, \tag{6.152}$$

where $\mathbf{r}_D$ is the droplet position, $\mathbf{v}_D$ is the droplet velocity vector, $m_D$ is the mass of the droplet, $\mathbf{F}_D$ is the drag force, and $\mathbf{F}_L$ is the lift force. Some discussion of the required closure relationships and further references can be found in [6].

## 6.5.2 TWO-FLUID MODEL

The *two-fluid model* is derived from the generic conservation principle given by Eq. (6.35), assuming that the considered mixture consists of just two components. The conservation equations for mass, momentum, and energy are obtained for each component separately by using the definitions of $\psi_k$, $\mathbf{J}_k$, and $\phi_k$, given in Table 6.1. The formulation of the conservation equations together with the derivation of the required constitutive relationships to close the system of equations is extensively discussed by, e.g., Ishii and Hibiki [113] and Drew and Passman [59]. We provide here a short summary of the model that can be applied to prediction of bubbly two-phase flows in fuel rod bundles. A more comprehensive discussion of the topic and further references can be found in [7, 237].

### Mass Conservation Equations

The mass conservation equations for both phases can be written as follows,

$$\frac{\partial \left( \alpha_k \overline{\rho_k}^X \right)}{\partial t} + \nabla \cdot \left( \alpha_k \overline{\rho_k}^X \overline{\mathbf{v}_k}^{X\rho} \right) = \Gamma_{ki} , \qquad (6.153)$$

with the interface mass transfer condition,

$$\sum_{k=1}^{2} \Gamma_{ki} = 0 . \qquad (6.154)$$

The interface mass transfer can be due to the thermodynamic non-equilibrium between two phases of the same species. For example, in a subcooled bubbly flow, bubbles condense and the corresponding interface mass transfer term is negative and equal to the mass of condensing vapor per unit volume and time. On the contrary, for saturated liquid droplets moving in a superheated vapor, the interface mass transfer term for vapor is positive and equal to the mass of evaporating droplets per unit volume and time.

### Momentum Conservation Equations

The momentum conservation equations for both phases are as follows,

$$\frac{\partial \left( \alpha_k \overline{\rho_k}^X \overline{\mathbf{v}_k}^{X\rho} \right)}{\partial t} + \nabla \cdot \left( \alpha_k \overline{\rho_k}^X \overline{\mathbf{v}_k}^{X\rho} \overline{\mathbf{v}_k}^{X\rho} \right) = -\alpha_k \nabla \overline{p_k}^X$$
$$+ \nabla \cdot \left[ \alpha_k \left( \overline{\boldsymbol{\tau}_k}^X + \boldsymbol{\tau}_k^t \right) \right] + \alpha_k \overline{\rho_k}^X \mathbf{g} + (\overline{p_k}^i - \overline{p_k}^X) \nabla \alpha_k \qquad (6.155)$$
$$+ (\overline{\mathbf{v}_k}^{i\rho} - \overline{\mathbf{v}_k}^{X\rho}) \Gamma_{ki} + \mathbf{M}_{ki} - \nabla \alpha_k \cdot \overline{\boldsymbol{\tau}_k}^i$$

In this equation $\overline{\boldsymbol{\tau}_k}^X$ is the mean shear stress tensor, $\boldsymbol{\tau}_k^t$ is the turbulent shear stress tensor, $\overline{p_k}^X$ is the mean pressure, $\overline{p_k}^i$ is the mean pressure at the interface, $\overline{\mathbf{v}_k}^{i\rho}$ is the mass-weighted mean velocity at the interface, $\overline{\boldsymbol{\tau}_k}^i$ is the mean shear stress at the interface, and $\mathbf{M}_{ki}$ represents all interfacial momentum sources.

**Energy Conservation Equations**

Thw energy conservation equations in terms of the specific enthalpies are given as,

$$\frac{\partial \left( \alpha_k \overline{\rho_k}^X \overline{i_k}^{X\rho} \right)}{\partial t} + \nabla \cdot \left( \alpha_k \overline{\rho_k}^X \overline{i_k}^{X\rho} \overline{v_k}^{X\rho} \right) = -\nabla \cdot \alpha_k \left[ \overline{q_k''}^X + \left( q_k'' \right)^t \right]$$
$$+ \left( \overline{i_k}^{i\rho} - \overline{i_k}^{X\rho} \right) \Gamma_{ki} + a_i \overline{q'''}^i + q_k''' \tag{6.156}$$

Here $\overline{i_k}^{X\rho}$ is the mass-weighted mean specific enthalpy of phase $k$, $\left( q_k'' \right)^t$ is the turbulent heat flux, $\overline{i_k}^{i\rho}$ is the mean specific enthalpy at the interface, $a_i$ is the interfacial area concentration, $\overline{q'''}^i$ is the mean interfacial heat flux, and $q_k'''$ is the volumetric heat source. All terms that result from mechanical energy dissipation are neglected in the equations, which is an acceptable approximation for processes dominated by heat transfer and phase changes.

### 6.5.3  TRANSPORTED LIQUID FILM MODEL

The *transported liquid film model* is suitable for annular two-phase flows with a thin liquid film, which is typically found just upstream of the dryout point in a heated channel. For such liquid films, it can be assumed that the flow in the direction normal to the wall can be disregarded and that the spatial gradients of flow variables tangential to the wall surface are insignificant compared to those in the direction normal to the wall. These assumptions suggest that advection can be addressed solely in the direction tangential to the wall. Consequently, the transport equations for the liquid film can be integrated in the direction normal to the wall, resulting in two-dimensional equations. All properties of the liquid film, which vary across the film thickness, are represented as depth-averaged quantities given as [146],

$$\overline{\phi} = \frac{1}{\delta} \int_0^\delta \phi \, dy, \tag{6.157}$$

where $\delta$ is the film thickness, $\phi$ is any film property, and $y$ is the distance normal from the wall surface. Omitting the bar to indicate the averaged quantities, the depth-averaged mass, momentum, and energy equations for two-dimensional liquid film are given as,

$$\frac{\partial (\rho \delta)}{\partial t} + \nabla_s \cdot (\rho \delta \mathbf{v}) = S_\delta, \tag{6.158}$$

$$\frac{\partial (\rho \delta \mathbf{v})}{\partial t} + \nabla_s \cdot (\rho \delta \mathbf{v} \mathbf{v}) = -\delta \nabla_s p + \mathbf{S}_v, \tag{6.159}$$

$$\frac{\partial (\rho \delta i)}{\partial t} + \nabla_s \cdot (\rho \delta i \mathbf{v}) = S_i, \tag{6.160}$$

where $\mathbf{v}$ is the mean film velocity vector in the surface, i is the mean film-specific enthalpy, $\nabla_s$ is the nabla operator tangential to the surface, $\rho$ is the density, $p$ is the pressure, and $S_\delta$, $\mathbf{S}_v$, and $S_i$ are the source terms in the mas, momentum, and energy equation, respectively. The formulation of these source terms is still an active research area and some examples can be found in the literature, e.g., [64, 145].

# PROBLEMS

### PROBLEM 6.1

A mixture of superheated vapor with 300 K superheat and saturated droplets at pressure 7 MPa is flowing in an adiabatic channel with mass fluxes 200 and 700 kg m$^{-2}$ s$^{-1}$, respectively. Calculate the actual and the thermal-equilibrium qualities for the mixture.

### PROBLEM 6.2

Explain the difference between the following two mean values of the quantity $\psi$: $\overline{\psi}^{X\rho}$ and $\overline{\psi}^{i\rho}$.

### PROBLEM 6.3

Consider the term $(\overline{\mathbf{v}_k}^{i\rho} - \overline{\mathbf{v}_k}^{X\rho})\Gamma_{ki}$ in the two-fluid momentum conservation given by Eq. (6.155) for two-phase flow of saturated droplets flowing in a stream of superheated vapor. Assume that droplets evaporate and move with lower speed than the vapor. What is the net momentum exchange between the phases? How the situation changes when the droplets move faster than the vapor?

### PROBLEM 6.4

A vertical tube with an inner diameter of 12 mm is uniformly heated with linear power $q' = 25$ kW m$^{-1}$ and cooled with water flowing upward at inlet pressure 7 MPa, inlet subcooling 10 K, and inlet mass flux 1500 kg m$^{-2}$ s$^{-1}$. Using the homogeneous equilibrium model, calculate the total friction pressure loss in the channel.

# 7 Convection Heat Transfer

Convective heat transfer involves the transfer of heat between different bodies due to the movement of fluid. In nuclear reactors, this type of heat transfer is a critical process that is responsible for the removal of heat from the reactor core. The circulating coolant in the primary circuit is continuously heated in the reactor core, where heat is transferred from the fuel cladding surface to the coolant. To keep the coolant and the core temperatures at constant and safe levels, the heat from the core has to be transferred at sufficient rates from the coolant to the secondary loop or to another heat sink.

The rate of convective heat transfer is observed to be proportional to the temperature difference and is conveniently expressed by *Newton's law of cooling*. The law states that the rate of convective heat transfer is governed by the temperature difference and by a *heat transfer coefficient* that is relatively independent of this temperature difference. The heat transfer coefficient, *h*, depends upon the physical properties of the fluid and the characteristic features of the fluid flow.

There are two main types of convective heat transfer: *natural convection*, in which the fluid flow is caused by buoyant forces, and *forced convection*, in which flow is caused by an outside force. Formulas and correlations that are available in many references to calculate heat transfer coefficients are always limited to specific conditions for which they have been derived. These conditions include the type of convection: either natural convection or forces convection, and the type of fluid flow: either laminar flow or turbulent flow. In some cases fluid flow is determined simultaneously by buoyancy forces and external forces and the corresponding convective heat transfer is referred to as *mixed convection* heat transfer.

In this chapter we discuss the convective heat transfer that is relevant to heat removal from a reactor core under various operational conditions. We incorporate specific conditions for different coolants, as considered in the context of water-cooled reactors, gas-cooled reactors, and liquid metal-cooled reactors.

## 7.1 NATURAL CONVECTION

Natural convection heat transfer is of paramount importance in the context of nuclear power reactor applications. From a safety perspective, its most significant attribute is its ability to facilitate heat removal and cool the reactor core. Consequently, it aids in keeping the temperature of the fuel and cladding within safe limits, averting overheating and potential damage, and ensuring the reactor's secure shutdown during emergencies. However, the rate of natural convection heat transfer is directly tied to the fluid flow rate induced by buoyancy forces and cannot be controlled. Therefore, a thorough understanding and effective utilization of natural convection heat transfer are essential for the safe and efficient operation of nuclear power reactors.

DOI: 10.1201/9781003255000-7

Natural convection of heat serves as the primary mechanism for heat removal in passive cooling systems, which are employed for core cooling during regular operation, transients, design basis accidents, and even severe accidents. Passive cooling is also implemented in other components such as steam generators and water cooling heat exchangers, which facilitate the transfer of heat from the reactor or steam generator to the cooling water. Furthermore, modern reactor designs incorporate specific passive safety systems, including passive safety injection systems, passive residual heat removal systems, and passive containment cooling systems.

## 7.1.1 NATURAL CONVECTION IN POOLS

One of the configurations for the study of natural convection consists of a heated vertical wall, with known height $H$, immersed in a large pool of fluid with temperature $T_\infty$. The complete governing equations for the steady, constant property, two-dimensional flow are

$$\frac{\partial u}{\partial x} + \frac{\partial v}{\partial y} = 0 , \tag{7.1}$$

$$\rho \left( u\frac{\partial u}{\partial x} + v\frac{\partial u}{\partial y} \right) = -\frac{\partial p}{\partial x} + \mu \nabla^2 u , \tag{7.2}$$

$$\rho \left( u\frac{\partial v}{\partial x} + v\frac{\partial v}{\partial y} \right) = -\frac{\partial p}{\partial y} + \mu \nabla^2 v - \rho g , \tag{7.3}$$

$$u\frac{\partial T}{\partial x} + v\frac{\partial T}{\partial y} = a\nabla^2 T . \tag{7.4}$$

Here $u$ and $v$ are velocity components in the horizontal and the vertical direction, respectively, $T$ is the temperature, $a$ is the thermal diffusivity, $p$ is the pressure, and $-\rho g$ is the body force term in the vertical momentum equation. Noting that the pressure gradient in the vertical direction is determined by the fluid density as,

$$\frac{\partial p}{\partial y} \simeq \frac{\mathrm{d}p_\infty}{\mathrm{d}y} = -\rho_\infty g , \tag{7.5}$$

the momentum equation in the vertical direction becomes,

$$\rho \left( u\frac{\partial v}{\partial x} + v\frac{\partial v}{\partial y} \right) = \mu \nabla^2 v + (\rho_\infty - \rho) g . \tag{7.6}$$

Thus we notice that when $\rho_\infty - \rho > 0$, there is a positive body force acting on the fluid in the vertical direction. This is the *buoyancy force* that is responsible for the fluid motion during natural convection. Since the density variation is due to temperature change, there is a coupling between the temperature field and the velocity field. Mathematically the coupling can be achieved using the *Boussinesq approximation* of the momentum equation in which the following linear relationship between the fluid density and the temperature is assumed,

$$\frac{\rho_\infty - \rho}{\rho_\infty} \simeq -\beta \left( T_\infty - T \right), \tag{7.7}$$

where $\beta$ is the volume expansion coefficient at constant pressure defined as,

$$\beta = -\frac{1}{\rho}\left(\frac{\partial\rho}{\partial T}\right)_p. \tag{7.8}$$

Substituting approximation (7.7) into Eq. (7.9) yields,

$$u\frac{\partial v}{\partial x}+v\frac{\partial v}{\partial y}=v\nabla^2 v+g\beta\left(T-T_\infty\right). \tag{7.9}$$

This equation represents the momentum balance in the vertical direction in which three kinds of forces are present: inertia forces on the left-hand side and friction and buoyancy forces on the right-hand side of the equation. An order-of-magnitude analysis indicates that for fluids with a high Prandtl number (that is with a high $v/a$ ratio) the fluid motion is dominated by the friction and buoyancy balance, and the Nusselt number varies as [13],

$$\mathrm{Nu}=\frac{hH}{\lambda}\sim\mathrm{Ra}_H^{1/4}, \tag{7.10}$$

where the *Rayleigh number* is defined as,

$$\mathrm{Ra}_H=\frac{g\beta\Delta T H^3}{av}. \tag{7.11}$$

For low Prandtl-number fluids, the analysis shows that the fluid motion is governed by the buoyancy and inertia balance and the Nusselt number varies as,

$$\mathrm{Nu}\sim\left(\mathrm{Ra}_H\mathrm{Pr}\right)^{1/4}. \tag{7.12}$$

The wall-averaged Nusselt number corresponding to the two Pr limits are [142],

$$\overline{\mathrm{Nu}_H}=\begin{cases}0.671\,\mathrm{Ra}_H^{1/4} & \text{as}\quad \mathrm{Pr}\to\infty\\[2mm]0.8\left(\mathrm{Ra}_H\,\mathrm{Pr}\right)^{1/4} & \text{as}\quad \mathrm{Pr}\to 0\end{cases}. \tag{7.13}$$

The boundary layer flow along a vertical wall remains laminar if the distance $y$ is small enough so that the following is satisfied,

$$\mathrm{Gr}_y\lesssim 10^9\quad\text{for}\quad 10^{-3}\le\mathrm{Pr}\le 10^3, \tag{7.14}$$

where the *Grashof number* is defined as,

$$\mathrm{Gr}_y=\frac{g\beta\Delta T y^3}{v^2}=\frac{\mathrm{Ra}_y}{\mathrm{Pr}}. \tag{7.15}$$

A correlation for wall-averaged Nusselt number that covers laminar, transition, and turbulent range is given as [38],

$$\overline{\mathrm{Nu}_y}=\left\{0.825+\frac{0.387\mathrm{Ra}_y^{1/6}}{[1+(0.492/\mathrm{Pr})^{9/16}]^{8/27}}\right\}^2. \tag{7.16}$$

The correlation is valid for the entire Rayleigh number range, $10^{-1} < \mathrm{Ra}_y < 10^{12}$, and for all Prandtl numbers. The physical properties used in $\overline{\mathrm{Nu}}_y$, $\mathrm{Ra}_y$ and $\mathrm{Pr}$ are evaluated at the film temperature $(T_w + T_\infty)/2$. Some additional correlations for natural convection applications are provided in Appendix C.

## 7.2 FORCED CONVECTION

The basic feature of forced convection heat transfer is that the velocity field is well known or can be found from given boundary conditions. For steady-state laminar flow of a constant-property fluid in a simple geometry, such as a circular tube or a concentric annulus, the velocity field can be found analytically. When the flow is turbulent, it is necessary to employ models to account for turbulent effects in friction and heat transfer processes. In §5 we discussed various approaches to predict velocity fields for fluid flow in channels. In this section we turn our attention to convective heat transfer associated with a known velocity field.

### 7.2.1 HEAT TRANSFER TO LAMINAR CHANNEL FLOW

When fluid enters a heated channel, two processes occur simultaneously. On the one hand, the velocity field develops along the channel due to the viscous force acting on the channel walls. As a result, the velocity field is divided into two regions, a core flow, in which the viscous effects are negligibly small, and the hydrodynamic boundary layer, dominated by the viscous effects. On the other hand, similar process takes place with the temperature field, where fluid layers in the wall proximity are heated and a thermal boundary layer is formed. Beyond the thermal boundary layer, a thermal core flow region exists with approximately uniform temperature distribution. After some length from the channel entrance, the velocity field becomes fully developed and the radial velocity profile does not change anymore. Even though due to heat transfer the fluid mean temperature will change along the channel, the temperature profile will become fully developed as well when channel is heated with uniform heat flux or uniform wall temperature is applied.

### Fully Developed Flow

As already mentioned, far enough from a channel entrance, the velocity profile is fully developed and can be described in terms of coordinates in the channel cross-section area. For such conditions, the energy conservation equation can be solved for the unknown temperature distribution in the channel.

#### Circular Channels

We will investigate heat transfer to fluid flowing in a circular tube with radius $R$, with uniform wall temperature $T_w$, and subject to the following conditions,

1. Steady-state conditions,
2. Axisymmetric flow,

3. Constant fluid properties,
4. Hydrodynamically fully developed flow,
5. Axial heat conduction negligible comparing to axial heat convection.

With these assumptions, the energy equation can be written as,

$$w\frac{\partial T}{\partial z} = a\left(\frac{\partial^2 T}{\partial r^2} + \frac{1}{r}\frac{\partial T}{\partial r}\right), \tag{7.17}$$

Where $T$ is the fluid temperature, $w$ is the axial velocity component, $z$ is the axial co-ordinate, $r$ is the radial coordinate, and $a$ is the fluid thermal diffusivity. The equation can be written in a dimensionless form by introducing the following variables,

$$\theta = \frac{T_w - T}{T_w - T_0}, \quad \eta = \frac{r}{R}, \quad \zeta = \frac{a}{2UR}\frac{z}{R} = \frac{2}{\text{Pe}}\frac{z}{D}. \tag{7.18}$$

Here $D = 2R$ is the tube diameter, $U$ is the area-averaged fluid velocity in the tube, $T_0$ is the inlet temperature, and Pe is the *Peclet number*:

$$\text{Pe} = \frac{UD}{a} = \frac{UD}{v}\frac{v}{a} = \text{Re Pr}, \tag{7.19}$$

where Re is the *Reynolds number* and Pr is the *Prandtl number*. After substitution of the dimensionless variables, the energy equation becomes,

$$\frac{\partial\theta^2}{\partial\eta^2} + \frac{1}{\eta}\frac{\partial\theta}{\partial\eta} = (1 - \eta^2)\frac{\partial\theta}{\partial\zeta}, \tag{7.20}$$

and the boundary conditions are as follows:

1. Given wall temperature $\theta = 0$ for $\eta = 1$,
2. Symmetry at the axis $\partial\Theta/\partial\eta = 0$ for $\eta = 0$,
3. Given inlet temperature $\theta = 1$ for $\zeta = 0$.

This classical problem was treated for the first time by Graetz and is recognized in the heat transfer literature as the **Graetz problem**. Applying the method of variable separation, the solution to the problem can be found in terms of infinite series, from which the following local value of the *Nusselt number* can be obtained,

$$\text{Nu}_z = \frac{h_z D}{\lambda} = \frac{\sum_{n=0}^{\infty} G_n \exp\left(-k_n^2 \zeta\right)}{2\sum_{n=0}^{\infty} \frac{G_n}{k_n^2} \exp\left(-k_n^2 \zeta\right)}. \tag{7.21}$$

Here $k_n$ are eigenvalues and $G_n$ are constants of the Graetz series solutions. Their nine first values are given in Table 7.1.

Taking $\zeta \to \infty$, Eq. (7.21) gives the following asymptotic value of the Nusselt number,

$$\text{Nu}_\infty = \frac{k_0^2}{2} \approx 3.66. \tag{7.22}$$

**TABLE 7.1**

**Values of $k_n$ and $G_n$ in Eq. (7.21) for $n = 0$ to 8**

| $n$ | $k_n$ | $G_n$ | $n$ | $k_n$ | $G_n$ | $n$ | $k_n$ | $G_n$ |
|---|---|---|---|---|---|---|---|---|
| 0 | 2.70436 | 0.748775 | 3 | 14.67108 | 0.415418 | 6 | 26.66866 | 0.339622 |
| 1 | 6.67903 | 0.543830 | 4 | 18.66990 | 0.382920 | 7 | 30.66832 | 0.324062 |
| 2 | 10.67338 | 0.462861 | 5 | 22.66914 | 0.358686 | 8 | 34.66807 | 0.311014 |

Defining the distance from the beginning of heated length to the place where $(\mathrm{Nu}_z - \mathrm{Nu}_\infty)/\mathrm{Nu}_\infty = 0.01$ as the *thermal entrance length*, $L_T$, Eq. (7.21) gives,

$$L_T/D = 0.055 \, \mathrm{Pe}. \tag{7.23}$$

This equation indicates that for constant Reynolds number, the thermal entrance length increases with increasing Prandtl number.

For practical calculations, the following approximations for the local Nusselt number are recommended,

$$\mathrm{Nu}_z = \begin{cases} 1.077\zeta_R^{-1/3} - 0.7 & \zeta_R \leq 0.01 \\ 3.657 + 6.874 \left(10^3 \zeta_R\right)^{-0.488} \exp\left(-57.2\zeta_R\right) & \zeta_R > 0.01 \end{cases}, \tag{7.24}$$

where $\zeta_R = (z/R)/\mathrm{Pe}$.

Similar analysis for a round tube heated with uniform heat flux $q_w''$ gives the following approximations for the local Nusselt number,

$$\mathrm{Nu}_z = \begin{cases} 1.302\zeta_R^{-1/3} - 1.0 & \zeta_R \leq 5 \times 10^{-5} \\ 1.302\zeta_R^{-1/3} - 0.5 & 5 \times 10^{-5} < \zeta_R \leq 1.5 \times 10^{-3} \\ 4.364 + 8.68 \left(10^3 \zeta_R\right)^{-0.506} \exp\left(-41.0\zeta_R\right) & \zeta_R > 1.5 \times 10^{-3} \end{cases}. \tag{7.25}$$

Here $\mathrm{Nu}_z = q_w'' D/\lambda \left[T_w(z) - T_m(z)\right]$, with $T_m$–mean temperature in a tube cross section and $T_w$–wall temperature. The corresponding asymptotic Nusselt number $\mathrm{Nu}_\infty$ and the thermal entrance length $L_T$ are obtained as,

$$\mathrm{Nu}_\infty = \frac{48}{11} \approx 4.36, \tag{7.26}$$

$$L_T/D = 0.07 \, \mathrm{Pe}. \tag{7.27}$$

## Non-Circular Channels

In the case of non-circular channels, the heat convection prediction is difficult due to complicated expressions that describe the lateral distribution of velocity. Such problems can be usually solved when adopting some simplifying assumptions.

**Thermally and Hydraulically Developing Flow**

In most cases of practical interest, the velocity and temperature fields develop simultaneously downstream of the channel entrance. A closed-form expression that covers both the entrance and fully developed regions is as follows [39],

$$\frac{\mathrm{Nu}_z}{4.364\left[1+(\mathrm{Gz}/29.6)^2\right]^{1/6}} =$$

$$\left[1+\left(\frac{\mathrm{Gz}/19.04}{\left[1+(\mathrm{Pr}/0.0207)^{2/3}\right]^{1/2}\left[1+(\mathrm{Gz}/29.6)^2\right]^{1/3}}\right)^{3/2}\right]^{1/3}, \qquad (7.28)$$

where $\mathrm{Nu}_z = q_w'' D/\lambda \left[T_w(z) - T_m(z)\right]$, with $T_m$–mean temperature in a tube cross section and $T_w$–wall temperature, and Gz$= \pi/(4\zeta_R)$ is the *Graetz number*.

The heat transfer and pressure drop expressions for thermally and hydraulically developing flow in channels with other cross-sectional shapes can be found in [198]. In general, however, it is worth noting that the Nusselt number in the entrance section obeys a relationship of the following type,

$$\mathrm{Nu}_z = \frac{\mathrm{const}}{\sqrt{\frac{z/D_h}{\mathrm{Pe}}}}. \qquad (7.29)$$

## 7.2.2 HEAT TRANSFER TO TURBULENT CHANNEL FLOW

Experiments show that for turbulent channel flow of fluid with Pr $= 1$ the velocity and temperature fields are similar. In this case, and in the case of turbulent flow along a plate, there is a simple relationship between the Nusselt number Nu and the Fanning friction factor $C_f$ as follows,

$$\mathrm{Nu} = \frac{C_f}{2}\mathrm{Re}, \quad \text{where} \quad C_f = \frac{2\tau_w}{\rho U^2}. \qquad (7.30)$$

This relationship, connecting the heat transfer coefficient and the friction factor, is called the *Reynolds analogy*.

For Pr $\neq 1$, which is the most frequent case, we need to consider separately the viscous sublayer and the turbulent core flow. In the turbulent region the Reynolds analogy is assumed, that is,

$$\nu_t = a_t, \qquad (7.31)$$

and because $a_t \gg a$ and $\nu_t \gg \nu$, we can assume that $\nu = a = 0$. We also assume that the temperature and the velocity have a linear distribution in the viscous sublayer, so the equations describing the heat and momentum transfer in the sublayer are given as,

$$\frac{\tau}{\rho} = \nu\frac{\mathrm{d}w}{\mathrm{d}y}, \qquad (7.32)$$

$$\frac{q''}{\rho c_p} = -a\frac{dT}{dy}.$$ (7.33)

Assuming that the viscous sublayer has thickness $\delta$ and the shear stress and the heat flux in the sublayer are constant and equal to the corresponding wall values, that is $\tau = \text{const} = \tau_w$ and $q'' = \text{const} = q_w''$, Eqs. (7.32) and (7.33) can be integrated as follows,

$$\int_0^{w_\delta} dw = \frac{\tau_w}{\rho}\int_0^\delta \frac{dy}{v}, \quad w_\delta = \frac{\tau_w}{\rho v}\delta$$ (7.34)

$$\int_{T_w}^{T_\delta} dT = -\frac{q_w''}{\rho c_p}\int_0^\delta \frac{dy}{a}, \quad T_w - T_\delta = \frac{q_w''}{\rho c_p a}\delta.$$ (7.35)

Similar integration can be performed in the turbulent core, assuming velocity variation from $w_\delta$ to $U$ and temperature variation from $T_\delta$ to $T_b$, where $U$ and $T_b$ is the mean velocity and the bulk temperature, respectively. As a result, the following relationship is obtained,

$$\frac{U - w_\delta}{T_\delta - T_b} = \frac{c_p \tau_w}{q_w''}.$$ (7.36)

Now we can combine Eqs. (7.32)–(7.36) to obtain,

$$T_w - T_b = \frac{q_w''}{c_p \tau_w}\left(U - w_\delta + w_\delta\frac{v}{a}\right) = \frac{q_w''}{c_p \tau_w}[w_\delta(\text{Pr} - 1) + U].$$ (7.37)

Substituting,

$$h = \frac{q_w''}{T_w - T_b} \quad \text{and} \quad C_f = \frac{2\tau_w}{\rho U^2},$$

the expression for $T_w - T_b$ takes the following form,

$$\frac{1}{h} = \frac{1}{c_p C_f \rho U^2/2}[w_\delta(\text{Pr} - 1) + U] = \frac{1}{c_p C_f \rho U/2}\left[\frac{w_\delta}{U}(\text{Pr} - 1) + 1\right].$$

Since the *Stanton number* is defined as

$$\text{St} = \frac{h}{c_p \rho U} = \frac{hD}{\lambda}\frac{\mu}{\rho U D}\frac{\lambda}{c_p \mu} = \frac{\text{Nu}}{\text{Re Pr}},$$ (7.38)

the following relationship, called the *Prandtl analogy*, is obtained,

$$\text{St} = \frac{C_f/2}{1 + \frac{w_\delta}{U}(\text{Pr} - 1)}.$$ (7.39)

Taking the thickness of the viscous sublayer equal to $y^+ = 5$ we get,

$$w_\delta = 5u_\tau = 5U\sqrt{C_f/2} \quad \text{and} \quad \frac{w_\delta}{U} = 5\sqrt{C_f/2}.$$

Using this result in Eq. (7.39), we obtain a new expression for the Stanton number,

$$\text{St} = \frac{C_f/2}{1 + 5\sqrt{C_f/2}(\text{Pr} - 1)}.$$ (7.40)

Prandtl analogy can be applied to various heat transfer problems with turbulent boundary layer, including flows in channels and flows over flat plates.

In Colburn's empirical correlation, the Stanton number is expressed in terms of the friction factor as follows [40],

$$St\,Pr^{2/3} \cong \frac{C_f}{2}.$$  (7.41)

In the special case of a pipe with smooth internal surface, the following formula can be derived,

$$Nu = 0.023\,Re^{0.8}\,Pr^{1/3},$$  (7.42)

which is valid for the following ranges of the Reynolds number: $2 \cdot 10^4 < Re < 10^6$, the Prandtl number: $0.7 < Pr < 160$, and the distance from the inlet: $L/D > 60$.

Another popular formula to predict heat transfer coefficient in tubes is a correlation due to Dittus and Boelter [56],

$$Nu = 0.023\,Re^{0.8}\,Pr^n,$$  (7.43)

which is valid for the following ranges of the Prandtl number: $0.7 \leq Pr \leq 120$, the Reynolds number: $2500 \leq Re \leq 1.24 \cdot 10^5$, and the distance from the inlet: $L/D > 60$. The Prandtl number exponent is $n = 0.4$ when the fluid is being heated and $n = 0.3$ when the fluid is being cooled.

A correlation that can be used both in constant heat flux and constant temperature applications was proposed by Gnielinski [77],

$$Nu = \frac{\frac{C_f}{2}\,(Re - 1000)\,Pr}{1 + 12.7\sqrt{\frac{C_f}{2}}\,(Pr^{2/3} - 1)}.$$  (7.44)

The correlation is accurate within $\pm 10\%$ in the range $0.5 < Pr < 2000$ and $2300 < Re < 5 \times 10^6$.

For the low Prandtl number applications, such as heat transfer to liquid metals, Notter and Sleicher proposed the following correlation [169],

$$Nu = \begin{cases} 6.3 + 0.0167\,Re^{0.85}\,Pr^{0.93} & \text{for } q_w'' = \text{const} \\ 4.8 + 0.0156\,Re^{0.85}\,Pr^{0.93} & \text{for } T_w = \text{const} \end{cases},$$  (7.45)

where all the properties are calculated at the mean temperature $T_m = (T_{in} + T_{out})/2$, $q_w''$ is the wall heat flux, and $T_w$ is the wall temperature. The correlation is valid for $0.004 < Pr < 0.1$ and $10^3 < Re < 10^6$.

## 7.3   HEAT TRANSFER TO SUPERCRITICAL FLUIDS

Supercritical fluids have physical properties that make them attractive for applications in a variety of fields such as thermal power engineering, chemical and biochemical reactions, wastewater treatment, and many more. Supercritical fluids, including supercritical water, are being investigated as potential coolants and working fluids for both fission and fusion reactors. In fission reactors, supercritical fluids can be used as coolants to improve the thermal efficiency and reduce the size of the reactor core. In addition to supercritical water, other supercritical fluids that are being considered for this application include carbon dioxide, nitrogen, and helium. In fusion reactors, supercritical fluids are being investigated as potential coolants and tritium breeding materials. Supercritical water is one of the fluids being considered for this application, along with other materials such as liquid metals and molten salts.

However, there are still significant challenges to overcome in the use of supercritical fluids in both fission and fusion reactors. These challenges include the need for materials that can withstand high temperatures and pressures, as well as the potential for corrosion and erosion of the reactor components. Additionally, the behavior of supercritical fluids in complex flow regimes needs to be better understood to design safe and efficient reactor systems. In particular, the significant departure of the heat transfer coefficient from the values commonly encountered for subcritical fluids need additional research.

Heat transfer to supercritical fluids plays an important role in many of the envisioned applications and thus proper understanding of the heat transfer characteristics is of significant importance. Although this type of heat transfer has been extensively studied in the past decades, there is still a lack of full understanding of the involved phenomena. In particular, due to significantly varying physical properties of supercritical fluids in the vicinity of a pseudocritical point, the intensity of heat transfer varies and can be either lower, higher, or at the same level as for constant-property subcritical fluids.

Heat transfer to supercritical fluids is generally more efficient than to other fluids, particularly in certain temperature and pressure ranges. Supercritical fluids have unique thermodynamic properties that make them very efficient at absorbing and transferring heat. For example, supercritical water has a high thermal conductivity, a high heat capacity, and a low viscosity, which makes it an excellent heat transfer medium. Since supercritical fluids have no surface tension, they can make direct contact with the heat source, allowing for more efficient heat transfer.

However, the efficiency of heat transfer to supercritical fluids can depend on several factors, including the temperature and pressure of the fluid, the heat transfer surface geometry, the heat flux, and the fluid flow rate. In general, three modes of heat transfer are distinguished: the deteriorated heat transfer, the enhanced heat transfer, and the normal heat transfer.

The deteriorated heat transfer is of particular interest in nuclear applications since its occurrence can lead to local high wall temperature spots and can potentially lead to wall damage. The deteriorated heat transfer is characterized by a decrease in the rate of heat transfer as the heat flux increases. The exact cause of heat transfer deterioration is not fully understood, and its explanation requires additional theoretical and experimental research.

In this section the thermal properties of supercritical fluids and the current methods to predict heat transfer to supercritical fluids are presented. The basic properties of water and carbon dioxide are shown, since these fluids are frequently considered as working fluids in thermal power engineering. We also make a short overview of the main results of experimental research and compare selected correlations and models with experimental results.

### 7.3.1 BASIC PROPERTIES OF SUPERCRITICAL FLUIDS

A supercritical fluid is such a fluid that exists at a temperature and pressure above its critical point, where it is not possible to distinguish between liquid and vapor phases, and thus there are no clear phase change phenomena as for fluids at subcritical pressures. Supercritical fluids have no surface tension because they are not subject to the vapor-liquid interface and no molecules have the attraction to the interior of the liquid. The physical properties of supercritical fluids vary significantly with pressure and temperature, especially along a pseudocritical curve, which in a phase diagram is a prolongation of the liquid-vapor phase change curve. Supercritical fluids can have very different properties than the regular fluids. For instance, supercritical water differs from regular water since it is non-polar and acidic. These property changes, along with high pressure and temperature conditions, are causing the development of new materials to be needed. Such materials will need to have improved corrosion and oxidation resistance enhanced strength and embrittlement resistance and advanced creep resistance.

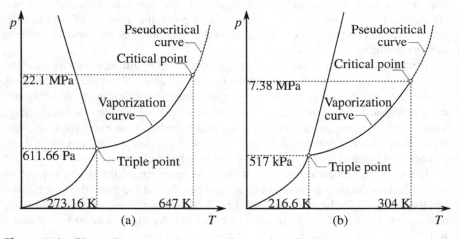

**Figure 7.1**    Phase diagrams: (a) - water, (b) - carbon dioxide.

### Phase Diagrams

The phase diagrams for water and carbon dioxide are shown in Fig. 7.1. The two characteristic points on the diagram are the *triple point*, at which solid, liquid and vapor phases coexist, and the *critical point*, beyond which the vaporization no longer takes place. Critical pressure and temperature values for selected fluids are given in Table 7.2.

**TABLE 7.2**
**Critical Properties of Selected Fluids**

| Fluid | Critical Temperature (K) | Critical Pressure (kPa) | Critical Density (kg/m$^3$) |
|---|---|---|---|
| Water (H$_2$O) | 647.096 | 22064.0 | 322 |
| Heavy Water (D$_2$O) | 643.847 | 21661.8 | 356 |
| Carbon Dioxide (CO$_2$) | 304.128 | 7377.3 | 469 |
| Oxygen (O$_2$) | 154.481 | 5043.3 | 436 |
| Nitrogen (N$_2$) | 126.192 | 3395.8 | 313 |
| Hydrogen (H$_2$) | 33.145 | 1296.4 | 31.3 |
| Helium (He) | 5.195 | 227.6 | 72.6 |

In the subcritical region, the liquid and vapor physical properties change across the vaporization curve in a discontinuous manner. Even in the supercritical region the properties change from "liquid-like" to "vapor-like" properties, however, this change is continuous. At certain locations in the supercritical region these changes are particularly severe. For example, specific heat and viscosity have local maxima, whereas density and viscosity experience steep change from liquid-like to vapor-like values. The points of such significant property changes are called pseudocritical points and they constitute the pseudocritical curve on the phase diagram, as shown in Fig. 7.1.

The pseudocritical curves for supercritical water and carbon dioxide are shown in Fig. 7.2. The curves have been determined as a collection of pressure and temperature points at which the specific heat has the maximum value. As can be seen, when the pressure increases about 50% above the critical pressure, the pseudocritical temperature increases only about 6%. Polynomial approximations of the curves in the pressure range $1 < p_R < 1.5$ are given by Eqs. (7.46) and (7.47),

$$\frac{T_{pc}}{T_{cr}} = 1 + 0.131(p_R - 1) - 0.0249(p_R - 1)^2 - 0.0110(p_R - 1)^3 \quad \text{for H}_2\text{O}, \quad (7.46)$$

$$\frac{T_{pc}}{T_{cr}} = 1 + 0.147(p_R - 1) - 0.0604(p_R - 1)^2 + 0.0335(p_R - 1)^3 \quad \text{for CO}_2, \quad (7.47)$$

where $p_R = p/p_{cr}$ is the reduced pressure.

## Variation of Supercritical Fluid Properties

Supercritical fluid properties strongly vary as a function of temperature and pressure. For a fixed pressure, the region in the vicinity of the corresponding pseudocritical temperature is a location of significant property changes. The density, viscosity, specific heat, and thermal conductivity variation in the vicinity of the pseudocritical temperature for water and carbon dioxide are shown in Figs. 7.3 and 7.4, respectively.

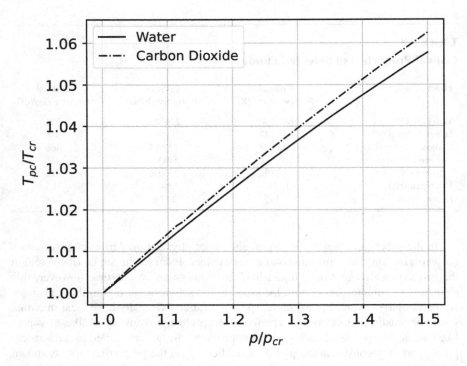

**Figure 7.2** Pseudocritical curves for water and carbon dioxide.

The property changes for the two supercritical fluids show some similarities. The density and the dynamic viscosity monotonically decrease with increasing temperature, with the highest drop in the vicinity of the pseudocritical temperature.

The specific heat capacity has a non-monotonic behavior, with a clear maximum at the pseudocritical temperature. For pressures close to the critical pressure the heat capacity varies in a range of several orders of magnitude for $0.96 < T/T_{pc} < 1.04$. This variation becomes less severe when pressure increases and departs from the critical pressure.

The thermal conductivity has a more complex variation with pressure and temperature, showing some similarities the the density and the specific heat behavior. For pressures close to the pseudocritical pressure, a clear maximum of the conductivity can be observed in the narrow temperature range around the pseudocritical temperature. With increasing pressure the maximum becomes less visible and eventually a monotonically decreasing thermal conductivity with increasing temperature is observed.

Variable thermal fluid properties influence the friction losses and heat transfer rates to supercritical fluids. In particular, when the pseudocritical point is located in the boundary layer, heat flux variations are causing local temperature variations, which lead to local property variations. As a result the linear relationship between

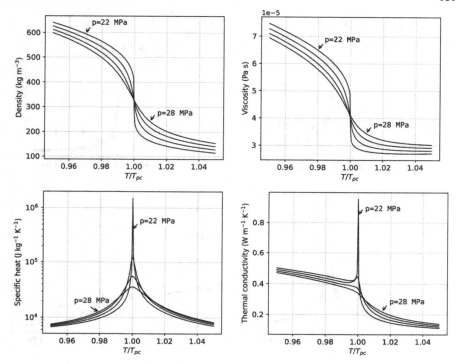

**Figure 7.3**  Variation of the physical properties of supercritical water in the pseud-ocritical region.

the heat flux and the wall-bulk temperature drop is significantly violated. This effect is further strengthened by high Prandtl number that characterizes supercritical fluids. For laminar and natural convection heat transfer, thermal boundary layer is much thinner than the velocity boundary layer, thus the fluid properties vary significantly throughout the velocity boundary layer.

### 7.3.2  WALL TEMPERATURE MEASUREMENTS

Since the conventional heat transfer correlations are not applicable to predict wall temperature in channels cooled with supercritical fluids, considerable number of investigations have been performed to measure the wall temperature during heat transfer to fluids at pressures above the critical pressure. These measurements have been performed under a wide range of operating conditions, such as mass flux, heat flux, system pressure, flow orientation against gravity, and tube inner diameter, in order to capture the influence of these factors on the heat transfer rates.

The variations of wall temperature during heat transfer to supercritical water flowing in tubes at various conditions are shown in Fig. 7.5(a–d). The figures show the difference between the measured wall temperature $T_w$ and the corresponding

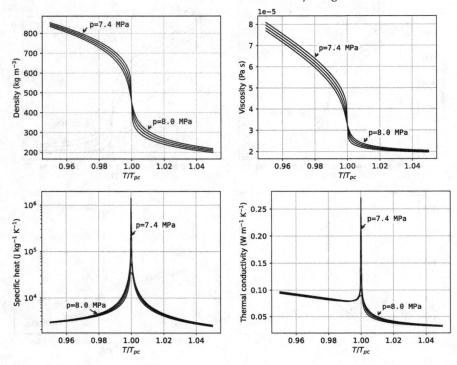

**Figure 7.4** Variation of the physical properties of supercritical carbon dioxide in the pseudocritical region.

pseudocritical temperature $T_{pc}$ as a function of a ratio of the bulk specific enthalpy $i_b$ and the specific enthalpy at the pseudocritical temperature $i_{bpc}$. Since axially-uniform heat flux was used in the experiments, $i_b/i_{bpc}$ ratio can be interpreted as a measure of a distance along the tube.

For $T_w - T_{pc} < 0$ a "liquid-like" fluid prevails in the tube cross-section, since the fluid temperature is lower than the pseudocritical temperature in the entire cross-section. For such conditions experimental data experience a rather small spread, similar to typical data obtained for subcritical fluids. However, for $T_w - T_{pc} > 0$ and for $i_b/i_{bpc} < 1$, the pseudocritical temperature is present in the thermal boundary layer causing significant property changes in that region. For such conditions a local heat transfer deterioration can occur, as is clearly visible in Fig. 7.5(a–c). For example, as shown in Fig. 7.5(a), increasing heat flux above 277.8 kW/m$^2$ causes wall temperature to locally increase up to 300 K above the pseudocritical temperature, even though the bulk temperature is below the pseudocritical temperature. Such significant temperature increase indicates a local heat transfer deterioration.

The influence of tube orientation on the measured wall temperature is shown in Fig. 7.5(d). It can be seen that at identical conditions, but various tube orientations, the measured wall temperature varies in a range of 20 K. These temperature

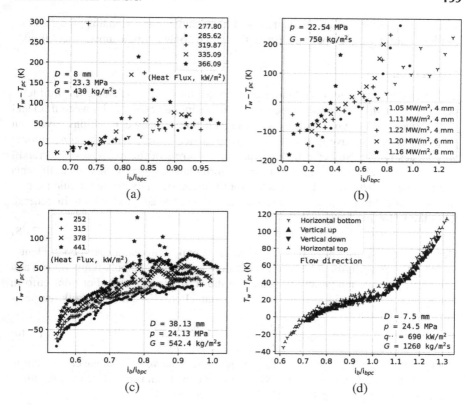

**Figure 7.5**  Measured wall temperature for heat transfer to supercritical water flow-ing in tubes: (a)–low mass flux conditions for vertical upflow [200], (b)–tube diame-ter influence at moderate mass flux for vertical upflow [76], (c)–vertical upflow in a large-diameter tube [141], (d)–influence of flow orientation at high mass flux [239].

variations are much less than the ones observed in Fig. 7.5(a–c), since the heat trans-fer deterioration does not take place. However, due to the local property changes the heat transfer coefficient changes as well. For example, for a horizontal flow, the fluid density at the top of a cross-section is less than the density at the bottom due to the buoyancy effects. The property variations in a tube cross-section are causing variations in the heat transfer coefficient.

### 7.3.3   WALL TEMPERATURE PREDICTIONS

Thermal-hydraulic designs and optimization of fuel assemblies require predictions of clad temperature at various flow and heat flux conditions. The predicted clad tem-perature at hot spots will provide information about thermal margins in the reac-tor core. One of the challenges is the accuracy of predictions, taking into account the variability of the operating conditions and the geometry details of core. In this section we present various available correlations and computational approaches that can be used for wall temperature prediction when cooled with a supercritical fluid.

## Correlations

Convective heat transfer correlations are very useful tools to calculate the wall temperature when the fluid bulk temperature and the applied heat flux between fluid and wall surface are known. Similarly as for regular fluids, many correlations have been proposed for convective heat transfer to supercritical fluids, using the conventional approaches for the Nusselt number expressed in terms of the Reynolds and Prandtl numbers. Main differences between various formulations are in values of coefficients and exponents that are used. To account for the effect of temperature on fluid properties, several definitions of the reference temperature are adopted, which, in general, takes the following form,

$$T_{ref} = c(T_w - T_b) + T_b. \qquad (7.48)$$

Here $T_w$ is the wall temperature, $T_b$ is the fluid bulk temperature, and coefficient $c$ varies between zero (the reference temperature is equal to the bulk temperature) and one (the reference temperature is equal to the wall temperature). For non-uniform temperature and velocity distributions in a channel cross-section, the bulk temperature is calculated as,

$$T_b = \frac{\langle \rho c_p u T \rangle_2}{\langle \rho c_p u \rangle_2} = \frac{\int_A \rho c_p u T \, dA}{\int_A \rho c_p u \, dA}, \qquad (7.49)$$

where $T$ and $u$ are local distributions of temperature and velocity in the cross section, respectively, and $\rho$ and $c_p$ are local values of the density and specific heat of the fluid. The averaging is performed over the cross-section area $A$.

Swenson et al. [213] correlated a wide range of their data for supercritical water as follows,

$$\mathrm{Nu}_w = 0.00459 \mathrm{Re}_w^{0.923} \mathrm{Pr}_{cw}^{0.613} \left( \frac{\rho_w}{\rho_b} \right)^{0.231}, \qquad (7.50)$$

where,

$$\mathrm{Pr}_{cw} \equiv \frac{i_w - i_b}{T_w - T_b} \frac{\mu_w}{\lambda_w}, \qquad (7.51)$$

is the Prandtl number that uses an averaged value of the specific heat given by the ratio of the specific enthalpy drop to the temperature drop between wall and bulk values. The Nusselt and Reynolds numbers are calculated with fluid properties determined at the wall temperature. For heat flux controlled systems, iterations are required to find the Nusselt number. The correlation is valid in the following range of parameters: pressure $p = 22.7$ to $41.3$ MPa, heat flux $q'' = 0.2$ to $2.0 \times 10^6$ W/m$^2$, mass flux $G = 200$ to $2000$ kg/m$^2$s, tube inner diameter $D = 9.4$ mm, fluid bulk temperature $T = 343$ to $848$ K, temperature drop $T_w - T_b = 6$ to $285$ K, and channel heated length $L = 1830$ mm.

Bishop et al. [18] proposed the following correlation,

$$\mathrm{Nu}_b = 0.0069 \mathrm{Re}_b^{0.90} \mathrm{Pr}_{cb}^{0.66} \left( \frac{\rho_w}{\rho_b} \right)^{0.43} \left( 1 + \frac{2.4}{x/D_h} \right), \qquad (7.52)$$

where,

$$\Pr_{cb} \equiv \frac{i_w - i_b}{T_w - T_b} \frac{\mu_b}{\lambda_b}, \tag{7.53}$$

is the Prandtl number that uses an averaged value of the specific heat and the remaining properties calculated at the bulk temperature. The correlation takes into account the entrance effect through the term $x/D_h$, where $x$ is the distance from the inlet and $D_h$ is the channel hydraulic diameter. The correlation is valid in the following range of parameters: pressure $p = 22.6$ to $27.5$ MPa, heat flux $q'' = 0.315$ to $3.47 \times 10^6$ W/m$^2$, mass flux $G = 678$ to $3660$ kg/m$^2$s, tube inner diameter $D = 2.5$ to $5.1$ mm, fluid bulk temperature $T = 567$ to $798$ K, and wall temperature $T_w = 625$ to $907$ K.

Krasnoshchekov and Protopopov [135] introduced a term to take into account the relationships between the wall, bulk, and pseudocritical temperatures. Using experimental data for carbon dioxide, they proposed the following correlation for the Nusselt number,

$$\mathrm{Nu}_b = \frac{C_f/2 \cdot \mathrm{Re}_b \mathrm{Pr}_b}{1.07 + 12.7 \left(C_f/2\right)^{1/2} \left(\mathrm{Pr}_b^{2/3} - 1\right)} \left(\frac{\rho_w}{\rho_b}\right)^{0.3} \left(\frac{c_{pa}}{c_{pb}}\right)^n, \tag{7.54}$$

where,

$$n = \begin{cases} n_1 & \text{for} \quad T_b \leq T_w \leq T_{pc} \quad \text{or} \quad 1.2 \cdot T_{pc} \leq T_b \leq T_w \\ n_2 & \text{for} \quad T_b \leq T_{pc} \leq T_w \\ n_3 & \text{for} \quad T_{pc} \leq T_b \leq 1.2 \cdot T_{pc} \quad \text{and} \quad T_b \leq T_w \end{cases}, \tag{7.55}$$

$$n_1 = 0.4 \tag{7.56}$$

$$n_2 = 0.4 + 0.2 \left(\frac{T_w}{T_{pc}} - 1\right) \tag{7.57}$$

$$n_3 = 0.4 + 0.2 \left(\frac{T_w}{T_{pc}} - 1\right) \left[1 - 5\left(\frac{T_b}{T_{pc}} - 1\right)\right] \tag{7.58}$$

$$C_f = \frac{1}{\left(3.64 \ln \mathrm{Re}_b - 3.28\right)^2}, \tag{7.59}$$

$$c_{pa} = \frac{i_w - i_b}{T_w - T_b}. \tag{7.60}$$

Here $T_{pc}$ is the pseudo-critical temperature, $C_f$ is the Fanning friction factor, and $c_{pb}$ is the specific heat capacity calculated at the bulk temperature. The correlation is valid in the following range of parameters: pressure $p/p_{cr} = 1.06$ to $1.33$, heat flux $q'' \leq 2.6$ MW/m$^2$, $\mathrm{Re}_b = 8 \cdot 10^4$ to $5 \cdot 10^5$, tube inner diameter $D = 4.1$ mm, and tube length $L = 2$ m.

Using the same expressions for the exponent $n$ and for the average specific heat capacity $c_{pa}$, Jackson proposed to reformulate the correlation given by Eq. (7.54) as follows [114],

$$\mathrm{Nu}_b = 0.0183 \mathrm{Re}_b^{0.82} \mathrm{Pr}_b^{0.5} \left(\frac{\rho_w}{\rho_b}\right)^{0.3} \left(\frac{c_{pa}}{c_{pb}}\right)^n. \tag{7.61}$$

This expression provides results in good agreement with the original expression given by Eq. (7.54).

Mokry et al. conducted a dimensional analysis to derive the general form of an empirical supercritical water heat transfer correlation. They employed recent experimental data for supercritical water and arrived at the following expression [164],

$$\mathrm{Nu}_b = 0.0061 \mathrm{Re}_b^{0.904} \mathrm{Pr}_{cb}^{0.684} \left( \frac{\rho_w}{\rho_b} \right)^{0.564}, \qquad (7.62)$$

where the definitions of variables are the same as in the Bishop et al. correlation given by Eq. (7.52). The experimental data have been obtained for supercritical water flowing upward in a 4-m long vertical tubes with 10 mm inner diameter, with system pressure around 24 MPa, inlet temperatures from 593 to 623 K, mass flux from 200 to 1500 kg/m$^2$s, and heat flux up to 1250 kW/m$^2$.

## Variation of Heat Transfer Coefficient

One of the main features of heat transfer to supercritical fluids is a significant variation of the heat transfer coefficient, and thus the wall temperature, with changing heat flux, specific enthalpy, system pressure, and flow orientation against the gravity. For conventional fluids, a variation of these parameters have a very limited influence on the heat transfer coefficient, especially under forced convection conditions. For supercritical fluids, the variations of the heat transfer coefficient can be very strong and difficult to predict using various correlations, as shown in Figs. 7.6 and 7.7. There are many reasons for this different behavior of regular and supercritical fluids, but the dominant one is due to the strong variation of properties of supercritical fluids with temperature in the vicinity of the pseudocritical point.

### Influence of Heat Flux

In general it is observed that the heat transfer coefficient decreases with increasing heat flux. This effect is shown in Fig. 7.6(a) for heat transfer to supercritical water in a vertical tube with 22.1 mm inner diameter. As can be seen, keeping constant pressure and mass flux, the heat transfer coefficient is significantly reduced when heat flux increases from 378 to 946 kW/m$^2$. This effect is particularly strong for $i_b / i_{bpc} \sim 0.7$.

### Influence of Mass Flux

Increasing mass flux in general leads to more efficient heat transfer, similar to regular fluids. However, this general trend can be reversed when buoyancy effects and significant property changes are involved. For example, increased mass flux can lead to such a change in these parameters that the overall heat transfer efficiency drops. This behavior has been confirmed experimentally by several researchers [240, 242].

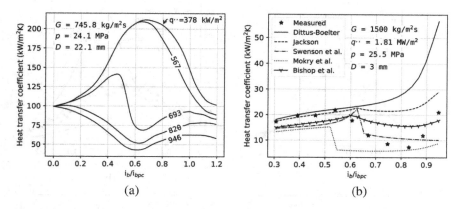

**Figure 7.6**   Heat transfer to supercritical water: (a)–variation of the heat transfer co-efficient with bulk enthalpy and heat flux [212], (b)–comparison of measured and calculated heat transfer coefficient in a deteriorated region (measurements from [173]).

## Influence of Pressure

The influence of pressure is rather weak compared to mass flux or heat flux, but it is particularly visible when the pressure is only slightly above the critical pressure. This can be explained by strong property variations with temperature in that pressure region. In general, heat transfer coefficient increases with decreasing pressure, when other parameters are kept constant. However, contradictory results have been reported as well [241].

## Influence of Flow Direction

The observed influence of flow direction against gravity on heat transfer coefficient is due to buoyancy effects. Depending on the flow direction, the buoyancy effect is either modifying the velocity distribution in a channel cross section or it causes a separation of hot and cold fluid layers. Examples of measured wall temperatures in a tube with various flow directions are shown in Fig. 7.5(d). For the given conditions, the measured wall temperature varies only slightly for vertical up and down flows. However, for horizontal flow, the heat transfer coefficient is slightly enhanced on the bottom of the tube in comparison with that on the top of the tube. In general, at relatively high ratio of heat flux to mass flux ($q''/G$), the heat transfer coefficient in downward flow is higher than that in upward flow. In addition, heat transfer deterioration is observed more frequently for upward flow than for downward flow. However, contradictory results are reported in the literature indicating that the effect of flow direction interferes with other effects and is rather complicated. As a result, prediction of the influence of flow direction on heat transfer coefficient is quite a challenging task, as illustrated in Fig. 7.7.

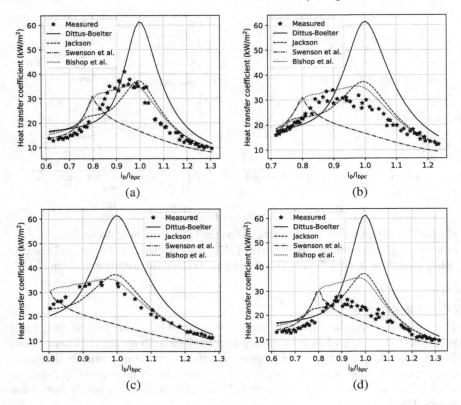

**Figure 7.7** Measured and calculated heat transfer coefficient for heat transfer to supercritical water at the following conditions: pressure 24.5 MPa, mass flux 1260 kg/m$^2$s, heat flux 698 kW/m$^2$, inner tube diameter 7.5 mm, and with various flow orientations: (a)–horizontal flow at the bottom of the tube, (b)–vertical up-flow, (c)–vertical downflow, (d)–horizontal flow at the top of the tube. Experimental data taken from [239].

## 7.4 CONVECTION HEAT TRANSFER IN ROD BUNDLES

The preceding results refer to convective heat transfer in simple ducts such as circular tubes. For ducts with more complex geometry, such as fuel rod bundles, specialized formulas have been developed.

Convective heat transfer for both upflow and downflow of water in fuel rod bundles at low Reynolds numbers (Re $< 10^4$) is of particular importance for systems with passive cooling. For such conditions the flow can be either laminar or turbulent. Experimental data show that the transition between these two regimes occurs at a specific Reynolds number that increases linearly with increasing $P/D_R$ ratio. For example, for a lattice with $P/D_R$ ratios of 1.25, 1.38, and 1.5, the transition from the laminar to turbulent flow occurs when the Reynolds number exceeds the following transition value [62, 63],

$$
\mathrm{Re}_T = \begin{cases} 10^4 \times (1.319\frac{P}{D_R} - 1.432) & \text{for triangular lattice} \\ 1.33 \cdot 10^4 \times (\frac{P}{D_R} - 1) & \text{for square lattice} \end{cases}, \qquad (7.63)
$$

where $P$ is the lattice pitch, $D_R$ is the rod diameter, and Re is based on the hydraulic equivalent diameter of the bundle, $D_h = 4A/P_w$, where $A$ is the bundle flow area and $P_w$ is the wetted perimeter.

### 7.4.1 CONVECTION AT HIGH REYNOLDS NUMBER

For convective heat transfer to gases and liquids flowing in rod bundles with square and triangle lattices, the following formula due to Weisman can be used [232],

$$
\mathrm{Nu} = C\,\mathrm{Re}^{0.8}\,\mathrm{Pr}^{1/3}, \qquad (7.64)
$$

where ,

$$
C = \begin{cases} 0.026\frac{P}{D_R} - 0.006 & \text{for triangular lattice with } 1.1 < \frac{P}{D_R} < 1.5 \\ 0.042\frac{P}{D_R} - 0.024 & \text{for square lattice with } 1.1 < \frac{P}{D_R} < 1.3 \end{cases}. \qquad (7.65)
$$

In these expressions $P$ is the lattice pitch and $D_R$ is the rod diameter. The Nusselt and Reynolds numbers are based on the equivalent hydraulic diameter of the bundle, $D_h = 4A/P_w$, where $A$ is the bundle flow area and $P_w$ is the wetted perimeter.

The most accurate of the correlations for the convective heat transfer to gases and liquids flowing in rod bundles is a formula due to Markoczy [153],

$$
\mathrm{Nu} = \mathrm{Nu}_{\mathrm{DB}}\left[1 + 0.91\,\mathrm{Re}^{-0.1}\,\mathrm{Pr}^{0.4}(1 - 2\,e^{-B})\right], \qquad (7.66)
$$

where,

$$
B = \begin{cases} \frac{2\sqrt{3}}{\pi}\left(\frac{P}{D_R}\right)^2 - 1 & \text{for triangular lattice} \\ \frac{4}{\pi}\left(\frac{P}{D_R}\right)^2 - 1 & \text{for square lattice} \end{cases}, \qquad (7.67)
$$

and $\mathrm{Nu}_{\mathrm{DB}}$ is calculated from the Dittus-Boelter correlation given by Eq. (7.43). The correlation is valid in the following ranges for the Reynolds number: $3 \cdot 10^3 < \mathrm{Re} < 10^6$, for the Prandtl number: $0.66 < \mathrm{Pr} < 5$, and for the pitch-to-diameter ratios: $1.02 < P/D_R < 2.5$.

For convective heat transfer to liquid metal flowing in a rod bundle the following correlation is applicable [20]:

$$
\mathrm{Nu} = \begin{cases} 24.14\,\log\left(-8.12 + 12.76s - 3.65s^2\right) & \text{for} \quad \mathrm{Pe} \leq 200 \\ 24.14\,\log\left(-8.12 + 12.76s - 3.65s^2\right) + \\ \quad 0.0174\left[1 - e^{-6(s-1)}\right](\mathrm{Pe} - 200)^{0.9} & \text{for} \quad 200 < \mathrm{Pe} \leq 2000 \end{cases}, \qquad (7.68)
$$

where the correlation is valid for $1.1 \leq s \leq 1.5$. In this formula $s = P/D_R$ is the pitch-to-diameter ratio and Pe is the Peclet number.

### 7.4.2 CONVECTION AT LOW REYNOLDS NUMBER

For water flowing in a uniformly heated vertical bundle with triangular lattice, the Nusselt number for combined convective heat transfer were best correlated by the following formulas [62],

$$\text{Nu}_C = \begin{cases} \left(\text{Nu}_F^3 + \text{Nu}_N^3\right)^{1/3} & \text{for combined upflow} \\ \left(\text{Nu}_F^2 - \text{Nu}_N^2\right)^{1/2} & \text{for combined downflow} \end{cases}, \tag{7.69}$$

where $\text{Nu}_F$ is the Nusselt number for the forced laminar upflow convection,

$$\text{Nu}_F = 0.511 \, \text{Re}^{0.46} \, \text{Pr}^{1/3}, \tag{7.70}$$

and $\text{Nu}_N$ is the Nusselt number for the natural laminar convection,

$$\text{Nu}_N = 0.272 \, \text{Ra}_q^{0.25}. \tag{7.71}$$

Here $\text{Ra}_q$ is the Rayleigh number based on the heat flux defined as,

$$\text{Ra}_q = \frac{g\beta q'' D_h^4}{\lambda a \nu}. \tag{7.72}$$

In this formula $q''$ is the heat flux, $\beta$ is the volumetric thermal expansion coefficient, $D_h$ is the hydraulic diameter, $\lambda$ is the thermal conductivity, $a$ is the thermal diffusivity, and $\nu$ is the kinematic viscosity. The correlation shows good agreement with experimental data obtained for $P/D_R = 1.38$, Re in a range from 1200 to $2.48 \cdot 10^4$ and Pr from 6.8 to 9.0 in the forced flow regime, and Re from 148 to 3800, $\text{Gr}_q$ from $1.3 \cdot 10^5$ to $3 \cdot 10^6$, and Ri from 0.01 to 9 in the combined convection regime. Here the heat-flux-based Grashof number, $\text{Gr}_q$, is defined as,

$$\text{Gr}_q = \frac{g\beta q'' D_h^4}{\lambda \nu^2} = \frac{\text{Ra}_q}{\text{Pr}}, \tag{7.73}$$

and the Richardson number, Ri, is defined as,

$$\text{Ri} = \frac{\text{Gr}_q}{\text{Re}^2}. \tag{7.74}$$

For a vertical bundle with a square lattice, the Nusselt number for combined convection can be found from the following formula [63],

$$\text{Nu}_C = \left(\text{Nu}_F^4 \pm \text{Nu}_N^4\right)^{1/4}, \tag{7.75}$$

where the plus sign should be used for the upflow conditions and the minus sign for the downflow conditions. This equation fits the experimental data for $P/D_R$ within $\pm 12\%$, and for $P/D_R$ 1.38 and 1.25 within $\pm 15\%$. The forced convection Nusselt number is given as,

$$\text{Nu}_F = \begin{cases} A \, \text{Re}^B \, \text{Pr}^{0.33} & \text{for laminar regime} \\ C \, \text{Re}^{0.8} \, \text{Pr}^{0.33} & \text{for turbulent regime} \end{cases}, \tag{7.76}$$

where the coefficients $A$, $B$, and $C$ are given as,

$$A = 2.97 - 1.76\frac{P}{D_R}, \quad B = 0.56\frac{P}{D_R} - 0.30, \quad C = 0.028\frac{P}{D_R} - 0.006. \tag{7.77}$$

The natural convection Nusselt number is correlated in terms of $\mathrm{Ra}_q$ as follows,

$$\mathrm{Nu}_N = \begin{cases} 0.178\,\mathrm{Ra}_q^{0.27} & \text{for } P/D_R = 1.5 \\[2mm] 0.057\,\mathrm{Ra}_q^{0.35} & \text{for } P/D_R = 1.25 \text{ and } 1.38 \end{cases} \tag{7.78}$$

In all dimensionless quantities, the heated equivalent diameter, defined as,

$$D_H = D_R \left[ \frac{4}{\pi} \left( \frac{P}{D_R} \right)^2 - 1 \right], \tag{7.79}$$

should be used as the characteristic length and the fluid physical properties should be determined at the coolant bulk temperature. The experimental data used in the derivation of the formula are as follows: $P/D_R$ ratios of 1.25, 1.38, and 1.5, Re varies from 250 to $3 \cdot 10^4$, Pr from 3 to 9, $\mathrm{Ra}_q$ from $5 \cdot 10^5$ to $3 \cdot 10^8$ for natural convection and from $10^7$ to $7 \cdot 10^8$ for combined convection, and Ri from 0.03 to 300. The natural convection data are correlated to within $\pm 10\%$ in terms of $\mathrm{Ra}_q$ and the combined convection data are correlated to within $\pm 15\%$. A comparison of a bundle with triangular lattice shows that for the same flow area per rod, the rod arrangement negligibly affects Nu in both forced and natural convection regimes.

## PROBLEMS

### PROBLEM 7.1

Calculate the wall-averaged heat transfer coefficient for a vertical wall with height equal to 1.5 m immersed in a pool of liquid sodium at a temperature equal to 450 K and at atmospheric pressure.

### PROBLEM 7.2

Due to a malfunction of the circulation pump, no forced convective cooling is available in a reactor core. Calculate the heat transfer coefficient between cladding and coolant for two cases: (1) when the core is filled with stagnant water at pressure 17.5 MPa and temperature 595 K, and (2) when the core is filled with stagnant water vapor at pressure 4.2 MPa and temperature 595 K. In both cases assume the core hydraulic diameter equal to 10 mm, the core height equal to 4.2 m, and the initial clad surface temperature equal to 645 K. What is the maximum heat flux on the cladding surface that can be retrieved by the coolant in each of the cases?

## PROBLEM 7.3

Calculate the convective heat transfer coefficient to water at 17.5 MPa pressure and 550 K temperature, flowing with a mass flux equal to 2450 kg m$^{-2}$ s$^{-1}$ in a fuel rod bundle with the rod diameter equal to 9.5 mm and the lattice pitch equal to 12.6 mm. Compare the heat transfer coefficients for the square and triangular lattices using the Markoczy correlation.

## PROBLEM 7.4

Supercritical water at pressure 23.3 MPa flows with a mass flux equal to 430 kg m$^{-2}$ s$^{-1}$ vertically upward in a smooth uniformly heated tube with an inner diameter equal to 8 mm. Calculate the tube wall temperature at the axial location where the ratio of the bulk specific enthalpy to the bulk specific enthalpy at the pseudocritical conditions is equal to 0.85. Compare results obtained for two values of the heat flux equal to 277.8 and 366.09 kW m$^{-2}$ and using the Dittus-Boelter and the Krasnoshchekov-Prototopov correlations. Compare the calculated temperature with the measured temperature presented in Fig. 7.5a.

## PROBLEM 7.5

Calculate the convective heat transfer coefficient to liquid lead at temperature 723 K flowing with a mean speed equal to 1.6 m s$^{-1}$ in a rod bundle with a square lattice. The lattice pitch is equal to 13.7 mm and the rod diameter is equal to 10.5 mm.

# 8 Boiling Heat Transfer

*Boiling* is a process of liquid evaporation in which the local liquid temperature exceeds the saturation temperature. It can take place for saturation temperatures that are contained between the temperature of the triple point and the temperature of the critical point of the liquid. *Boiling heat transfer* provides a very efficient mechanism for heat removal from a heated surface. High heat transfer rates are achieved through the evaporation process combined with a localized motion of the liquid-vapor interface. For a wide range of the heat flux magnitude, the temperature of the heated surface remains at a level close to the saturation temperature of the evaporating liquid. This feature is particularly important for systems where the heated wall temperature should not exceed a prescribed safety limit value.

In this chapter we discuss the various boiling heat transfer modes that can be encountered in nuclear applications. A short presentation of characteristic features of boiling heat transfer and an introduction of important concepts and definitions are given in §8.1. Sections §8.2, §8.3, and §8.4 are devoted to such topics as the onset of nucleate boiling, the bubble nucleation and growth, and the different modes of heat transfer that occur during nucleate boiling. Overviews of governing phenomena, correlations, and models for the two most common boiling modes, namely pool boiling and flow boiling, are given in sections §8.5 and §8.6 respectively.

## 8.1 GENERAL BOILING CHARACTERISTICS

Boiling heat transfer phenomena include many aspects of thermodynamics and fluid flow. One of the main characteristic features of boiling is the phase change during which heat, referred to as the *latent heat*, is absorbed at constant pressure. In general the latent heat is supplied to the boiling liquid through surface of a solid body, but it can be supplied in the whole volume of the liquid as internal heat sources, or to the liquid-vapor interface through thermal radiation that is not absorbed by the vapor, but is absorbed by the liquid.

Evaporation during *nucleate boiling* heat transfer takes place at specific locations called *nucleation sites*. The nucleation sites are in general located at a heated solid surface, where bubbles form over cavities with preexisting inert gas or vapor. This type of nucleation is prevailing in engineering systems and is thus of primary interest in this book. However, it should be mentioned that nucleation sites can also be found within the liquid, far from any solid surface. Such nucleation sites can typically occur in two forms. The first form is manifest as microscopic voids that constitute the nuclei necessary for the growth of microscopic bubbles, which can be created due to the thermal motions of liquid molecules. This is termed *homogeneous nucleation*. The second form exists as micron-sized bubbles of contaminant gas, suspended particles, or even high-energy particle radiation. In water, microbubbles of air seem to persist almost indefinitely and are almost impossible to remove completely. When

DOI: 10.1201/9781003255000-8

nucleation occurs at solid surfaces or at suspended contaminant particles in the liquid, it is termed *heterogeneous nucleation*.

## 8.1.1  BOILING SURFACE FEATURES

Experimental investigations show that boiling surface features play an important role in the boiling heat transfer phenomena. Interest in surface conditions, such as wettability and roughness, has been increasing in response to the development of new, more sophisticated experimental techniques and due to the introduction of manufactured surfaces using micro/nano technologies. Employing proper fluid and boiling surface fabrication, it has been possible to significantly increase the boiling heat transfer rates and CHF.

### Surface Roughness

Surface roughness commonly has a scale of microns and is expected to affect the heat transfer phenomena through influencing the interface and contact line movement. A parameter that is used to characterize the surface roughness is the **arithmetic average roughness**, $R_a \equiv \int_0^l |h(z)| \, dz / l$, defined as the average of the absolute height value along the sampling distance $l$. The geometric features of the rough surface determine the boiling parameters of the surface such as the *nucleation site density* and the *active nucleation sites* present on the heated surface.

### Surface Wettability and Contact Angle

Wettability is a property of a given material that describes a liquid ability to adhere to its surface. Liquid contact angle (usually called just contact angle) is used to determine the wettability of materials. If the contact angle is smaller than $90°$, the material is hydrophilic, otherwise is hydrophobic.

For a given gas-liquid-solid combination of materials, the contact angle is influenced by several factors, such as the interface movement, the solid surface roughness, the pressure, and the temperature. The equilibrium contact angle of a liquid drop on an ideal solid surface, which is smooth, non-deformable, and chemically homogeneous, can be described by the classical Young equation, given as

$$\cos \theta = \frac{\sigma_{sg} - \sigma_{sl}}{\sigma_{lg}}, \tag{8.1}$$

where $\sigma_{sg}$, $\sigma_{sl}$, and $\sigma_{lg}$, represent the solid-gas, solid-liquid, and liquid-gas interfacial tension, respectively.

When the gas-liquid interface is moving along the solid surface, a continuous range of contact angle values occurs. The maximum contact angle is referred to as the *advancing contact angle* ($\theta_A$) and the minimum contact angle is referred to as the *receding contact angle* ($\theta_R$). The difference between the advancing and receding contact angles ($\theta_A - \theta_R$) is often referred to as the *contact angle hysteresis*. It has been derived theoretically [214] and confirmed experimentally [33] that the equilibrium contact angle can be calculated from $\theta_A$ and $\theta_R$ as follows,

$$\theta = \arccos \left( \frac{r_A \cos \theta_A + r_R \cos \theta_R}{r_A + r_R} \right), \tag{8.2}$$

where,

$$r_A = \left( \frac{\sin^3 \theta_A}{2 - 3\cos\theta_A + \cos^3 \theta_A} \right)^{1/3}, \tag{8.3}$$

$$r_R = \left( \frac{\sin^3 \theta_R}{2 - 3\cos\theta_R + \cos^3 \theta_R} \right)^{1/3}. \tag{8.4}$$

For liquid moving quickly over a surface, the contact angle can be altered from its value at rest. The advancing contact angle (when liquid advances over the previously dry surface) will increase with speed, and the receding contact angle (when liquid recedes from a previously wet surface) will decrease.

Experimental data indicate a rather weak influence of pressure on the contact angle. However, the temperature effect is considered the dominant influencing factor. A classical, semi-empirical model of the temperature effect is as follows:

$$\cos\theta = 1 + C(T_{pc} - T)^{a/(b-a)}, \tag{8.5}$$

where $T$ and $T_{pc}$ are the system temperature and the pseudo-critical temperature at which the contact angle reduces to zero, respectively, $C$ is an integral constant, and $a$ and $b$ are two constants derived from a balance of intermolecular forces, which are assumed not to vary with temperature [1].

The temperature-dependent contact angle of water on an SS304 surface can be found from the following correlation:

$$\theta = \theta_0 \tanh\left[ 2.503 \left( \frac{T_{cr} - T}{T_{cr} - T_0} \right)^{1.223} \right], \tag{8.6}$$

where $\theta_0$, $T_0$, $T_{cr}$, $T$ are the contact angle at room temperature, room temperature, critical point temperature, and system temperature, respectively. This correlation can be used for temperatures up to 250°C, and it agrees with experimental data within ±6 percent [209].

## 8.1.2 THERMAL BOUNDARY LAYER

The temperature distribution in the thermal boundary layer affects both the onset of nucleate boiling and the efficiency of nucleate boiling heat transfer. The thickness of the thermal boundary layer is the distance across a boundary layer from the wall surface to a point where the fluid temperature has essentially reached the freestream temperature. This quantity plays an important role in the boiling heat transfer.

For free convection, the thickness of the thermal boundary layer can be found from the following expression [120],

$$\delta_T = 7.14 \left( \frac{\mu_l a_l}{\rho_l g \beta_l (T_w - T_{sat})} \right)^{1/3}, \tag{8.7}$$

where $a_l$ is the liquid thermal diffusivity and $\beta_l$ is the isobaric thermal expansion coefficient of the liquid.

For laminar flow over a flat plate at zero incidence, the thermal boundary layer thickness is given as,

$$\delta_T = 5.0\sqrt{\frac{vz}{U}}\text{Pr}^{-1/3}, \tag{8.8}$$

where Pr is the Prandtl number, $U$ is the freestream velocity, $v$ is the kinematic viscosity, and $z$ is the distance downstream from the start of the boundary layer.

For turbulent flow, the thermal boundary layer thickness can be obtained from the following expression,

$$\delta_T \approx 0.37z/\text{Re}_z^{1/5}, \tag{8.9}$$

where $\text{Re}_z = Uz/v$ is the Reynolds number.

Because the thermal boundary layer is usually very thin in pool nucleate boiling, the temperature profile is assumed to be linear and the thickness can be estimated as,

$$\delta_T \approx \frac{\lambda_l \Delta T_w}{q_w''}, \tag{8.10}$$

where $\Delta T_w = T_w - T_{\text{sat}}$ is the wall superheat and $q_w''$ is the wall heat flux.

When the thermal boundary layer is re-build after a bubble departure, its thickness can be derived by considering a transient conduction heat transfer in the liquid, and the following expression can be obtained,

$$\delta_T = \sqrt{\pi a_l \Delta t_W}, \tag{8.11}$$

where $a_l$ is the thermal diffusivity of liquid and $\Delta t_W$ is the bubble waiting time.

### 8.1.3  BOILING CURVE

Regimes in boiling heat transfer between a solid wall at temperature $T_w$ and a liquid with a far-field temperature $T_l \leq T_{\text{sat}}$ are frequently presented on a $q''$–$\Delta T_w$ plane, where $q''$ is a heat flux and $\Delta T_w \equiv T_w - T_{\text{sat}}$ is a *wall superheat*. The liquid saturation temperature $T_{\text{sat}} = T_{\text{sat}}(p)$ is determined by the prevailing system pressure $p$.

If heat flux is controlled on a heated surface, the heat flux is an independent variable and its value uniquely determines the corresponding wall superheat,

$$\Delta T_w = \frac{q''}{h_{2\phi}(q'')}, \tag{8.12}$$

where $h_{2\phi}(q'')$ is a boiling (two-phase) heat transfer coefficient. When the heat flux rises, it also leads to an increase in the wall superheat. However, this relationship is not linear, since, in general, the boiling heat transfer coefficient is a function of the heat flux magnitude.

In some boiling systems the wall temperature, rather than the heat flux, is determined. For example, if a wall separates two fluids and one of the fluids is evaporating whereas the other fluid is cooled in the single-phase heat transfer regime, the wall temperature (and thus the wall superheat) is determined by fluid temperatures and the heat transfer conditions. For such systems, the heat transfer rates (or the heat flux magnitudes) depend on the wall superheat as follows,

$$q'' = h_{2\phi}\,(\Delta T_w)\,\Delta T_w. \tag{8.13}$$

Again the relationship is nonlinear since the boiling heat transfer coefficient depends on the wall superheat. Traditionally, expression (8.13) is used to represent the $q''$–$\Delta T_w$ relationship, both for the heat-flux-controlled and the temperature-controlled boiling heat transfer. The relationship is often presented on a $\log q''$–$\log \Delta T_w$ plane and is called the *boiling curve*.

For low values of the wall superheat, natural convection heat transfer prevails. When the wall superheat exceeds a certain threshold value that is characteristic of the onset of nucleate boiling, vapor bubbles start appearing on the heated wall. At this point, the motion of liquid is additionally enhanced by bubble growth and detachments. As a result, heat transfer efficiency suddenly increases from a rather small value that corresponds to natural convection heat transfer to a much higher value that is typical of boiling heat transfer. In systems where wall temperature is controlled, the transition from natural convection to boiling heat transfer is accompanied by a sudden increase of the heat flux at the heated surface, as demonstrated by points $B$ and $B''$ on the boiling curve shown in Fig. 8.1. For a heat-flux controlled system, the wall superheat is suddenly reduced as indicated by points $B$ and $B'$ on the boiling curve.

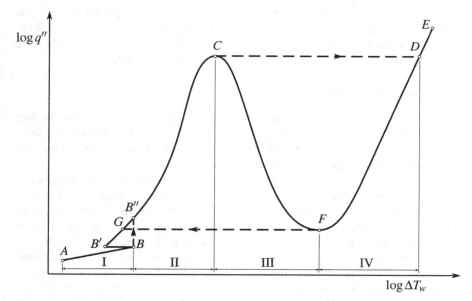

**Figure 8.1** A boiling curve: I–convection, II–nucleate boiling, III–transition boiling, IV–film boiling, $C$–critical heat flux, $F$–minimum heat flux that sustains stable film boiling.

As wall superheat or heat flux is increased in region II in Fig. 8.1, nucleate boiling heat flux reaches a peak value at point $C$. At that point, a boiling crisis, also called burnout or critical heat flux (CHF) occurs. With heat-flux-controlled heating the path $C$-$D$-$E$ is followed, along which a sudden increase of the wall superheat between points $C$ and $D$ takes place. The increase of the wall superheat is caused by a significant drop of the heat transfer coefficient as pre-CHF boiling heat transfer transits into the post-CHF boiling heat transfer regime. If wall temperature is controlled and increased between points $C$ and $D$, the corresponding wall heat flux will follow the path $C$-$F$-$D$. In the transition boiling region, the liquid touches the surface intermittently, and the heat transfer process may be alternating between nucleate and film boiling.

Operation along the curve $D$-$F$ is in the film boiling regime, characterized by a persistent vapor layer that coats the surface, thereby inhibiting direct contact between the liquid and the surface. This curve is followed when high heat flux at point $D$ is reduced beyond point $F$. The dashed line $F$-$G$ shows a sudden reduction of the wall superheat when film boiling at point $F$ reverts to nucleate boiling at point $G$. Here point $F$ represents a condition when the minimum heat flux is achieved that can sustain stable film boiling.

The qualitative overview of the boiling curve and the related boiling phenomena presented in this section is primarily concerned with the boiling of saturated liquid in an extensive ambient. Two parameters that strongly influence boiling phenomena are liquid subcooling and forced convection. For example, the maximum heat flux is strongly influenced by both these parameters. However, the general shape of the boiling curve and the various boiling heat transfer regimes are quite similar to those presented in the present section. In the following sections, a more comprehensive explanation of how these parameters affect the boiling is provided.

## 8.1.4  BOILING MODES

There are two major modes of boiling heat transfer: *pool boiling* and *flow boiling*. The pool-boiling mode corresponds to a free-convection heat transfer to a stationary liquid contained in a large pool. The three basic regimes of pool boiling include nucleate boiling, transition boiling, and film boiling. Each of these regimes can exist at either subcooled or saturated conditions and occurs over a range of wall superheats.

The flow-boiling mode corresponds to a forced-convection heat transfer to two-phase mixture flowing in a channel. Similarly to pool boiling, the flow-boiling mode manifests itself with various boiling regimes, depending on the local conditions in the channel. With increasing thermodynamic quality of the mixture, flow boiling gradually transits from subcooled nucleate boiling regime to saturated nucleate boiling, evaporation of liquid film, film boiling, and dry-wall mist flow boiling regime. At certain conditions, typical for loss-of-coolant accidents, the transition boiling can be established as well. The various boiling regimes in pool boiling and flow boiling are discussed in more detail in §8.5 and §8.6, respectively.

## 8.2 ONSET OF NUCLEATE BOILING

Boiling that manifests itself with *homogeneous nucleation* in a superheated bulk liquid is called *homogeneous boiling*. In clean pure systems the liquid temperature can significantly exceed the saturation temperature, thus, the liquid can be highly superheated before the nucleation is initiated. Such homogeneous nucleation can take quite violent forms and is associated with a sudden increase of the volume of the vapor-liquid mixture.

In engineering systems with submerged heated walls, the wall surfaces and neighboring liquid layers have the same temperature. When the surface temperature exceeds the liquid local saturation temperature with a certain margin, a bubble nucleation process is initiated. This phenomenon is known as an *onset of nucleate boiling* or an *incipience of boiling*. Since the bubble nucleation and boiling are confined to the wall surface, they are called *heterogeneous nucleation* and *heterogeneous boiling*, respectively. Here attention is focused on heterogeneous boiling, which in continuation is simply referred to as boiling.

### 8.2.1 PHASE EQUILIBRIUM

When discussing processes that take place at constant temperature and volume or constant temperature and pressure (for example chemical reactions or phase change), it is convenient to introduce two new thermodynamic functions. The first one is the *Helmholtz free energy* defined as,

$$\mathscr{F} \equiv \mathscr{U} - TS, \tag{8.14}$$

and the second is the *Gibbs free energy*, given as,

$$\mathscr{G} \equiv \mathscr{I} - TS, \tag{8.15}$$

where $\mathscr{U}$ is the thermal energy, $\mathscr{I}$ is the enthalpy, $S$ is the entropy, and $T$ is the temperature of the system.

A system (or a part of it), which has in its whole volume the same physical properties and is described with the same equation of state, is called a *phase*. From the phase definition it is clear that the value of any extensive thermodynamic function for the whole multiphase system is a sum of values of this function for each phase. Thus, if a system consists of water and vapor, the Helmholtz free energy of this multiphase system is $\mathscr{F}_{sys} = \mathscr{F}_l + \mathscr{F}_v$, where $\mathscr{F}_l$ and $\mathscr{F}_v$ is the Helmholtz free energy of the water and the vapor phase, respectively. To thermodynamic functions of each of the phases apply all relationships that are valid for single-phase systems. In particular, the *chemical potential* of phase $k$ is obtained as a partial derivative of the Gibbs free energy as follows,

$$\mu_k = \left(\frac{\partial \mathscr{G}_k}{\partial \mathscr{N}}\right)_{p,T}. \tag{8.16}$$

Here $\mathscr{N}$ is the number of atoms or molecules of the substance in phase $k$, $p$ is the pressure and $T$ is the temperature of the multiphase system.

If a multiphase system containing phases $k = 1, 2, 3, \ldots$ is at equilibrium, that is there are no phase-change processes taking place in the system, the following condition for each phase in the system has to be satisfied,

$$\mu_1 = \mu_2 = \mu_3 = \ldots \quad . \tag{8.17}$$

Consider a system that contains a mixture of two phases (liquid and gas) of the same species, separated by a sharp dividing surface, the interface. The total volume $V_m$ and the total number of molecules of the species $\mathcal{N}_m$ are assumed constant and given as,

$$V_l + V_v = V_m = \text{const}, \tag{8.18}$$

$$\mathcal{N}_l + \mathcal{N}_v = \mathcal{N}_m = \text{const}. \tag{8.19}$$

The first law of thermodynamics applies to each phase $k = l, v$ as follows,

$$d\mathcal{U}_k = T_k dS_k - p_k dV_k + \mu_k d\mathcal{N}_k. \tag{8.20}$$

In view of Eq. (8.19), the mixture interface has no mass, but it has energy $\mathcal{U}_i$, entropy $S_i$, and surface area $A_i$. The first law of thermodynamics for the interface can be written as,

$$\mathcal{U}_i = T_i S_i + \sigma A_i, \tag{8.21}$$

where $T_i$ is the interface temperature and $\sigma$ is the surface tension. The interface Helmholtz free energy is thus,

$$\mathcal{F}_i = \mathcal{U}_i - T_i S_i = \sigma A_i. \tag{8.22}$$

The mixture Helmholtz free energy can be found as a sum of the free energy of the phases and the interface,

$$\mathcal{F}_m = \mathcal{F}_l + \mathcal{F}_v + \mathcal{F}_i. \tag{8.23}$$

At phase equilibrium we should have $d\mathcal{F}_m = 0$, and using Eqs. (8.20) and (8.22) we get,

$$dF_m = -p_l dV_l - p_v dV_v - S_l dT_l - S_v dT_v + \mu_l d\mathcal{N}_l + \mu_v d\mathcal{N}_v + \sigma dA_i = 0. \tag{8.24}$$

Invoking conditions given by Eqs. (8.18) and (8.19), we have $dV_l = -dV_v$ and $d\mathcal{N}_l = -d\mathcal{N}_v$ and the phase equilibrium relationship is as follows,

$$-(p_v - p_l) dV_v + (\mu_v - \mu_l) d\mathcal{N}_v + \sigma dA_i - S_l dT_l - S_v dT_v = 0. \tag{8.25}$$

For a special case of approaching the phase equilibrium with $\mu_l = \mu_v$ and $dT_l = dT_v = 0$, the following additional condition has to be satisfied,

$$p_v - p_l = \sigma \frac{dA_i}{dV_v}. \tag{8.26}$$

Assuming that the system under consideration consists of a bubble with a radius $r_B$ submerged in liquid, and that the bubble radius changes from $r_B$ to $r_B + dr$, we have

$dA_i = 8\pi r_B dr$ and $dV_g = 4\pi r_B^2 dr$. Thus, the vapor will be in equilibrium with the liquid when,

$$p_v - p_l = \frac{2\sigma}{r_B}. \tag{8.27}$$

The equation shows that the gas pressure at the interface is higher than the liquid pressure by factor $2\sigma/r_B$. For water at atmospheric pressure and a vapor bubble with radius $r_B = 100$ $\mu$m the pressure difference is about 590 Pa. Since for water the surface tension decreases with increasing pressure, the pressure difference decreases to 52 Pa, when the pressure of the liquid-bubble system increases to 15 MPa.

The saturation pressure given for most fluids in property tables is defined at a flat interface. Consider a spherical liquid-vapor interface in a capillary tube, connected to a flat surface, at which the liquid and vapor have the same pressure $p_\infty$. Assuming a vertically-upward directed tube in which the vapor-liquid interface is at height $h$ above the flat surface (see Fig. 8.2), the liquid and vapor pressure at the interface in the capillary tube can be related to $p_\infty$ as follows,

$$p_l = p_\infty - \rho_l g h, \quad p_v = p_\infty - \rho_v g h. \tag{8.28}$$

Here $\rho_l$ and $\rho_v$ are densities of liquid and vapor, respectively. Division of the equations yields,

$$\frac{p_\infty - p_v}{p_\infty - p_l} = \frac{\rho_v}{\rho_l}. \tag{8.29}$$

Then eliminating either $p_l$ or $p_v$ with Eq. (8.27) gives the relationship between the pressure at the curved and flat interfaces as,

$$p_v = p_\infty - \frac{\rho_v}{\rho_l - \rho_v}\frac{2\sigma}{r_B}, \tag{8.30}$$

$$p_l = p_\infty - \frac{\rho_l}{\rho_l - \rho_v}\frac{2\sigma}{r_B}. \tag{8.31}$$

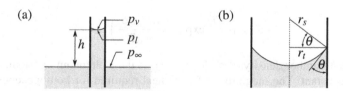

**Figure 8.2**   (a) Liquid and vapor pressures in a capillary tube, (b) relationship between the surface radius $r_s$, the capillary tube radius $r_t$, and the wetting angle $\theta$.

Both phases are superheated at a curved interface since their pressures are less than the saturation pressure $p_\infty$.

We can calculate the liquid superheat $\Delta T_{\text{sup},l}$ corresponding to $\Delta p_l = p_\infty - p_l$ using the *Clausius-Clapeyron relation*,

$$\frac{dT}{dp} = \frac{T \upsilon_{fg}}{i_{fg}}. \tag{8.32}$$

Here $v_{fg} \equiv 1/\rho_g - 1/\rho_l$ is a specific volume change at the saturation curve and $i_{fg}$ is the latent heat. Integration of the equation yields,

$$\Delta T_{\text{sup},l} \equiv T_l - T_{\text{sat}} = \int_{T_{\text{sat}}}^{T_l} dT = \int_{p_l}^{p_\infty} \frac{T v_{fg}}{i_{fg}} dp. \tag{8.33}$$

The integration requires an expression for the integrand as a function of pressure. Various approximations can be used for such relationships. In the simplest case, assuming,

$$\frac{T v_{fg}}{i_{fg}} = \frac{T_{\text{sat}} v_{fg}}{i_{fg}} = \text{const}, \tag{8.34}$$

we get,

$$T_l - T_{\text{sat}} = \frac{T_{\text{sat}} v_{fg}}{i_{fg}} (p_\infty - p_l) = \frac{T_{\text{sat}} v_{fg}}{i_{fg}} \frac{\rho_l}{\rho_l - \rho_v} \frac{2\sigma}{r_B}, \tag{8.35}$$

where Eq. (8.31) was used for the pressure difference in the liquid phase.

## 8.2.2 HOMOGENEOUS NUCLEATION

Homogeneous nucleation has been treated either on the ground of the thermodynamic equilibrium theory or as a result of statistical density fluctuations. The kinetic view of homogeneous nucleation suggests that due to density fluctuations in the liquid, there is the probability that a sufficient number of molecules with greater than average energy can form a vapor bubble with a certain equilibrium radius $r_{BE}$. Taking liquid temperature $T_l$ and using Eq. (8.35), the following equilibrium radius of the bubble is found,

$$r_{BE} = \frac{2\sigma}{T_l - T_{\text{sat}}} \frac{T_{\text{sat}} v_{fg}}{i_{fg}} \frac{\rho_l}{\rho_l - \rho_v}, \tag{8.36}$$

where $T_{\text{sat}}$ is a saturation temperature at a flat interface. The rate of formation of liquid clusters with $T_l > T_{\text{sat}}$ is given as [42],

$$J = n \frac{k_B T_l}{h} \exp\left(-\frac{4\pi r_{BE}^2 \sigma}{3 k_B T_i}\right), \tag{8.37}$$

where n is the number density of molecules, $k_B$ is the Boltzmann constant, and h is the Planck constant. The magnitude of superheat required for homogeneous nucleation is obtained by combining equations (8.36) with (8.37),

$$T_l - T_{\text{sat}} = \frac{T_{\text{sat}} v_{fg}}{i_{fg}} \frac{\rho_l}{\rho_l - \rho_v} \left[\frac{16\pi\sigma^3}{3 k_B T_l \ln(n k_B T_l / hJ)}\right]^{1/2}. \tag{8.38}$$

## 8.2.3 HETEROGENEOUS NUCLEATION ON FLAT SOLID SURFACE

For heterogeneous nucleation at solid surfaces, the required liquid superheat is different from that for homogeneous nucleation, and it was shown that it can be obtained from the following modified form of Eq. (8.38) [42],

$$T_l - T_{sat} = \frac{T_{sat} v_{fg}}{i_{fg}} \frac{\rho_l}{\rho_l - \rho_v} \left[ \frac{16\pi\sigma^3 f(\theta)}{3k_B T_l \ln(nk_B T_l / hJ)} \right]^{1/2}, \quad (8.39)$$

where $f(\theta)$ is a function of the contact angle. For flat solid surface the function is as follows [42],

$$f(\theta) = \frac{1}{4}\left(2 + 3\cos\theta - \cos^3\theta\right), \quad (8.40)$$

and for a conical cavity with half-angle $\phi$ the function is given as [134],

$$f(\theta) = \frac{1}{4}\left[2 - 3\sin(\theta - \phi) + \sin^3(\theta - \phi)\right]. \quad (8.41)$$

These equations are valid for pure liquids and solids. Chemical impurities or dissolved gasses tend to reduce the superheats below those predicted.

### 8.2.4 ONSET OF NUCLEATE POOL BOILING

Due to a cyclic character of the ebullition process, the liquid layer adjacent to the wall is heated transiently. When relatively cold liquid surrounds a cavity, the embryo bubble will not grow. Only when the liquid is superheated as indicated by Eqs. (8.35) and (8.39), the bubble growth will be possible, provided that the bubble radius is $r_b$ or greater. However, the bubble size is limited by the decreasing liquid superheat when moving away from the heated wall. When the bubble is too large, the vapor will start condensing at the bubble cap. Thus, there is a certain size range in which the bubble embryo will grow to make a cavity into an active nucleation site.

Hsu [93] postulated the following relationships between the height of the embryo bubble $y_B$, the radius of the bubble embryo $r_B$, and the mouth radius of the cavity $r_c$,

$$y_B = C_1 r_c = (1 + \cos\theta)\frac{r_c}{\sin\theta}, \quad r_B = C_2 r_c = \frac{r_c}{\sin\theta}, \quad (8.42)$$

where $\theta$ is the contact angle. Assuming further a linear temperature distribution in liquid as a function of the distance $y$ from the wall,

$$T(y) = \left(1 - \frac{y}{\delta}\right)(T_w - T_\infty) + T_\infty \quad (8.43)$$

the range of the active nucleation sites is found as,

$$\{r_{c,min}, r_{c,max}\} = \frac{\delta(T_w - T_{sat})}{2C_1(T_w - T_\infty)}\left[1 \pm \sqrt{1 - \frac{8C_1}{C_2}\frac{(T_w - T_\infty)T_{sat}\sigma}{(T_w - T_{sat})^2\delta\rho_v i_{fg}}}\right]. \quad (8.44)$$

To use the equation, the limiting thermal boundary layer thickness $\delta$ is needed. For forced convective boiling this thickness was proposed to be equivalent to the dimensional thickness of the viscous sublayer $y^+ = 7$ [88].

To account for the roughness effect on the wall superheat at the onset of nucleate boiling, the following expression is proposed [231],

$$\Delta T_w = \frac{q''R_a}{\lambda_f} + \frac{2\sigma}{mR_a},$$  (8.45)

where $m$ is the slope of the vapor pressure curve and $R_a$ is the surface roughness.

### 8.2.5 ONSET OF NUCLEATE FLOW BOILING

Nucleation in forced convection is suppressed when the limiting thermal boundary layer $\delta$ is thinned by increases in the bulk velocity. Bergles and Rohsenow [17] modified the Hsu model and used the relationship $-\lambda(dT/dy) = h(T_w - T_b)$ for the liquid temperature profile to replace $\delta$. Their resulting onset of boiling criterion was obtained (once rewriting the equation in SI units) from curve fitting as follows,

$$q'' = 0.0018p^{1.156}(1.8\Delta T_w)^{2.83/p^{0.0234}}.$$  (8.46)

Here $q''$ is the wall heat flux in W/m$^2$, $p$ is pressure in Pa, and $\Delta T_w$ is the wall superheat in K. The correlation is valid for water over a pressure range of 0.1–13.8 MPa.

Sato and Matsumura [192] took a similar approach as Hsu and derived the following condition for the onset of nucleate boiling,

$$q'' = \frac{\lambda_l i_{fg}\rho_v\Delta T_w^2}{8\sigma T_{\text{sat}}}.$$  (8.47)

This correlation was further modified by Davis and Andersson [50] to take into account the effect of the contact angle $\theta$ on the bubble shape,

$$q'' = \frac{\lambda_l i_{fg}\rho_v\Delta T_w^2}{8(1+\cos\theta)\sigma T_{\text{sat}}}.$$  (8.48)

Basu et al. [10] introduced a suppression factor to take into account the possible flooding of some cavities and preventing them to become active nucleation sites. This effect is particularly strong for hydrophilic surfaces and is described by the following equation,

$$q'' = \frac{F^2\lambda_l i_{fg}\rho_v\Delta T_w^2}{2\sigma T_{\text{sat}}},$$  (8.49)

where

$$F = 1 - \exp\left[-\left(\frac{\pi\theta}{180}\right)^3 - 0.5\left(\frac{\pi\theta}{180}\right)\right],$$  (8.50)

This empirical correlation was obtained for various heated surfaces with the contact angle in a range 1–85°.

## 8.3 BUBBLE NUCLEATION AND GROWTH

The theory of bubble nucleation and growth is fundamental to understanding boiling heat transfer and identifying potential pathways to boiling crisis. In this section, we discuss some of the most crucial aspects of this field.

### 8.3.1 ACTIVE NUCLEATION SITE DENSITY

A general nucleation site activation theory provides the following expression for the *active nucleation site density* [73, 74],

$$N_a'' = N_0'' \exp\left\{\left[\frac{-16\pi\sigma^3 M^2 N_A}{3\rho_f^2 R_u^3 [\ln(p_\infty/p_g)]^2)}\right]\left(\frac{1}{T_w}\right)^3 \phi\right\}, \tag{8.51}$$

where $N_0''$ (sites/m$^2$) and $\phi$ are constants, $M$–molar mass, $N_A$–Avogadro's constant, $R_u$–universal gas constant, and $T_w$–wall temperature in K. In photographic investigation of saturated nucleate boiling with several different fluids it was observed that the relationship for the active nucleation site density can be simplified as [4],

$$N_a'' = N_0'' \exp\left[\frac{-3.305 \cdot 10^9}{T_w^3}\right]. \tag{8.52}$$

**Effect of Surface Roughness**

The effect of surface roughness on the active nucleation site density is proposed to be calculated as [15],

$$N_a'' = 218.8 \mathrm{Pr}^{1.63} \frac{1}{\gamma} \delta_r^{-0.4} \Delta T_w^3, \tag{8.53}$$

where $\gamma$ is the surface/liquid interaction parameter,

$$\gamma = \left(\frac{\lambda_w \rho_w c_{p,w}}{\lambda_f \rho_f c_{p,f}}\right)^{1/2}, \tag{8.54}$$

and $\delta_r$ is a dimensionless surface roughness parameter,

$$\delta_r = 14.5 - 4.5\left(\frac{R_a p}{\sigma}\right) + 0.4\left(\frac{R_a p}{\sigma}\right)^2. \tag{8.55}$$

Here $R_a$ is the *arithmetic average roughness* in a range of 0.02–1.17 $\mu$m, $p$ is the pressure in Pa, Pr is the Prandtl number, and $\Delta T_w$ is the wall superheat in K.

### 8.3.2 GROWING BUBBLE CHARACTERISTICS

Typical modeling assumptions and dimensions of a growing bubble are shown in Fig. 8.3. It is assumed that a growing bubble has the shape of a spherical cap that makes an angle $\theta$ with the heated surface. The macroscale bubble dimensions are shown in Fig. 8.3a. Assuming that the bubble radius is $r_B$, the bubble volume $V_B$ and interface area $A_B$ can be calculated from the following expressions,

$$V_B = \frac{\pi r_B^3}{3} (1+\cos\theta)^2 (2-\cos\theta), \tag{8.56}$$

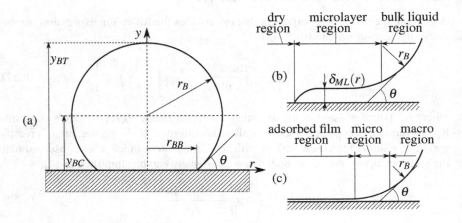

**Figure 8.3**  Bubble growing models: (a) - macroscale bubble dimensions, (b) - microlayer model, (c) - three-phase contact line model.

$$A_B = 2\pi r_B^2 (1+\cos\theta), \tag{8.57}$$

and the area beneath the bubble, also called the bubble base area, is found as,

$$A_{BB} = \pi r_{BB}^2 = \pi r_B^2 \sin^2\theta, \tag{8.58}$$

where $r_{BB} = r_B \sin\theta$ is the radius of the bubble base. The bubble center is located at a distance $y_{BC} = r_B \cos\theta$ from the heated surface, and the distance from the bubble top to the surface is $y_{BT} = r_B (1+\cos\theta)$. The bubble base surface is not completely dry, as schematically shown in Fig. 8.3b. At certain conditions during bubble growth, a liquid microlayer is created on the base surface. Due to evaporation the thickness of the microlayer varies with the distance from the base center. When the microlayer evaporates completely, a dry region is created. This dry region is initiated at the center of the bubble base and its radius increases during the bubble growth. After bubble departure, the dry region is rewetted with liquid that fills the volume released by the bubble.

Depending on local conditions during bubble growth, the microlayer is not created, as illustrated in Fig. 8.3c. In this case the bubble base surface is covered with a thin, non-evaporating adsorbed liquid film. Most intensive evaporation, significantly contributing to the bubble growth, takes place in the micro-region located in the vicinity of the three-phase (solid, liquid, and vapor) contact line.

### 8.3.3  BUBBLE GROWTH RATE

A cyclic bubble growth and departure from a single nucleation site during nucleate boiling is called an *ebullition cycle*. The ebullition cycle is characterized by several parameters, such as the frequency of bubble departures, the size of the departing bubbles, and the bubble growth time. All these parameters have been intensively investigated, both analytically and experimentally.

During growth, the bubble radius increases due to the increasing volume of the vapor phase generated on the bubble interface. At the same time, complex patterns of wall surface temperature and heat flux can be observed beneath the bubble base, as illustrated in Fig. 8.4. The asymptotic growth rate of a bubble in a uniformly

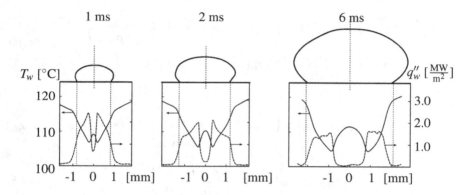

**Figure 8.4**  Bubble growth on a heated surface with corresponding instantaneous temperature and heat flux distributions (retrieved from [215]).

superheated infinite liquid can be found as [68, 178, 227],

$$r_B(t) \cong \left(\frac{12}{\pi}\right)^{1/2} \mathrm{Ja}_{\mathrm{sup}} \sqrt{a_l t}, \tag{8.59}$$

in which the Jakob number applies to the liquid superheating:

$$\mathrm{Ja}_{\mathrm{sup}} = \frac{\rho_l c_{p,l}(T_l - T_{\mathrm{sat}})}{\rho_v i_{fg}}. \tag{8.60}$$

In a similar manner, bubble growth rate due to microlayer evaporation is found as [227],

$$r_B(t) \cong b \left(\frac{12}{\pi}\right)^{1/2} \mathrm{Ja}(t) \sqrt{a_l t}, \tag{8.61}$$

where the time-dependent Jakob number is defined as,

$$\mathrm{Ja}(t) = \frac{\rho_l c_{p,l} \Delta T_w}{\rho_v i_{fg}} \exp\left[-\left(\frac{t}{\Delta t_G}\right)^{1/2}\right]. \tag{8.62}$$

Here $b$ is a time-independent bubble growth parameter, $\Delta t_G$ is the bubble growth time, and $\Delta T_w$ is the wall superheat. The bubble growth parameter varies from $b = 0.794$ for a hemispherical bubble to $b = 1$ for a sphere.

## Influence of Subcooling

The bulk liquid subcooling influences the main bubble dynamics parameters. In highly subcooled conditions bubbles start shrinking due to condensation before departing from the wall. Under such conditions, the bubble departure diameter is smaller and the growing time is shorter in comparison to the less subcooled conditions. Taking into account the bubble evaporation and condensation processes, the bubble growth model can be expressed as [123],

$$
\frac{dr_B(t)}{dt} = \frac{5\alpha_1(t)}{4}\mathrm{Pr}^{-1/2}\mathrm{Ja}\left(\frac{a_l}{t}\right)^{1/2} + \alpha_2(t)\sqrt{\frac{3}{\pi}}\lambda_l\Delta T_w\left(\frac{1}{a_l^{1/2}\rho_v i_{fg}}\right)t^{-1/2}
$$
$$
- [1 - \alpha_2(t)]\frac{\Delta T_{sub}\lambda_l}{d_B\rho_v i_{fg}}\left(2 + 0.6\mathrm{Gr}^{1/4}\mathrm{Pr}^{1/3}\right) \tag{8.63}
$$

where,

$$
\alpha_1(t) = \begin{cases} 1 - \left(\frac{t}{0.7\Delta t_G}\right)^3 & \text{for } 0 \le t \le 0.7\Delta t_G \\ 0 & \text{otherwise} \end{cases} \tag{8.64}
$$

$$
\alpha_2(t) = \min\left[1, 0.11\left(\frac{t}{0.7\Delta t_G}\right)^{-1/2}\right], \tag{8.65}
$$

$$
\mathrm{Gr} = \frac{g\beta\Delta T_{sub}d_B}{v_l^2}. \tag{8.66}
$$

Here $\Delta t_G$ is the bubble growth time, $\Delta T_w = T_w - T_{sat}$ is the wall superheat, $d_B = 2r_B$ is the instantaneous bubble diameter, Ja is the Jakob number based on the liquid subcooling, $\Delta T_{sub} = T_{sat} - T_l$, and Pr is the liquid Prandtl number.

## Influence of Bulk Flow

Compared to pool nucleate boiling, the number of investigations of bubble dynamics in flow boiling is rather limited. This is mainly due to the complexity of the governing phenomena that include the turbulent bulk convection and hydrodynamic interactions between the liquid phase and the vapor bubbles. The bubble growth rates are influenced by various heat transfer mechanisms in a complex manner due to non-uniform temperature distributions in the flow field. Different heat transfer mechanisms that contribute to the bubble size change include evaporation of the microlayer beneath the bubble, heat transfer from the superheated thermal boundary layer, and condensation at the bubble cap where the bubble is exposed to subcooled liquid layers.

Several analytical models for calculating bubble growth rates have been proposed in the literature. Plesset and Zwick [178] and Foster and Zuber [68] models were formulated assuming heat conduction from a superheated liquid layer surrounding the growing bubble. The expressions for the bubble radius as a function of time have the following general form,

$$r_B(t) = \frac{C}{\sqrt{\pi}} \mathrm{Ja_{sup}} \sqrt{a_l t}, \tag{8.67}$$

where $C$ is a constant equal to $2\sqrt{3}$ or in a range from 2 to $2\sqrt{3}$, respectively, and $\mathrm{Ja_{sup}}$ is the Jakob number given as,

$$\mathrm{Ja_{sup}} = \frac{\rho_l c_{p,l}(T_l - T_{sat})}{\rho_v i_{fg}}. \tag{8.68}$$

Mikic et al. [158] developed an analytical expression for a bubble growth under combined inertia and heat conduction controlled conditions as follows,

$$r_B(t) = \frac{2}{3} \frac{B^2}{A} \left[ (t^+ + 1)^{3/2} - (t^+)^{3/2} - 1 \right], \tag{8.69}$$

where $t^+ = A^2 t / B^2$, and

$$A = \left[ b \frac{i_{fg} \rho_v \Delta T_{sup}}{\rho_l T_{sat}} \right]^{\frac{1}{2}}, \tag{8.70}$$

$$B = \left[ \frac{12}{\pi} a_l \right]^{\frac{1}{2}} \mathrm{Ja_{sup}}. \tag{8.71}$$

For bubble growth in an infinite medium $b = 2/3$ and for bubble growth on a surface $b = \pi/7$.

Cooper and Lloyd [47] observed that the rates of growth of bubbles were of the same order as the rates of evaporation from the microlayers. Using a simplified hydrodynamic theory, they showed that the thickness of the microlayer was $0.8\sqrt{vt}$, and experimental observations indicated thicknesses in the range 0.5 to $1.0\sqrt{vt}$. Using simplifying assumptions, a closed analytic form was obtained as follows,

$$r_B(t) = \frac{2}{C} \mathrm{Ja_{sup}} \mathrm{Pr}^{-1/2} \sqrt{a_l t}, \tag{8.72}$$

with $C = 0.8$ and $\mathrm{Pr} = \mu_l c_{p,l} / \lambda_l$.

### 8.3.4 BUBBLE DEPARTURE FREQUENCY

The bubble departure frequency, $f_{BD}$, can be calculated from the bubble waiting time, $\Delta t_W$ and bubble growth time, $\Delta t_G$, as follows,

$$f_{BD} = \frac{1}{\Delta t_W + \Delta t_G}. \tag{8.73}$$

The waiting time during ebullition cycle is the time needed to rebuild the thermal layer following the bubble departure and to create conditions for the inception of the next bubble at the same nucleation site. The bubble growth time is the time required for bubble growth from inception to the departure size. Thus, the complete ebullition cycle time is $\Delta t_W + \Delta t_G$ and corresponds to a characteristic time scale of nucleation boiling heat transfer.

Using the criterion of bubble nucleation and the potential flow theory, the following expression for the bubble waiting time was proposed [84],

$$\Delta t_W = \frac{9}{16\pi a_l} \left\{ \frac{2(T_w - T_b)r_c}{T_w - T_{\text{sat}}\left[1 + (2\sigma/r_c\rho_v i_{fg})\right]} \right\}^2, \tag{8.74}$$

where $a_l$ is the thermal diffusivity of liquid and $r_c$ is the cavity radius. In the derivation of the equation, it was proposed that the waiting time corresponds to the time needed for heating of a thermal layer with a thickness equal to $2r_c/3$.

During the waiting period, the temperature in the thermal boundary layer increases due to the transient conduction of heat from the wall surface to the liquid. Using a one-dimensional model of heat conduction in a semi-infinite body, the liquid temperature can be expressed in terms of time and the distance from the wall surface. Assuming constant wall temperature $T_w$ and initial liquid temperature equal to the bulk temperature $T_b$, the solution takes the following form,

$$T_l(y,t) = T_w - (T_w - T_b) \times \text{erf}\left(\frac{y}{2\sqrt{a_l t}}\right), \tag{8.75}$$

where $y$ is the distance from the wall and $a_l = \lambda_l/(\rho_l c_{p,l})$ is the thermal diffusivity of liquid. The corresponding wall heat flux is given as,

$$q_w'' = \frac{\lambda_l(T_w - T_l)}{\sqrt{\pi a_l t}}. \tag{8.76}$$

The above relations for the temperature field and the wall heat flux can be used to determine the rate of heat transfer and resulting vaporization rate at the bubble interface. To evaluate the waiting time, we consider the temperature at the top of a hemispherical bubble of radius $r_c$ and postulate that the temperature at this location must exceed the equilibrium superheat for the bubble to grow. As a result, the following relation for the waiting time is obtained [157],

$$\Delta t_W = \frac{1}{4a_l} \left\{ \frac{r_c}{\text{erfc}^{-1}\left[\frac{T_{\text{sat}} - T_b}{T_w - T_b} + \frac{2\sigma T_{\text{sat}}(v_v - v_l)}{(T_w - T_b)i_{fg}r_c}\right]} \right\}^2. \tag{8.77}$$

### 8.3.5 BUBBLE DEPARTURE DIAMETER

The bubble departure diameter is an important nucleate boiling parameter since it is strongly related to the heat transfer characteristics. Together with the frequency of bubble formation and the nucleation site density it determines the nucleate boiling heat flux and the evaporation rate. Therefore, the bubble departure diameter has been intensively investigated, both experimentally and theoretically, and a variety of expressions and models have been proposed.

The earliest correlation has been proposed by Fritz in 1935, who utilized the theory of capillarity to get an equilibrium bubble shape and formulated a balance equation for the gravity and surface tension forces. A solution of the equation yielded the following bubble diameter at departure [72],

$$d_{BD} = 0.0208\theta \left[ \frac{\sigma}{g\left(\rho_l - \rho_v\right)} \right]^{1/2}, \tag{8.78}$$

where $\sigma$ is the surface tension and $\theta$ is the contact angle measured in degrees. The departure bubble diameter calculated from this equation agrees well with experimental data for water at atmospheric pressure. For water at pressures up to 14.2 MPa, the following correlation can be used [131],

$$d_{BD} = 2.5 \times 10^{-4}\theta \left( \frac{\rho_l - \rho_v}{\rho_l} \right)^{0.9} \left[ \frac{\sigma}{g\left(\rho_l - \rho_v\right)} \right]^{1/2}. \tag{8.79}$$

The effect of pressure and wall superheat $\Delta T_w = T_w - T_{\text{sat}}$ is taken into account in the following expression [41],

$$d_{BD} = \theta \left[ \frac{\sigma}{g\left(\rho_l - \rho_v\right)} \right]^{1/2} \frac{\rho_l c_{p,l} \Delta T_w}{\rho_v i_{fg}}. \tag{8.80}$$

As can be seen, this equation predicts that the bubble departure diameter is directly proportional to the wall superheat. However, such a trend is not confirmed by experimental data.

When the effect of only system pressure is considered, the bubble departure diameter can be found as [43]

$$d_{BD} = C \left[ \frac{\sigma}{g\left(\rho_l - \rho_v\right)} \right]^{1/2} \left( \frac{\rho_l c_{p,l} T_{\text{sat}}}{\rho_v i_{fg}} \right)^{5/4}, \tag{8.81}$$

where $C = 1.5 \times 10^{-4}$ for water and $4.65 \times 10^{-4}$ for fluids other than water.

### Influence of Bulk Flow

During forced convection nucleate boiling, when a bubble grows large enough at a nucleation cavity, it first departs from the cavity location, then slides along the heated surface, and finally lifts off at some distance downstream. Experiments indicate a systematic influence of the flow mass flux, bulk liquid subcooling, and applied wall heat flux on this process. In particular, the bubble departure size, the bubble lift-off size and the time scales of these processes are strongly dependent on the above-mentioned parameters.

Various approaches to determine the bubble departure size have been applied. A force-balance approach is based on consideration of the various forces acting on a bubble prior to and during the departure. Thus this approach is a direct extension of the method used for the calculation of the bubble departure size during pool boiling, where the balance between surface tension force, buoyancy force, and drag force are typically considered. During force convection boiling, additional forces are included such as, for example, the unsteady force due to bubble growth and the shear lift force.

The bubble energy balance approach constitutes another method to determine the bubble departure diameter. In this approach, a differential energy balance equation is formulated for a single bubble. The equation includes various evaporation and condensation processes contributing to the bubble size change. Initially the bubble is growing due to the prevailing evaporation rates from the microlayer and the superheated thermal boundary layer. With increasing bubble size, the bubble cap enters the subcooled liquid layers where vapor condensation takes place. At certain time instant the bubble gets its maximum size and it starts shrinking afterwords. This time instant is considered as the bubble departure time and the corresponding maximum bubble size is considered as the bubble departure size.

### Correlations

In its simplest form, the influence of bulk subcooling on the bubble departure diameter can be given as follows [136],

$$d_{BD} = 6 \times 10^{-4} e^{-\Delta T_{sub}/45}$$

where $\Delta T_{sub}$ is the liquid bulk subcooling in K and $d_{BD}$ is the bubble departure diameter in m.

### Energy Balance Approach

For subcooled nucleate flow boiling of water, bubble growth rate, average bubble diameter at departure, and average growth time can be calculated from semi-empirical correlations derived from a heat transfer controlled bubble model. Inertia-controlled bubble growth can be neglected since its duration is very short, as determined experimentally. Assuming that the microlayer evaporation is the main contribution to the bubble growth and that the bubble growth within the liquid superheated layer can be neglected since it is small, the energy balance for a bubble can be written as [225],

$$\frac{\pi}{6}\rho_v i_{fg} \frac{dd_B^3}{dt} = q''_{ML}\frac{\pi d_B^2}{4}\left(1 - \frac{d_{BDP}^2}{d_B^2}\right) - h_c\Delta T_{sub}\frac{\pi d_B^2}{2}, \tag{8.82}$$

with initial condition $d_B = 0$ at $t = 0$. Here $d_B$ is the bubble diameter, $d_{BDP}$–diameter of the dry patch beneath the bubble, $q''_{ML}$–heat flux in the microlayer region, $h_c$–heat transfer coefficient during condensation at the bubble cap, and $\Delta T_{sub}$–subcooling of the liquid. Applying closure relationships and using experimental data to determine correlation parameters, the following expression for the average bubble departure diameter has been obtained,

$$d_{BD} = \frac{2.42 \times 10^{-5} p^{0.709}}{\sqrt{b\Phi}} a, \tag{8.83}$$

where,

$$a = \frac{(q''_w - h_c \Delta T_{sub})^{1/3} \lambda_l}{2 C^{1/3} i_{fg} \rho_v \sqrt{\frac{\pi \lambda_l}{\rho_l c_{p,l}}}} \sqrt{\frac{\lambda_w \rho_w c_{p,w}}{\lambda_l \rho_l c_{p,l}}},$$

$$b = \frac{\Delta T_{sub}}{2 (1 - \rho_v/\rho_l)},$$

$$C = \frac{i_{fg} \mu_l \left[ \frac{c_{p,l}}{0.013 \, i_{fg} Pr^{1.7}} \right]^{1/3}}{\sqrt{\frac{\sigma}{(\rho_l - \rho_v) g}}},$$

$$\dot{\Phi} = \begin{cases} (U_l/0.61)^{0.47} & \text{for} \quad U_l > 0.61 \text{ m/s} \\ 1 & \text{for} \quad U_l \leq 0.61 \text{ m/s} \end{cases}.$$

The experimental range of the correlation is:

$$0.1 < p < 17.7 \text{ MPa}$$
$$0.47 < q''_w < 10.64 \text{ MW/m}^2$$
$$3.0 < \Delta T_{sub} < 86 \text{ K}$$
$$0.08 < d_{BD} < 1.24 \text{ mm}$$

## Force Balance Approach

It can be assumed that bubble departure and lift-off are governed by the gravity force, the adhesion force, and the hydrodynamic forces exerted by flowing liquid on the bubble surface. Several contributing mechanisms are taken into account such as the surface tension, the drag in the flow direction, the liquid-bubble interaction due to asymmetric growth of the bubble attached to the wall, and the gravity.

The wall tangential and perpendicular components of the surface tension force are given as [127],

$$F_s^{\parallel} = -1.25 d_{BB} \sigma \frac{\pi (\theta_A - \theta_R)}{\pi^2 - (\theta_A - \theta_R)^2} (\sin \theta_A + \sin \theta_R), \tag{8.84}$$

$$F_s^{\perp} = -d_{BB} \sigma \frac{\pi}{\theta_A - \theta_R} (\cos \theta_R - \cos \theta_A), \tag{8.85}$$

where $d_{BB}$ is the bubble base diameter, $\theta_A$ is the advancing contact angle and $\theta_R$ is the receding contact angle.

The steady drag force acting on a bubble attached to the wall surface can be estimated from an expression derived for the steady unbounded uniform flow over a spherical bubble given as [155],

$$F_D^{\parallel} = C_D 6 \pi \mu_l v_l^{\parallel} r_B, \tag{8.86}$$

where the drag force coefficient is given by the following correlation valid for $0 \leq Re \leq 1000$,

$$C_D = \frac{2}{3} + \left[ \left( \frac{12}{Re_B} \right)^{0.65} + 0.862 \right]^{-1.54}, \tag{8.87}$$

and the bubble Reynolds number is defined as

$$\text{Re}_B = \frac{\rho_l v_l^{\parallel} d_B}{\mu_l}, \quad d_B = 2r_B. \tag{8.88}$$

Here $v_l^{\parallel}$ is the liquid velocity component parallel to the wall surface at the bubble center location. It can be assumed that the time-averaged liquid velocity near the wall follows the turbulent single-phase logarithmic law of the wall.

During the bubble growth process, the bubble is distorted and inclined in the flow direction by an angle $\theta_I$ measured from the direction normal to the wall surface. As a result, a growth force, also called unsteady drag force is created with the following tangential and normal components [127],

$$F_{du}^{\parallel} = -\rho_l \pi r_B^2 \left( \ddot{r}_B r_B + \frac{3}{2} \dot{r}_B^2 \right) \sin \theta_I, \tag{8.89}$$

$$F_{du}^{\perp} = -\rho_l \pi r_B^2 \left( \ddot{r}_B r_B + \frac{3}{2} \dot{r}_B^2 \right) \cos \theta_I, \tag{8.90}$$

where $r_B$ is the time-dependent radius of the growing bubble. The parallel growth force component $F_{du}^{\parallel}$ acts in the direction opposite to the flow, whereas the normal growth force component $F_{du}^{\perp}$ acts towards the wall, preventing the bubble from departing and lifting from the nucleation cavity. To calculate the growth force it is necessary to know the bubble growth rate and the bubble inclination angle $\theta_I$. The bubble growth rate can be estimated from expressions given in the present section. The bubble inclination angle is in general not known and it has to be determined from experimental data.

Shear lift force is given as,

$$F_L^{\perp} = C_L \frac{\rho_l \left( v_l^{\parallel} \right)^2}{2} \pi r_B^2. \tag{8.91}$$

Here $v_l^{\parallel}$ is the liquid velocity component parallel to the wall surface and $C_L$ is the lift force coefficient. The liquid velocity can be found from a law of the wall for single-phase flow, using bubble radius $r_B$ as the wall distance. The lift force coefficient for a bubble attached to the wall is not known and has to be estimated from expressions developed for shear lift force on a spherical bubble in an unbounded flow field. Over a large range of Reynolds numbers, this coefficient can be estimated from the following correlation [127, 156],

$$C_L = 3.877 G_s^{1/2} \left( \text{Re}_B^{-2} + 0.014 G_s^2 \right)^{1/4}, \tag{8.92}$$

where

$$G_s = \left| \frac{dv_l^{\parallel}}{dn} \right| \frac{r_B}{v_l^{\parallel}} \tag{8.93}$$

is the dimensionless shear rate of the oncoming flow, $n$ is the direction normal to the wall surface and the bubble Reynolds number $Re_B$ is given by Eq. (8.88).

The buoyancy force acting on a bubble attached to a wall with an inclination angle $\varphi$ has both the normal and parallel components as follows,

$$F_b^{\parallel} = V_B \left(\rho_l - \rho_v\right) g \sin\varphi, \quad F_b^{\perp} = V_B \left(\rho_l - \rho_v\right) g \cos\varphi, \tag{8.94}$$

where $V_B$ is the bubble volume.

The hydrodynamic force is caused by the dynamic pressure due to the liquid flowing around a stationary bubble and can be estimated by considering an inviscid flow over a sphere in an unbounded flow field. This force is acting in the direction normal to the wall and can be estimated as [127],

$$F_h^{\perp} = \frac{9}{8} \rho_l \left(v_l^{\parallel}\right)^2 \frac{\pi d_{BB}^2}{4}. \tag{8.95}$$

The additional force that is considered is due to the pressure difference inside and outside of the bubble at a reference point over the contact area. This pressure difference is determined by the Young-Laplace equation, Eq. (8.27), and thus depends on the local radius of curvature of the bubble. Assuming that this radius is approximately equal to $5r_B$, the contact pressure force can be estimated as [127],

$$F_{cp}^{\perp} = \frac{2\sigma}{5r_B} \frac{\pi d_{BB}^2}{4}. \tag{8.96}$$

The total normal force and total tangential force acting on a bubble attached to the nucleation cavity determine conditions for the bubble's departure and lift off from the wall. In particular, when the total tangential force exceeds zero, the bubble will depart from the nucleation cavity and slide along the wall in the flow direction. Similarly, when the total force in the normal direction exceeds zero, the bubble will lift off from the wall surface. However, should the total normal force exceed zero once the tangential force is still zero or negative, the bubble will lift off directly from the nucleation cavity without first sliding.

The force balance approach requires closure relationships for the advancing, receding, and inclination angles. These angles are not known in general and need to be determined experimentally. In addition, these angles are continuously changing from the point of inception until the point of bubble departure. Since this evolution process is not well known, the angles are taken to be constant and equal to their mean values at the point of departure. As a result, the force balance model will predict that initially the total normal and tangential forces are negative until the point of departure and lift-off.

Another important and uncertain parameter that needs closure is the bubble base diameter. It is assumed that detaching bubbles have a constant and non-zero base diameter. In reality, this diameter varies during the departure process and can approach zero due to the necking phenomenon. Since the bubble base diameter is rather difficult to measure and its evolution during the bubble departure is still not well known, a constant value or a constant fraction of the bubble departure diameter is assumed.

## 8.4 HEAT TRANSFER DURING NUCLEATE BOILING

In this section, we delve into the different modes of heat transfer that occur during nucleate boiling. Specifically, we provide a detailed discussion on topics like the partitioning of heat flux and the different mechanisms of heat transfer during the ebullition cycle.

### 8.4.1 HEAT FLUX PARTITIONING

To analyze the boiling process, it is convenient to introduce a *boiling unit area* $A_{bu}$ and *boiling unit time interval* $\Delta t_{bu}$. For nucleate boiling with known active nucleation site density $N_a''$ and bubble departure frequency $f_{BD}$, these quantities can be approximated as,

$$A_{bu} = \frac{1}{N_a''}, \tag{8.97}$$

$$\Delta t_{bu} = \frac{1}{f_{BD}}. \tag{8.98}$$

We assume next that the time and space averaged parameters over the boiling unit area and time represent the boiling parameters in the macroscale. In particular, if $Q_{bu}$ is the heat transferred over the boiling unit area and during unit time intervals, the macroscale wall heat flux $q_w''$ can be found as,

$$q_w''(\mathbf{r}_w, t) = \frac{Q_{bu}}{A_{bu}\Delta t_{bu}} = \frac{1}{A_{bu}\Delta t_{bu}} \int_{t-\Delta t_{bu}}^{t} \int_{A_{bu}} q_{inst}''(\mathbf{r}_w + \mathbf{s}, t')dA(\mathbf{s})dt'. \tag{8.99}$$

Here $\mathbf{r}_w$ is the location of the boiling unit area centroid on the boiling surface and $\mathbf{s}$ is the location of the differential area $dA$ within the boiling unit area.

The instantaneous wall heat flux $q_{inst}''$ has a complex spatiotemporal microstructure resulting from the various contributing heat transfer mechanisms, such as convection, micro-convection, microlayer evaporation, and dry spot rewetting. Usually these mechanisms prevail over specific fractions of the boiling unit area and time. These fractions can overlap either in time or space. For example, microlayer evaporation and dry spot rewetting prevail over the same (more or less) fraction of a boiling surface, but these processes occur during different time instants. Thus, for heat transfer mechanism $i$ occurring over an area $A_{bu,i}(t)$ during a time interval $\Delta t_{bu,i}$, where a prevailing local instantaneous heat flux is $q_{inst,i}''$, the transferred heat fraction can be found as,

$$Q_{bu,i} = \int_{\Delta t_{bu,i}} \int_{A_{bu,i}(t)} q_{inst,i}'' dA dt. \tag{8.100}$$

It should be noted that in general $A_{bu,i}$ changes with time. For example, the microlayer evaporation area gradually increases when the bubble is growing. Since the total heat transferred over the boiling unit area and time is a sum of all participating mechanisms, the macroscopic wall heat flux can be obtained as,

$$q_w'' = \frac{1}{A_{bu}\Delta t_{bu}} \sum_i Q_{bu,i} = \frac{1}{A_{bu}\Delta t_{bu}} \sum_i \int_{\Delta t_{bu,i}} \int_{A_{bu,i}(t)} q_{inst,i}'' dA dt, \tag{8.101}$$

where summation includes all contributing boiling heat transfer mechanisms. We now introduce an instantaneous area-averaged heat flux,

$$\langle q''_{inst,i}\rangle_2 \equiv \frac{1}{A_{bu,i}(t)} \int_{A_{bu,i}(t)} q''_{inst,i}\,\mathrm{d}A, \tag{8.102}$$

to express the macroscopic wall heat flux as follows,

$$q''_w = \frac{1}{A_{bu}\Delta t_{bu}} \sum_i \int_{\Delta t_{bu,i}} A_{bu,i}(t)\langle q''_{inst,i}\rangle_2\,\mathrm{d}t. \tag{8.103}$$

Let us now introduce the following time average area-weighted heat flux,

$$\overline{\langle q''_{inst,i}\rangle_2}^A \equiv \frac{\int_{\Delta t_{bu,i}} A_{bu,i}(t)\langle q''_{inst,i}\rangle_2\,\mathrm{d}t}{\int_{\Delta t_{bu,i}} A_{bu,i}(t)\,\mathrm{d}t}$$
$$= \frac{1}{\overline{A_{bu,i}(t)}\Delta t_{bu,i}} \int_{\Delta t_{bu,i}} A_{bu,i}(t)\langle q''_{inst,i}\rangle_2\,\mathrm{d}t \tag{8.104}$$

Here

$$\overline{A_{bu,i}(t)} \equiv \frac{1}{\Delta t_{bu,i}} \int_{\Delta t_{bu,i}} A_{bu,i}(t)\,\mathrm{d}t \tag{8.105}$$

is the time average area on which mechanism $i$ prevails during the boiling unit time. Using these definitions, the macroscopic wall heat flux becomes,

$$q''_w = \frac{\sum_i \overline{A_{bu,i}(t)}\Delta t_{bu,i}\overline{\langle q''_{inst,i}\rangle_2}^A}{A_{bu}\Delta t_{bu}} = \sum_i \overline{\varepsilon_{bu,i}}\tau_{bu,i}\overline{\langle q''_{inst,i}\rangle_2}^A. \tag{8.106}$$

The derived equation represents the principle of the wall heat flux partitioning based on the time average area fraction,

$$\overline{\varepsilon_{bu,i}} \equiv \frac{\overline{A_{bu,i}(t)}}{A_{bu}}, \tag{8.107}$$

and the time fraction for mechanism $i$

$$\tau_{bu,i} \equiv \frac{\Delta t_{bu,i}}{\Delta t_{bu}}. \tag{8.108}$$

The sum of products of area and time fractions has to satisfy a normalization condition. Assuming that all considered boiling mechanisms have a constant and same heat flux equal to the macroscopic wall heat flux, the following condition is obtained,

$$\sum_i \overline{\varepsilon_{bu,i}}\tau_{bu,i} = 1. \tag{8.109}$$

Thus, in summary, the task of heat flux partitioning during boiling includes the following steps:

1. Identification of contributing boiling heat transfer mechanisms during a boiling unit time interval $\Delta t_{bu}$ and over a boiling unit area $A_{bu}$.
2. Determination of the time average area fraction $\overline{\varepsilon_{bu,i}}$ for each of the mechanisms.
3. Determination of the time fraction of the boiling unit time $\tau_{bu,i}$ during which each of the mechanisms prevails.
4. Determination of the time average area-weighted instantaneous heat flux $\overline{\langle q''_{inst,i}\rangle_2}^A$ pertinent to each of the boiling heat transfer mechanisms.

Two general types of heat flux partitioning are commonly applied. A *complete heat flux partitioning* takes into account all heat transfer mechanisms contributing to the generation of the vapor phase and the increase of the liquid phase enthalpy. In this case mechanisms such as microlayer evaporation, contact line evaporation, rewetting, convection, and transient conduction are considered. In the *incomplete heat flux partitioning* the evaporation heat flux is found from the energy balance based on the departing bubble size, rather than from a consideration of the various contributing evaporation mechanisms.

## 8.4.2 HEAT TRANSFER MECHANISMS

At low heat fluxes, isolated bubbles are formed at active nucleation sites. Such bubbles do not interact with each other since the distance that separates them is large enough to exclude any mutual disturbances. As the wall heat flux increases, the number of active nucleation sites increases and neighboring isolated bubbles are close enough to interact. For both isolated and coalescing bubble regimes, the wall heat transfer can be partitioned into several heat transfer mechanisms. These mechanisms include evaporation of microlayer beneath a growing bubble, contact line heat transfer, rewetting heat transfer, post-rewetting transient conduction, and convection heat transfer. The relative importance of the various heat transfer mechanisms is still a subject of intensive experimental and theoretical investigations.

Experiments indicate that the isolated bubble growth is caused by the evaporation of a liquid microlayer beneath the bubble or evaporation in the vicinity of the contact line. The third contributing mechanism is the evaporation of superheated liquid in the thermal boundary layer. The microlayer model is based on the assumption that the bubble growth is so fast that a thin liquid film is trapped between the bubble and the heated wall. The thickness of the liquid film ranges from a few to several tens of $\mu$m and decreases with the square root of time due to evaporation. The evaporation of the film is caused by heat conduction through the microlayer from the heated wall surface to the bubble-microlayer interface. Generally, it is understood that the evaporation of the microlayer contributes significantly to the total evaporation rate linked with the ebullition cycle.

At certain conditions no microlayer is deposited on the heated surface during bubble growth and evaporation in the vicinity of the three-phase contact line becomes a dominating heat transfer mechanism. A thin adsorbed film with a thickness not greater than a few nm covers the heater surface beneath the bubble. Due to significant intermolecular forces, this film is not evaporating. However, in the transition between the adsorbed film and bulk liquid, the liquid film is very thin and poses a rather small heat transfer resistance. As a result, high evaporation rates can be expected in this region, significantly contributing to the overall heat transfer and bubble growth [67].

## Microlayer Evaporation

When the vapor bubble grows fast enough, a thin liquid microlayer is created beneath the bubble and the heated surface. The thickness of the microlayer changes with the distance from the nucleation site and with time. Its maximum thickness can be calculated from the following correlation [47],

$$\delta_{ML,max} = 0.8\sqrt{v_l \Delta t_G}, \tag{8.110}$$

where $\Delta t_G$ is the bubble growth time. Initial microlayer thickness for a bubble whose interface moves with speed $U_i$ can be found from the following expression [67],

$$\delta_{ML,0} = \left(\frac{v_l U_i}{g}\right)^{1/2} f(\text{Ca}), \tag{8.111}$$

where

$$f(\text{Ca}) \approx \begin{cases} 0.93\,\text{Ca}^{1/6} & \text{for Ca} \ll 1 \\ 1 & \text{for Ca} \gg 1 \end{cases} \tag{8.112}$$

and $\text{Ca} = v_l \rho_l U_i / \sigma$ is the capillary number. Assuming that the heat is conducted through the microlayer in the direction normal to the surface, the local instantaneous wall heat flux can be calculated as,

$$q''_{inst,ML} = \frac{\lambda_l (T_w - T_{sat})}{\delta_{ML}}, \tag{8.113}$$

where $\delta_{ML}$ is the local thickness of the microlayer.

The initial radial distribution of the microlayer thickness has been measured experimentally using laser interferometry. The measured data suggest the following expression [215],

$$\delta_{ML,0} = ar^b, \tag{8.114}$$

where $a$ and $b$ are parameters that vary with fluid and heater properties, $r$ is the radial distance from the nucleation site, and $\delta_{ML,0}$ is the initial thickness of the microlayer. The measured thickness has been compared with a simple microlayer evaporation model in which it is assumed that the entire microlayer is depleted due to evaporation only and no advection of liquid takes place. Under such circumstances, at each radial location $r$ the following is valid,

$$\delta_{ML,evap}(r) = \int_0^{t_{evap}(r)} \frac{q''_{inst,ML}(r,t)}{(i_{fg} - c_{p,l}\Delta T_{ML})\rho_l}\,dt. \tag{8.115}$$

Here it is assumed that liquid in the microlayer is superheated and its temperature is $T = T_{sat} + \Delta T_{ML}$.

The relative importance of the microlayer evaporation to the bubble growth can be investigated experimentally by comparing the bubble volume increase with the volume of vapor generated due to microlayer evaporation. Experiments indicate that the microlayer evaporation contributes between 20 and 50 percent to the overall bubble growth. The rest of the bubble volume is due to the evaporation of superheated liquid in the thermal boundary layer. This rather wide variation of experimental results is attributed to differences in the thermal conductivity of substrates used in experiments. When the thermal conductivity of the substrate is high, heat is effectively conducted into the microlayer region facilitating evaporation. On the contrary, for low-conductivity substrates, the evaporation of the microlayer is limited by higher thermal resistance and lower heat flow to the microlayer region [215].

## Contact Line Evaporation

If the bubble growth is sufficiently slow, a microlayer underneath the growing bubble is not created, leading to the so-called *contact line evaporation* regime. In this regime, evaporation takes place in the vicinity of the three-phase contact line (see Fig. 8.3). In the modeling of contact line evaporation, the influence of attractive intermolecular forces between the wall molecules and the molecules at the liquid-vapor interface is taken into account. These intermolecular forces pose an additional resistance against evaporation that can be expressed as a shift of the phase equilibrium to higher interfacial equilibrium temperatures. As a result, an apparent dry region beneath the bubble is covered by a thin adsorbed film with a thickness in the order of a few nm. The film in this region, referred to as the adsorbed film region, can not be evaporated due to intermolecular forces [67].

Overall heat transfer is significantly reduced when the microlayer is absent and the contact line evaporation dominates in the nucleate-boiling process. Due to that, it is important to determine under which conditions a transition from the microlayer to the contact line regime occurs. Theoretical and experimental investigations suggest that parameters such as the Jakob number based on the wall superheat, the contact angle, the liquid viscosity, and the surface tension play a dominant role in the transition. However, it is still not clear what is the influence of these parameters. In particular, in some sources, it is claimed that the microlayer regime dominates for low Jakob number values [67, 125], whereas opposite conclusions are drawn from direct numerical simulations of nucleate boiling in the microlayer regime [226]. According to the latter, the microlayer regime will dominate when the following criterion is satisfied,

$$\text{Ja} > 3.23 \cdot 10^{-8} \times \frac{(\theta - 5)^3}{\text{Ca}}. \tag{8.116}$$

This regime will exist always when $\text{Ja} > 70$. Here Ja and Ca are the Jakob number and the capillary number, respectively, defined as,

$$\text{Ja} = \frac{\rho_l c_{p,l}(T_w - T_{\text{sat}})}{\rho_v i_{fg}}, \tag{8.117}$$

$$\text{Ca} = \frac{\mu_l u_{BG}}{\sigma}, \tag{8.118}$$

where $u_{BG}$ is the bubble growth speed. The criterion is valid for the contact angle in the following range $15° < \theta < 90°$ and for $2.2 \cdot 10^{-4} < CaJa < 1.2 \cdot 10^{-2}$. In terms of the bubble growth speed, the criterion can be expressed as,

$$u_{BG} > 1.81 \cdot 10^{-4} \times (\theta - 5)^{3/2} \left( \frac{a_l \sigma}{\delta_T \mu_l} \right)^{1/2}, \tag{8.119}$$

where $\delta_T$ is the thickness of the thermal boundary layer in which the liquid is superheated.

## Rewetting

Rewetting takes place when liquid enters the dry spot area beneath the departing bubble. Rewetting front moves from outer boundaries of the dry region towards its center. The rewetting heat flux is significantly lower than the heat flux during the microlayer evaporation and has a peak value in the vicinity of the rewetting front. Since the area on which the rewetting heat transfer takes place is quite small, the contribution of this mechanism to overall heat transfer is rather insignificant.

The instantaneous heat flux during rewetting can be determined using the transient conduction theory, which is based on the model of heat transfer to a semi-infinite body [157],

$$q_w''(t) = \frac{\lambda_l (T_w - T_b)}{\sqrt{\pi a_l t}}. \tag{8.120}$$

In this equation it is assumed that $T_w$ and $T_b$ represent constant temperatures for the wall and bulk liquid, respectively, and $t$ is the time after the bulk liquid covers the entire wall surface. Experiments show that Eq. (8.120) gives too high heat flux values compared to the measured data. The main reason for the discrepancy is the finite speed with which the rewetting takes place, whereas Eq. (8.120) is valid for an immediate rewetting of the entire dry region. In addition, the mean initial temperature of the liquid in the wall vicinity is not equal to $T_b$ but it is somewhat higher and in a range between the bulk temperature and the wall temperature. Finally, the wall surface temperature is not constant during rewetting due to transient conduction in the solid body. Taking these effects into account, the instantaneous heat flux during rewetting can be estimated as,

$$q_{w,eff}''(t) = \frac{\lambda_l (T_{w,eff} - T_{l,eff})}{\sqrt{\pi a_l t}} = \frac{\lambda_l \Delta T_{w,eff}}{\sqrt{\pi a_l t}}, \tag{8.121}$$

where $T_{l,eff}$ is the effective liquid temperature in the vicinity of the rewetted wall surface, $T_{w,eff}$ is the effective wall surface temperature and $\Delta T_{w,eff}$ is the corresponding effective temperature difference. Using a model of two connected semi-infinite bodies, the wall surface effective temperature can be estimated as,

$$T_{w,eff} = \frac{T_w \sqrt{\lambda_s c_{p,s} \rho_s} + T_b \sqrt{\lambda_l c_{p,l} \rho_l}}{\sqrt{\lambda_s c_{p,s} \rho_s} + \sqrt{\lambda_l c_{p,l} \rho_l}}, \tag{8.122}$$

where index $s$ refers to the solid wall material and $T_w$, $T_b$ are constant wall and liquid bulk temperatures before rewetting, respectively.

Assuming that the instantaneous rewetting area is that area of the initially dry region, which is covered with liquid, the time-averaged rewetting area can be found as,

$$\overline{A_{bu,R}} = \frac{1}{\Delta t_{bu,R}} \int_0^{\Delta t_{bu,R}} \pi \left[ r_{DR}^2 - r_R^2(t) \right] dt = \frac{2}{3} \pi r_{DR}^2. \tag{8.123}$$

Here $r_{DR}$ is the radius of the dry region, $r_R(t) = r_{DR} - u_R \cdot t$ is the radius of the rewetting front at time $t$, $u_R$ is the rewetting front speed and $\Delta t_{bu,R} = r_{DR}/u_R$ is the rewetting time, that is, the time when the rewetting front reaches the dry region center at $r = 0$.

The rewetting heat flux varies both with the radial position and time. Assuming a model based on Eq. (8.121) and applying coordinates that are moving with the rewetting front speed, the instantaneous heat flux distribution in the rewetting region is given as,

$$q_{inst,R}''(r,t) = \frac{\lambda_l \Delta T_{w,eff}}{\sqrt{\pi a_l \left( t - \frac{r_{DR}-r}{u_R} \right)}}, \quad \text{for} \quad r_{DR} - u_R \cdot t < r < r_{DR}. \tag{8.124}$$

The instantaneous area-averaged rewetting heat flux is found in Eq. (8.102) as follows,

$$\begin{aligned}
\langle q_{inst,R}'' \rangle 2 &= \frac{2}{r_{DR}^2 - (r_{DR} - u_R \cdot t)^2} \int_{r_{DR}-u_R \cdot t}^{r_{DR}} r q_{inst,R}''(r,t) dr \\
&= \frac{4}{3} \frac{\lambda_l \Delta T_{w,eff}}{\sqrt{\pi a_l t}} \frac{2u_R \cdot t - 3r_{DR}}{u_R \cdot t - 2r_{DR}}
\end{aligned} \tag{8.125}$$

As can be seen, the area-averaged rewetting heat flux decreases with time and at the end of the rewetting process, for $t = \Delta t_{bu,R}$ it reaches the minimum value given as,

$$\langle q_{inst,Rmin}'' \rangle 2 = \frac{4}{3} \frac{\lambda_l \Delta T_{w,eff}}{\sqrt{\pi a_l \Delta t_{bu,R}}}. \tag{8.126}$$

The time average area-weighted heat flux during rewetting is given by Eq. (8.104) and can be calculated from the following expression,

$$\overline{\langle q_{inst,R}'' \rangle 2}^A = \frac{1}{A_{bu,R}(t)\Delta t_{bu,R}} \int_{\Delta t_{bu,R}} A_{bu,R}(t) \langle q_{inst,R}'' \rangle 2 dt. \tag{8.127}$$

The remaining characteristic parameters for the rewetting process include the time-averaged area fraction,

$$\overline{\varepsilon_{bu,R}} = \frac{\overline{A_{bu,R}}}{A_{bu}} = \frac{2}{3} \pi r_{DR}^2 N_a'', \tag{8.128}$$

and the rewetting time fraction,

$$\tau_{bu,R} = \frac{\Delta_{bu,R}}{\Delta_{bu}} = \frac{r_{DR}}{u_R} f_{BD}. \tag{8.129}$$

## Transient Heat Conduction

After rewetting of the dry region, the liquid temperature in the boundary layer is still lower than during the bubble growth, since the warmer liquid that is lost during bubble growth and departure is replaced by colder liquid from the bulk. In general there are three mechanisms of the warm liquid loss from the thermal boundary layer: the evaporation of liquid in the microlayer and in the thermal boundary layer, the liquid displacement from the thermal boundary layer due to the bubble growth, and the liquid entrainment from the thermal boundary layer by the departing and coalescing bubbles. All these mechanisms combined cause a significant temperature reduction in the thermal boundary layer in a certain region around the bubble and promote transient heat conduction from the wall surface to the liquid.

The instantaneous heat flux during transient conduction can be calculated from Eq. (8.120). However, experimental data indicate that the measured heat flux is lower than this equation would suggest, thus, the following approximation could be used,

$$q''_{inst,TC}(t) = \frac{\lambda_l C (T_w - T_b)}{\sqrt{\pi a_l t}}, \tag{8.130}$$

where $C < 1$ is a modeling parameter derived from experimental data. For nucleate boiling of FC-72, this parameter was found to be in a range from 0.25 to 0.4 [162]. For nucleate boiling of water, the parameter was greater than 0.5 but still sufficiently smaller than unity [215].

Assuming that the transient conduction heat transfer takes place over a certain constant area $A_{bu,TC}$, the time average area-weighted heat flux during that process can be calculated as,

$$\overline{\langle q''_{inst,TC} \rangle 2}^A = \frac{2 \lambda_l C (T_w - T_b)}{\sqrt{\pi a_l \Delta t_{TC}}}, \tag{8.131}$$

with the time fraction given as,

$$\tau_{bu,TC} = \Delta t_{TC} f_{BD}. \tag{8.132}$$

In these expressions it is assumed that the transient heat conduction prevails during the time $\Delta t_{TC}$.

The time-average area for transient heat conduction is usually specified as being proportional to the heater area covered by hemispherical bubbles with a diameter equal to the departure diameter, multiplied by an *influence factor K*,

$$A_{bu,TC} = K \pi r_{BD}^2, \tag{8.133}$$

where $r_{BD}$ is the bubble departure radius. The value of $K$ varies significantly in the literature from 4 [136] to 5–8 [51]. Recent experimental data indicate that the influence factor can be as low as 0.5 [3]. Using the area for the transient conduction given by Eq. (8.133), the time-averaged area fraction for transient conduction becomes,

$$\overline{\varepsilon_{bu,TC}} = K \pi r_{BD}^2 N_a'', \tag{8.134}$$

where $N_a''$ is the active nucleation site density. For high heat flux conditions, a correction method for the overlapping area of influence should be applied [3].

## Convective Heat Transfer

The convective heat transfer occurs in the region that is not influenced by bubble growth and departure. For pool boiling, the instantaneous heat flux can be calculated from relationships that are developed for specific conditions, such as natural convection from a horizontal or vertical plate. Some examples of such relationships are provided in §7.1.

When a wall-resolved approach is used in the calculation of the convective heat flux, a pure heat transfer by conduction in the wall vicinity can be assumed as follows,

$$q''_{inst,NC} = \rho_l a_l c_{p,l} \frac{dT}{dn},\qquad(8.135)$$

where $NC$ stands for natural convection, $a_l$ is the thermal diffusivity of liquid and $dT/dn$ is the temperature gradient in the direction normal to the surface.

Since the convective heat transfer is not influenced by bubble growth and departure, the time fraction of that process is equal to unity, that is,

$$\tau_{bu,NC} = 1.\qquad(8.136)$$

The time averaged area for convective heat transfer can be found from Eq. (8.137) as follows,

$$\overline{\varepsilon_{bu,NC}} = 1 - \sum_{i \neq NC} \overline{\varepsilon_{bu,i}} \tau_{bu,i}.\qquad(8.137)$$

## Transient Micro-Convection

The transient micro-convection is caused by liquid motion induced by the bubble growth, coalescence, and departure. The enhancement of convective heat transfer is rather limited during the motion of isolated bubbles since they do not significantly agitate the surrounding liquid. However, the interaction between coalescing bubbles appears to greatly enhance convective heat transfer, since the bubbles are promoting warm liquid outflow and cold liquid inflow in the narrow space between them [215].

## 8.5  POOL BOILING

Pool boiling can occur in LWRs under various accidental scenarios. A *small-break LOCA* in PWRs involves the depressurization of an originally subcooled system, while heat from the fuel rods is still transferred to the reactor coolant. After the reactor coolant pumps stop operating, coolant flow is determined by various natural circulation mechanisms. For very small break areas (less than approximately $4.6 \cdot 10^{-3}$ m$^2$) single-phase natural circulation is the dominant mechanism causing water to circulate through the core and the steam generator tubes. For larger break areas the flow pattern will change from single-phase natural circulation to circulation caused by boiling in the core and condensation in the steam generator tubes (*reflux boiling*).

Under such conditions, it is reasonable to assume, as a conservative measure, a pool boiling heat transfer mode in the core. Early in the small break LOCA depressurization, forced convection heat transfer prevails at all core elevations. After forced circulation is interrupted due to the coastdown of the reactor coolant pumps, heat transfer below the two-phase mixture surface is predominantly by nucleate boiling with some transition boiling or even stable film boiling near the surface.

A *reflood* phase of LOCA is associated with the emergency cooling in the LWR core when fuel rods are quenched during refilling water in a reactor vessel. The zircaloy or chromium-coated (ATF) fuel behavior under quenching is examined under transient pool boiling heat transfer conditions. Since a portion of the fuel bundle is uncovered, a quenching front is established with a liquid region below the front and a two-phase mixture ahead of it. For predicting quench front movement, precursory cooling due to boiling is important. Various pool boiling heat transfer regimes can be considered for that purpose.

One of the most important safety concerns when designing a LWR core is preventing or eliminating a boiling crisis during a *reactivity-initiated accident* (RIA). Among the various scenarios considered for RIA, the most penalizing in a PWR consists of a control rod ejection at hot standby condition and a control rod drop accident at cold zero power conditions in a BWR. Thus RIA is associated with an abrupt and unexpected insertion of reactivity that results in an unwanted surge in reactor power during which fuel and core structural components can experience rapid temperature increase. Fuel performance analysis under RIA conditions shows that CHF is one of the top contributors to uncertainties out of many crucial fuel-to-coolant heat transfer parameters. During RIA a CHF is induced when fuel assembly with stagnant coolant is submitted to a significant power pulse. The influence of such parameters as the degree of subcooling, pressure, surface characteristics (contact angle, oxide layer formation, surface activation), and rapid heating effects in pool boiling have to be considered.

### 8.5.1   NUCLEATE POOL BOILING

Nucleate pool boiling is a very efficient heat removal process in which vapor bubbles form at nucleation sites on a surface to be cooled and heat is transferred from the surface to quiescent liquid. The bubble nucleation and growth and involved heat transfer micro-processes have been discussed in §8.3 and §8.4, respectively. In this section, we present useful methods to predict a heat transfer coefficient during nucleate pool boiling.

Many correlations of heat transfer coefficient in saturated nucleate pool boiling were originally derived in a form similar to the forced convective heat transfer such as $Nu = c \times Re^{n_1} \times Pr^{n_2} \times ....$ The dimensionless groups Nu, Re, and others were defined to reflect the important phenomena underlying boiling. Experimental data show that the main parameters influencing nucleate pool boiling heat transfer include thermophysical properties of the surface material, the interaction between the solid, liquid, and vapor phases at the interface, and the surface microgeometry.

One of the first correlations for the nucleate pool boiling, which is applicable to a wide spectrum of liquid-solid pairs, was proposed by Rohsenow and is given as follows [188],

$$\frac{c_{p,l}\Delta T_w}{i_{fg}} = C_{sf}\left[\frac{q_w''}{\mu_l i_{fg}}\sqrt{\frac{\sigma}{g(\rho_l - \rho_v)}}\right]^m\left(\frac{c_{p,l}\mu_l}{\lambda_l}\right)^n, \tag{8.138}$$

where $C_{sf}$ is a constant that depends on the solid-fluid combination. This correlation with $C_{sf} = 0.013$, $m = 0.33$, and $n = 1.7$ was shown to correlate well for water with nickel and stainless steel in a pressure range of 0.1 to 16.8 MPa. The values of correlation parameters and applicability ranges for other solid-fluid combinations are provided in [177].

For solid-fluid pairs with known surface roughness $R_a$ (in $\mu$m) and contact angle $\theta$, and for which the $C_{sf}$ constant is not known, the following revised form of Eq. (8.138) can be used to estimate heat transfer rates [148],

$$q_w'' = 5.185 \cdot 10^5 C_s \frac{\lambda_l^{3.03}}{(i_{fg}\mu_l)^{2.03}}\sqrt{\frac{g(\rho_l - \rho_v)}{\sigma}}\Delta T_w^{3.03}, \tag{8.139}$$

where,

$$C_s = (1 - \cos\theta)^{0.5}\left[1 + \frac{5.45}{(R_a - 3.5)^2 + 2.61}\right]\gamma^{-0.04}, \quad \theta = \max(\theta, 15°), \tag{8.140}$$

and $\gamma$ is the influence parameter of heating surface material given as,

$$\gamma = \sqrt{\frac{\lambda_s\rho_s c_{p,s}}{\lambda_l\rho_l c_{p,l}}}. \tag{8.141}$$

The SI units should be used in the correlation (except for $R_a$). The correlation validity range based on the used experimental data is limited to pressure from 4 to 200 kPa, contact angle from 1 to 90°, surface roughness from 0.004 to 2.22 $\mu$m, and the wall superheat from 6 to 60 K.

The heat transfer coefficient for various working fluids with molecular weight $M$ and the maximum peak height of surface roughness $R_p$ can be found as [46],

$$h = \frac{55 p_R^{(0.12 - 0.2\log R_p)}(q_w'')^{0.67}}{[\log(1/p_R)]^{0.55} M^{0.5}}. \tag{8.142}$$

Here the heat transfer coefficient $h$ is in W/(m$^2$K), heat flux $q_w''$ in W/m$^2$, and peak roughness $R_p$ in $\mu$m.

Forster and Zuber proposed the following correlation for the heat flux in terms of the wall superheat and physical properties of liquid and vapor [69]

$$q_w'' = 0.00122\left(\frac{\lambda_l^{0.79}c_{p,l}^{0.45}\rho_l^{0.49}}{\sigma^{0.5}\mu_l^{0.29}i_{fg}^{0.24}\rho_v^{0.24}}\right)\Delta T_w^{1.24}[p_{sat}(T_w) - p_l], \tag{8.143}$$

where $\Delta T_w = T_w - T_{sat}$ is the wall superheat in K, $\lambda_l$ - liquid thermal conductivity in kW/(m K), $c_{p,l}$ - liquid specific heat in kJ/(kg K), $\rho_l$, $\rho_v$ - liquid and vapor density in kg/m$^3$, respectively, $p_{sat}(T_w)$, $p_l$ - saturation pressure at wall temperature and liquid pressure in Pa, respectively, $\sigma$ - surface tension in N/m, $\mu_l$ - liquid dynamic viscosity in Ns/m$^2$, $i_{fg}$ - latent heat in kJ/kg with the resulting value of wall heat flux $q''_w$ in kW/m$^2$.

### Effect of Subcooling

In subcooled pool boiling the bulk liquid temperature far from the boiling surface is below the saturation temperature. To make the bubble nucleation possible, a layer of superheated liquid must be present in the vicinity of the boiling surface. When the condition for the onset of nucleate boiling is satisfied, bubbles start growing from nucleation sites. With increasing heat flux, bubbles coalesce to a vapor layer located between the subcooled bulk liquid and the superheated liquid layer attached to the heated wall. Experiments show that void fraction increases rapidly from zero to unity when moving from the liquid layer on the heated surface to the vapor layers hovering above the surface. Similarly, as in saturated pool boiling, large vapor masses are formed and detached from the heated surface. Despite the similarities in boiling characteristics in saturated and subcooled boiling, with increasing subcooling the heat transfer coefficient decreases and the vapor masses detaching from the wall collapse at higher rates. With high enough subcooling and a heat flux significantly below the critical heat flux, the boiling heat transfer will eventually cease and natural convection heat transfer will prevail.

### Effect of Surface Orientation

In a study of pool boiling of water at atmospheric pressure performed by Nishikawa et al. [168] it was shown that, above a certain threshold heat flux level, the pool boiling curves for all surface orientations are virtually identical. It appears that this threshold heat flux value corresponds to the transition from the isolated bubble regime to the regime of slugs and columns when the heat flux reaches a value given by the following correlation [163]

$$q''_{w,tr} = 0.11\rho_v i_{fg}\theta^{1/2}\left(\frac{\sigma g}{\rho_l - \rho_v}\right)^{1/4}, \tag{8.144}$$

where $\theta$ is the contact angle given in degrees. The correlation is consistent with experimental data when the contact angle is between 35° and 85° [150].

At low heat flux levels, for a given fixed value of the heat flux, the wall superheat monotonically decreases with increasing inclination angle. The wall superheat is the highest for the upward-facing surface (inclination angle equal to 0°), whereas the downward-facing surface (inclination angle equal to 180°) results in the lowest superheat.

### 8.5.2   TRANSITION POOL BOILING

Transition boiling heat transfer is an intermediate heat transfer regime between the nucleate boiling crisis and the minimum film boiling (region III between points $C$ and

$F$ in Fig. 8.1). The main feature of boiling heat transfer in this range is that the heat transfer rates decrease with increasing wall superheat, which manifests itself with a negative slope of the boiling curve $(dq''_w/d(\Delta T_w) < 0)$. This type of boiling can only exist in temperature-controlled heat transfer systems. For heat-flux-controlled systems, the region III of the boiling curve cannot be realized.

Experimental observations reveal that during transition boiling heat transfer, the heating surface is intermittently wetted by the liquid and covered by a vapor film. The mean time fraction of the heating surface in contact with each phase depends on the wall superheat, fluid properties, and the wall material properties. Above a certain limiting value of the wall superheat a stable film boiling is established. The surface temperature that corresponds to this limiting wall superheat is frequently referred to as the minimum film boiling temperature or the Leidenfrost temperature.

Assuming that $h_{fb}$ is the heat transfer coefficient for the stable film boiling and $q''_{mfb}$ is the minimum heat flux that will sustain film boiling, the minimum film boiling temperature $T_{mfb}$ can be estimated as,

$$T_{mfb} = T_{sat} + \frac{q''_{mfb}}{h_{fb}}. \tag{8.145}$$

Using such arguments, Berenson obtained the following expression for the minimum film boiling temperature,

$$T_{mfb} = T_{sat} +$$

$$0.127 \frac{\rho_v i_{fg}}{\lambda_v} \left[ \frac{g(\rho_l - \rho_v)}{\rho_l + \rho_v} \right]^{2/3} \left[ \frac{\sigma}{g(\rho_l - \rho_v)} \right]^{1/2} \left[ \frac{\mu_v}{g(\rho_l - \rho_v)} \right]^{1/3}. \tag{8.146}$$

Henry [87] developed a model of the Leidenfrost phenomenon that included the effects of transient wetting and subsequent liquid microlayer evaporation and arrived at the following expressions:

$$\frac{T_{mfb,H} - T_{mfb}}{T_{mfb} - T_l} = 0.42 \left[ \sqrt{\frac{\lambda_l \rho_l c_{p,l}}{\lambda_s \rho_s c_{p,s}}} \left( \frac{i_{fg}}{c_{p,s}(T_{mfb} - T_{sat})} \right) \right]^{0.6}, \tag{8.147}$$

where $T_{mfb,H}$ is the minimum film boiling temperature obtained from the Henry model and $T_{mfb}$ is given by Eq. (8.146).

Ramilison and Lienhard performed experiments to investigate the influence of the contact angle and arrived at the following correlation for the minimum film boiling temperature [28, 182],

$$T_{mfb} = T_{sat} + 0.97(T_{hn} - T_{sat}) \exp(-0.0006\theta_A^{1.8}), \tag{8.148}$$

where $\theta_A$ is the advancing contact angle in degrees and $T_{hn}$ is the homogeneous nucleation temperature derived by Lienhard from the corresponding state theory and given as [149]

$$T_{hn} = \left[ 0.932 + 0.077 \left( \frac{T_{sat}}{T_{cr}} \right)^9 \right] T_{cr}, \tag{8.149}$$

where $T_{cr}$ is the critical temperature.

Based on own experimental data and data obtained by Berentson, Ramilison and Lienhard recommended the following correlation for predicting the heat transfer in the transition boiling heat transfer regime [182]

$$\text{Bi}^* = 3.74 \cdot 10^{-6} K (\text{Ja}^*)^2, \tag{8.150}$$

where

$$\text{Bi}^* \equiv \frac{(q_w'' - q_{fb}'') \sqrt{a_s \tau}}{\lambda_s \Delta T_w K}, \tag{8.151}$$

$$\text{Ja}^* \equiv \frac{\rho_s c_{p,s} (T_{\text{mfb}} - T_w)}{\rho_v i_{fg}}, \tag{8.152}$$

$$\tau \equiv \left[ \frac{\sigma}{g^3 (\rho_l - \rho_v)} \right]^{1/4}, \tag{8.153}$$

$$K \equiv \frac{\lambda_l / a_l^{1/2}}{\lambda_l / a_l^{1/2} + \lambda_s / a_s^{1/2}}. \tag{8.154}$$

Here variables with index $s$ refer to the solid heater material, $q_{fb}''$ is the heat flux predicted for film boiling at the given wall superheat, and $\tau$ is the characteristic period of the Taylor wave at the interface.

The heat flux in the region close to the departure from nucleate boiling is slightly dependent on the surface condition of the heater, whereas the film boiling heat flux is independent of the surface condition of the heater. The minimum film boiling heat fluxes for relatively large contact angles can be calculated from the following correlation proposed by Berenson [16],

$$q_{\text{min}}'' = 0.09 \rho_v i_{fg} \sqrt[4]{\frac{g(\rho_l - \rho_v)}{(\rho_l + \rho_v)^2}}. \tag{8.155}$$

### 8.5.3  FILM POOL BOILING

At wall surface temperature above the Leidenfrost temperature, the bulk liquid and the heating surface are separated by a stable vapor film. This boiling regime, indicated as region IV in Fig. 8.1, is known as film boiling.

Heat transfer rates and the minimum heat flux during film boiling are known to depend on the orientation and shape of the boiling surface, the type (regular or chaotic) of bubble motion in the fluid bulk, and the flow regime (laminar or turbulent) of vapor in the film. For horizontal plates with regular motion of bubbles and laminar the flow of vapor in the film, the minimum heat flux is shown to be as follows [16]:

$$q_{\text{min}}'' = 0.09 \rho_v i_{fg}' \left[ \frac{g(\rho_l - \rho_v)}{\rho_l + \rho_v} \right]^{1/2} \left[ \frac{\sigma}{g(\rho_l - \rho_v)} \right]^{1/4}, \tag{8.156}$$

where $i'_{fg}$ is the effective latent heat that includes the sensible heat accumulated in the superheated vapor in the film. The equation for the Nusselt number becomes,

$$\mathrm{Nu}_f = 0.425 \mathrm{Ra}_f^{1/4} \left[ \frac{i'_{fg}}{c_{p,v}\Delta T_w} \right]_f^{1/4}, \tag{8.157}$$

where subscript $f$ indicates that the fluid properties should be calculated at the film temperature $T_f = (T_w + T_{sat})/2$. The Nusselt (Nu) and Rayleigh (Ra) numbers are defined as follows:

$$\mathrm{Nu} = \frac{hL_c}{\lambda_v}, \tag{8.158}$$

$$\mathrm{Ra} = \mathrm{GrPr}_v = \left[ \frac{L_c^3 \rho_v (\rho_l - \rho_v) g}{\mu_v^2} \right] \left( \frac{c_{p,v}\mu_v}{\lambda_v} \right). \tag{8.159}$$

Here $L_c$ is the capillary length given by the following formula,

$$L_c = \left[ \frac{\sigma}{g(\rho_l - \rho_v)} \right]^{1/2}. \tag{8.160}$$

For film boiling on the vertical plate, the correlation becomes [94],

$$\mathrm{Nu}_f = 0.943 \mathrm{Ra}_f^{1/2} \left[ \frac{i_{fg}\left(1 + 0.34 c_{p,v}\Delta T_w / i_{fg}\right)^2}{c_{p,v}\Delta T_w} \right]_f^{1/2}, \tag{8.161}$$

where Nu and Ra use a characteristic length $L$ equal to the vertical distance from the bottom of the plate.

Typically heat transfer coefficient for film boiling in a pool is expressed in terms of the Nusselt number (Nu), which is correlated with the Rayleigh number (Ra) and the Prandtl number (Pr). More recent models are based on the dimensional analysis. For vertical heated flat plates immersed in subcooled liquid pools the Nusselt number is correlated to the Rayleigh number and the superheat Jacob number ($\mathrm{Ja}_{sup}$) that incorporate the plate superheat. A correlation that takes into consideration the effect of physical properties of the vapor film, liquid subcooling $\Delta T_{sub} = T_{sat} - T_b$, and wall superheat $\Delta T_w = T_w - T_{sat}$ is given as follows,

$$\mathrm{Nu} = C\,\mathrm{Ra}^{0.43}\left(1 + 4.5\frac{\mathrm{Ja}_{sub}}{\mathrm{Ja}_{sup}^{1.8}}\right), \tag{8.162}$$

where

$$\mathrm{Ja}_{sub} = \frac{c_{p,f}\Delta T_{sub}}{i_{fg}}, \tag{8.163}$$

$$\mathrm{Ra} = \frac{g\beta\Delta T_w D^3}{\nu_v a_v}, \tag{8.164}$$

$$\mathrm{Ja}_{sup} = \frac{c_{p,v}\Delta T_w}{i_{fg}}, \tag{8.165}$$

and $C$ is a material-dependent constant equal to 0.045, 0.055, and 0.068 for the stainless steel, zirconium, and Inconnel-600. The correlation agrees within $\pm 40$ percent with experimental data obtained for vertical rods with an outer diameter of 9.5 mm and a length of 0.245 m. All samples included naturally formed oxide layers [61].

## 8.6 FLOW BOILING

Flow boiling is one of the most effective heat transfer mechanisms in which liquid evaporation at a heated surface is combined with the liquid circulation over the surface. When this process takes place in heated channels, various boiling and two-phase flow regimes occur. The boiling first occurs at a certain location in the channel with sufficient wall superheat, where conditions for the onset of nucleate boiling are met. At relatively low flow quality, bubble nucleation at the wall occurs and a bubbly two-phase flow regime prevails. At high qualities and mass flow rates, the flow regime is normally annular with a liquid film covering channel walls. When the wall superheat is sufficient, nucleate boiling is still present within the liquid film. With thinning liquid film and augmenting heat convection in the film due to high flow velocity, the nucleate boiling may be suppressed, in which case heat transfer is only by convection through the liquid film and evaporation occurs only at its interface.

In comparison to pool boiling, heat transfer in flow boiling is less dependent on heat flux, but it is rather strongly influenced by the local flow quality and the mass flux. Thus, both nucleate boiling and convective heat transfer must be taken into account to predict heat transfer data. At low flow qualities and high heat fluxes, heat transfer is dominated by the nucleate boiling while convection dominates at high mass fluxes and high-quality two-phase flows. For intermediate conditions, both mechanisms are contributing to overall heat transfer and are equally important.

Under specific limiting conditions within a boiling channel, elaborated further in §9, the mechanism of boiling heat transfer undergoes a significant transformation due to the onset of a boiling crisis. The boiling heat transfer that prevails downstream of the boiling crisis point is frequently referred to as the post-critical heat flux (post-CHF) regime. In this heat transfer regime, the heated surface temperature becomes very high and the wall superheat can reach several hundreds of kelvins. Under certain extreme conditions, the temperature of the wall can surpass the melting point of the wall material, resulting in damage to the wall.

The main reason for the high wall temperature in the post-CHF regime is that the wall is covered by the vapor phase rather than the liquid phase. Due to the low thermal conductivity of the vapor, the heat transfer coefficient, and thus the heat transfer rate, is significantly reduced.

There are two major paths that can lead to the post-CHF regime in a boiling channel. The first path is when the channel power increases above a certain limiting value, referred to as the critical power. This type of situation may occur in a water-cooled nuclear reactor that exceeds thermal safety limits and operates at too high a power level. The second path is when coolant flow in the channel drops below a certain minimum value necessary for heat removal. This situation may occur during a loss of coolant accident (LOCA) in a water-cooled nuclear reactor.

### 8.6.1  NUCLEATE FLOW BOILING

The nucleate flow boiling has many similarities with nucleate pool boiling. The ebullition process and the roles of various heat transfer mechanisms such as the transient conduction, the microlayer evaporation, and the microconvection are similar to those in pool boiling. However, the liquid circulation over the heated surface influences the thickness and formation of the thermal boundary layer and gives rise to hydrodynamic forces which interact with bubbles during departure and lift off from the wall. These effects have to be taken into account to discern the specific features of nucleate flow boiling.

Two general groups of models are under development. One group is based on the bubble departure analysis and comparing it with the saturated boiling [211]. The other group is using the heat flux partitioning method that yields the wall temperature and evaporation rate. An example of the latter is the *Chen correlation* that is applicable to convective boiling heat transfer in water-cooled channels for a wide range of parameters [32]. In this approach, the heat transfer rate per unit area $q''$ is found as:

$$q'' = h(T_w - T_{sat}) , \tag{8.166}$$

where $h$ is the heat transfer coefficient, $T_w$ is the heated surface temperature, and $T_{sat}$ is the saturation temperature. The fundamental assumption made by Chen is that the heat transfer coefficient is a superposition of two effects, one resulting from microscopic phenomena caused by nucleate boiling ($h_{nuc}$), and the other resulting from the bulk convective heat transfer contribution ($h_{bulk}$):

$$h = h_{nuc} + h_{bulk} . \tag{8.167}$$

The two heat transfer coefficients are given as follows:

$$h_{bulk} = 0.023 \left( \frac{\lambda_f}{D_h} \right) \mathrm{Re}^{0.8} \mathrm{Pr}^{0.4} \cdot F , \tag{8.168}$$

$$h_{nuc} = 0.00122 \left[ \frac{\lambda_f^{0.79} c_{p,f}^{0.45} \rho_f^{0.49}}{\sigma^{0.5} \mu_f^{0.29} i_{fg}^{0.24} \rho_g^{0.24}} \right] \Delta T_w^{0.24} \left( p_{sat}(T_w) - p_f \right)^{0.75} \cdot S , \tag{8.169}$$

where

$$F = \begin{cases} 1 & X_{tt}^{-1} \leq 0.1 \\ 2.35 \left( 0.213 + \frac{1}{X_{tt}} \right)^{0.736} & X_{tt}^{-1} > 0.1 \end{cases} , \tag{8.170}$$

$$S = \left( 1 + 2.56 \cdot 10^{-6} F^{1.463} \cdot \mathrm{Re}^{1.17} \right)^{-1} , \tag{8.171}$$

$$\mathrm{Re} = \frac{G(1-x)D_h}{\mu} , \tag{8.172}$$

$$X_{tt} = \left( \frac{1-x}{x} \right)^{0.9} \left( \frac{\rho_g}{\rho_f} \right)^{0.5} \left( \frac{\mu_f}{\mu_g} \right)^{0.1} . \tag{8.173}$$

The application of the correlation requires iteration according to the following scheme:

1. For a given mass flux, fluid properties, local heat flux and quality: guess the wall superheat $\Delta T_w$.
2. Find the parameter $X_{tt}$ from Eq. (8.173).
3. Find the Reynolds number Re from Eq. (8.172).
4. Find the parameter $F$ from Eq. (8.170).
5. Find the parameter $S$ from Eq. (8.171).
6. Find the partial heat transfer coefficient $h_{bulk}$ from Eq. (8.168).
7. Find the partial heat transfer coefficient $h_{nuc}$ from Eq. (8.169).
8. Find the overall heat transfer coefficient $h$ from Eq. (8.167).
9. Find the new value of the wall superheat $\Delta T_w$ from Eq. (8.166).
10. Continue repeating steps 7 through 9 until convergence is achieved and the old and new wall superheat values are equal to each other within a specified convergence limit.

### 8.6.2 POST-CHF FLOW BOILING

Depending on the void content in a boiling channel, the post-CHF boiling regime can be either inverted annular flow boiling (IAFB), inverted slug film boiling (ISFB), or dispersed flow film boiling (DFFB). In this section we discuss the IAFB and DFFB regimes since they are most often considered in nuclear reactor safety analyses. In addition, we consider the transition flow boiling that may potentially arise during the reflooding phase of various loss of coolant accident scenarios.

### Inverted Annular Flow Boiling

The inverted annular flow boiling heat transfer prevails in heated channels at low qualities and subcooled conditions. This heat transfer regime follows the occurrence of the departure from nucleate boiling (DNB) that leads to the creation of a vapor film flowing along the heated wall. The core of the flow contains a subcooled or saturated liquid phase. During various postulated scenarios of the loss of coolant accident, the inverted annular flow boiling regime occurs just downstream of the quenching front, where the wall surface temperature is very high and exceeds the Leidenfrost temperature.

Bromley [24] performed film boiling experiments on a horizontal tube and derived a correlation for the natural convection heat transfer coefficient as follows,

$$h = C \left[ \frac{g\rho_v(\rho_l - \rho_v)i'_{fg}\lambda_v^3}{\mu_v D(T_w - T_{sat})} \right]^{1/4}, \tag{8.174}$$

where $i'_{fg}$ is calculated as the enthalpy difference between the vapor at its arithmetic average temperature and liquid at saturation temperature, $D$ is the tube diameter, and $C$ is a constant to be specified. The correlation is applicable for $U/(gD) < 1.0$, where $U$ is the area-averaged liquid velocity. For forced convection, when $U/(gD) > 2.0$, the following correlation is proposed [25],

$$h = 2.7\sqrt{\frac{U\lambda_v\rho_f i'_{fg}}{D(T_w - T_{sat})}}. \tag{8.175}$$

## Dispersed Flow Film Boiling

The *dispersed flow film boiling* prevails in a heated channel downstream of dryout of the liquid film in annular flow. Since the mist two-phase flow exists in this region, this boiling regime is frequently referred to as the *mist flow evaporation*. In BWR applications, the preferred nomenclature is the **post-dryout heat transfer**.

The simplest correlation approach to predict wall temperature in the post-dryout region is based on assumptions that thermodynamic equilibrium exists between the phases and that the heat transfer is dominated by turbulent vapor convection. These correlations are similar to the single-phase convective heat transfer correlations and calculate the local heat transfer coefficient as a function of local mass flux, equilibrium quality, and saturated vapor properties. Using this approach, Dougall and Rohsenow proposed the following correlation [58],

$$h_v = 0.023\frac{\lambda_v}{D}\left[\frac{GD}{\mu_v}\left(x_e + \frac{\rho_v}{\rho_l}(1 - x_e)\right)\right]^{0.8}\left(\frac{c_{p,v}\mu_v}{\lambda_v}\right)^{0.4}, \tag{8.176}$$

where $h_v$ is the heat transfer coefficient, $\lambda_v$ is the vapor thermal conductivity, $D$ is the tube diameter, $G$ is the mass flux, $\mu_v$ is the vapor dynamic viscosity, $x_e$ is the equilibrium quality, $\rho_v$, $\rho_l$ is the vapor and liquid density, respectively, and $c_{p,v}$ is the vapor specific heat capacity. All physical properties should be taken at the system saturation temperature. For post-dryout in tubes and annuli at reactor operating conditions, Groeneveld developed the following correlation [79],

$$h_v = a\frac{\lambda_v}{D_h}\left[\frac{GD_h}{\mu_v}\left(x + \frac{\rho_v}{\rho_l}(1 - x)\right)\right]^b\left(\frac{c_{p,v}\mu_v}{\lambda_v}\right)_w^c Y^d\left(\frac{q''_w}{3.152}\right)^e, \tag{8.177}$$

where $x$ is the flow quality, $D_h$ is the hydraulic diameter, $q''_w$ is the wall heat flux in W/m$^2$, and $Y$ is a correction factor defined as,

$$Y = 1.0 - 0.1\left(\frac{\rho_l - \rho_g}{\rho_g}\right)^{0.4}(1 - x_e)^{0.4}. \tag{8.178}$$

The best-fit values of constants $a$–$e$ in Eq. (8.177) are provided in Table 8.1 and the range of data on which the correlation is based is given in Table 8.2. The last term in Eq. (8.177) represents the influence of the heat flux on the heat transfer coefficient. In general, this term only slightly improves the agreement of the correlation with the experimental data and can be dropped by setting $e = 0$. It should be noted that the wall temperature should be used to calculate the vapor Prandtl number. Thus, to calculate the heat transfer coefficient when the wall temperature is not known and the heat flux is given, iterations will be necessary.

**TABLE 8.1**

**Constants $a$–$e$ in Eq. (8.177) [79]**

| Geometry | $a$ | $b$ | $c$ | $d$ | $e$ |
|---|---|---|---|---|---|
| Tubes | $1.85 \cdot 10^{-4}$ | 1.00 | 1.57 | -1.12 | 0.131 |
| | $1.09 \cdot 10^{-3}$ | 0.989 | 1.41 | -1.15 | 0 |
| Annuli | $1.30 \cdot 10^{-2}$ | 0.664 | 1.68 | -1.12 | 0.133 |
| | $5.20 \cdot 10^{-2}$ | 0.688 | 1.26 | -1.06 | 0 |
| Tubes and annuli | $7.75 \cdot 10^{-4}$ | 0.902 | 1.47 | -1.54 | 0.112 |
| | $3.27 \cdot 10^{-3}$ | 0.901 | 1.32 | -1.50 | 0 |

**TABLE 8.2**

**Range of Data on which Correlation Given by Eq. (8.177) is Based [79]**

| Parameter | Tube | Annulus |
|---|---|---|
| Flow direction | vertical and horizontal | vertical |
| Hydraulic diameter $D_h$, mm | 2.5–25 | 1.5–6.3 |
| Pressure $p$, MPa | 6.8–21.5 | 3.4–10 |
| Mass flux $G$, kg/m$^2$s | 70–530 | 80–410 |
| Flow quality $x$ | 0.10–0.90 | 0.10–0.90 |
| Heat flux $q_w''$, kW/m$^2$ | 120–2100 | 450–2250 |
| $Nu_v = h_v D_h / \lambda_v$ | 95–1770 | 160–640 |
| $(GD_h/\mu_v)[x+(1-x)\rho_v/\rho_l]$ | $6.6 \cdot 10^4$–$1.3 \cdot 10^6$ | $1.0 \cdot 10^5$–$3.9 \cdot 10^5$ |
| $Pr_{v,w} = (c_{p,v}\mu_v/\lambda_v)_w$ | 0.88–2.21 | 0.91–1.22 |
| $Y$ | 0.706–0.976 | 0.610–0.963 |

The correlations provided by Eqs. (8.176) and (8.177) are developed under the assumption that the vapor is at the saturation temperature. However, experimental data show that vapor can be significantly superheated during post-dryout heat transfer.

## Transition Flow Boiling

The transition flow boiling can be encountered in systems with the wall temperature control and forced flow of coolant. Such conditions would exist in the transition boiling region during the reflooding phase of large break loss of coolant accident scenarios for light water reactors. The transition boiling region is located just downstream of the quench front, where the wall surface temperature exceeds the CHF temperature, but is lower than the minimum stable film boiling temperature.

## PROBLEMS

### PROBLEM 8.1

Explain the main steps that need to be taken to perform heat flux partitioning for nucleate boiling heat transfer.

### PROBLEM 8.2

Describe the various heat transfer mechanisms that contribute to the overall nucleate boiling heat transfer.

### PROBLEM 8.3

Find the wall superheat and the nucleate boiling heat transfer coefficient in a fuel rod bundle with a square lattice with a rod diameter of 10 mm and a rod pitch of 12.7 mm. Assume a flow of saturated water and steam mixture at a pressure 7 MPa, with mass flux equal to 1250 kg m$^{-2}$ s$^{-1}$, and the thermodynamic equilibrium quality equal to 0.25. The local heat flux is equal to 750 kW m$^{-2}$. Use the isolated subchannel model and the Chen correlation.

# 9 Boiling Crisis

A sudden deterioration of heat transfer between a heated wall surface and neighboring liquid layers, accompanied by the creation of a persistent vapor layer on the surface and a significant increase of the wall temperature is called a *boiling crisis*. The local heat flux at which the boiling crisis occurs is termed a *critical heat flux* (CHF).

The boiling crisis phenomenon was initially identified during early studies of water pool boiling on a submerged heated wire. This led to the development of a *boiling curve*, which represents the relationship between heat flux and wall superheat, and it was characterized by clearly defined maximum and minimum heat flux values [171].

A comprehensive examination of the boiling crisis phenomenon commenced immediately following the initiation of the civil and military power reactor program, which was based on water-cooled reactor systems. Both experimental and theoretical investigations uncovered the existence of various modes of boiling crisis in heated channels with forced convective heat transfer. As many experiments resulted in the heated element melting or rupturing when a boiling crisis occurred, it became common practice to refer to the corresponding heat flux as "burnout heat flux", or simply *burnout*. However, this terminology was rightly criticized for being misleading, prompting a search for more precise definitions of the terms used in this field [187, 45].

In this book, we adhere to a nomenclature that has gained widespread acceptance in recent years. The literature typically describes two types of boiling crises. The first, which occurs at low bulk quality in a heated channel where nucleate boiling heat transfer prevails, is commonly known as the *departure from nucleate boiling* (DNB). The second type of boiling crisis, which occurs at the high bulk quality and during annular two-phase flow, is known as *dryout*.

A variety of methods are employed to predict the onset of a boiling crisis in heat transfer equipment. With the recent advancements in computational fluid dynamics and numerical heat transfer techniques, there is a growing interest in utilizing mechanistic models for physical processes, including the phenomena of boiling crises. However, in many systems with specific safety requirements, it is crucial to not only accurately predict but also estimate the uncertainties of these predictions. Therefore, correlation-based predictions continue to be the preferred choice for safety analysis applications.

In this chapter, we introduce the various types of correlations available for predicting boiling crises. We specifically discuss how these correlations should be chosen and implemented in calculations. We consider both pool boiling and flow boiling applications.

DOI: 10.1201/9781003255000-9

## 9.1 PRINCIPAL MECHANISMS AND EFFECTS

The boiling crisis is a complex phenomenon that occurs due to the interaction of numerous mechanisms. We discussed some of these mechanisms in §8. The examination of local mechanisms that govern the boiling crisis necessitates the use of sophisticated instrumentation to capture the distribution of local parameters near the boiling crisis point. This method is currently employed in experimental investigations focusing on nucleate boiling and the boiling crisis within pool boiling scenarios. However, applying such an approach to complex geometries, such as a fuel rod assembly, can be challenging or impractical.

The location of the boiling crisis point in a rod bundle is not predetermined and depends on the flow and heat transfer conditions. Therefore, to gather essential experimental information, the instrumentation should cover a significant portion of the test section. However, such a vast experimental database would only be beneficial if accompanied by the development of mechanistic and locally based prediction models. Although this type of approach is the ultimate goal in the field of boiling crisis research, its practical usage in nuclear power safety applications is still rather limited.

Global effects in boiling crisis phenomena are easier to capture and are useful for the development of empirical correlations. These effects are investigated by recording the occurrence of the boiling crisis when changing the test section geometry and orientation, as well as flow and heat transfer conditions. Specifically, when investigating the boiling crisis in heated channels, the influences of coolant mass flux, system pressure, inlet subcooling, and spatial power distribution are examined.

### 9.1.1 POOL BOILING CRISIS

The pool boiling crisis is influenced by a multitude of processes that operate on various scales. These range from the nanometer scale, which pertains to the characteristics of the boiling surface, to the micrometer scale, which involves the hydrodynamics of the microlayer beneath nucleating bubbles, and finally to the system scale, which encompasses the hydrodynamics of the two-phase mixture. Kutateladze noted the similarity between the boiling crisis condition and the flooding phenomena in distillation columns and used dimensional analysis arguments to derive a relationship for the maximum heat flux as follows: $q''_{max} \sim C \rho_g^{1/2} i_{fg} \left[ g \left( \rho_f - \rho_g \right) \sigma \right]^{1/4}$, and concluded that $C$ was equal to 0.16 [137]. This result was later derived theoretically by Zuber who considered Taylor wave motion and Helmholtz instability of vapor columns over a heated surface, finding that $C = \pi/24 \simeq 0.131$ [244]. According to this theory, a boiling crisis is caused by the hydrodynamic limit when the flow of liquid towards a heated surface is no longer possible. Recent investigations show, however, that the onset of boiling crises is caused by the creation of irreversible dry spots underneath growing bubbles [122, 216, 217].

In pool boiling, the three primary global effects encompass system pressure, fluid subcooling, and the orientation of the heated wall. System pressure has a significant impact on fluid properties such as the density ratio, latent heat, and surface tension. Consequently, pressure plays a pivotal role in the boiling crisis, influencing factors such as heat flux, bubble dynamics, and the overall stability of the boiling process. In the case of boiling water, the critical heat flux initially increases and then decreases as the pressure increases towards the critical point.

Subcooling of the ambient pool below saturation temperature typically results in a shift of the boiling curve towards higher levels of heat flux. The maximum heat flux is strongly impacted by subcooling. As the vapor departs from the region near the heated surface and ascends through the subcooled pool, it tends to condense. This condensation facilitates the flow of liquid towards the surface, making the boiling process more efficient.

Experimental findings suggest that the inclination angle of the heated surface significantly affects nucleate boiling heat transfer under low heat flux conditions. However, under high heat flux conditions, the effect of orientation is not substantial.

Visual observations have led to the categorization of pool boiling into three regions: upward facing ($0° \leq \theta \leq 90°$), vertical and near-downward facing ($90° \leq \theta \leq 165°$), and downward facing ($165° \leq \theta \leq 180°$). In the upward-facing region, vapor is generated and detaches easily in the vertical direction from the heated surface. In the vertical and near-downward region, the vapor grows and drifts upward along the surface in a wavy shape due to the vapor flow. When the heater faces downward, the vapor is trapped and covers the entire heated surface.

The boiling crisis heat flux decreases as the surface orientation shifts from horizontal facing upward to downward, according to the vapor behaviors at various inclination angles. The vapor velocity influences the distance between the two wavelength peaks of the vapor, with faster vapor flow resulting in a shorter wavelength. Moreover, the location of the boiling crisis is influenced by the surface orientation due to the behavior of the vapor.

## 9.1.2 FLOW BOILING CRISIS

The incorporation of a forced-convection effect typically enhances the single-phase heat transfer coefficient beyond that of natural convection alone, leading to an upward shift in the single-phase segment of the boiling curve. Furthermore, the maximum pool boiling heat flux is generally significantly increased by the introduction of a forced convection effect. In general, the boiling crisis phenomenon in a heated channel is influenced by several global parameters such as:

- coolant mass flow rate $G$,
- coolant pressure $p$,
- coolant quality at channel inlet $x_{in}$,
- heat flux $q''$,
- channel flow area $A$,
- channel heated perimeter $P_H$,
- channel unheated perimeter $P_{UH}$,
- channel heated length $L$.

The impact of these parameters on the conditions of the boiling crisis has been thoroughly examined across a variety of test sections. In the following subsections, we discuss some well-established findings related to uniformly-heated circular tubes.

### Mass Flux

A relationship between the critical heat flux and the mass flux is shown in Fig. 9.1. The four data sets shown in the figure represent critical quality values that were measured in circular tubes with inner diameter $D = 8$ mm and heated length $L$ varying from 1 to 5 m. The tubes were cooled with water at 7 MPa pressure and with inlet quality $x_{in} = -0.035$ [208].

For the mass flux higher than 1000 kg m$^{-2}$ s$^{-1}$ and tube lengths equal to 2 m or greater, the critical heat flux almost linearly increases with the mass flux with $\Delta q''_{cr}/\Delta G \approx const$. From Eq. (9.53) we can infer that under these conditions, with both system pressure and inlet quality being constant, the critical quality also remains approximately constant.

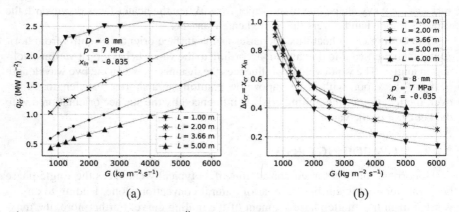

(a)                                      (b)

**Figure 9.1** Critical heat flux $q''_{cr}$ (a) and critical quality increase $\Delta x_{cr} = x_{cr} - x_{in}$ (b) as a function of the mass flux $G$ and the tube length $L$ in a round tube with inner diameter $D = 8$ mm, cooled with water at pressure $p = 7$ MPa and with inlet subcooling $x_{in} = -0.035$. Data from [208].

### System Pressure

The influence of the system pressure on the critical heat flux in $D = 8$ mm and $L = 3.66$ m tubes are shown in Fig. 9.2. For all cases the inlet subcooling is kept almost constant in a range $-0.065 < x_{in} < -0.025$. As shown in the figure, the critical heat flux and the increase of the critical quality strongly depend on the system pressure and both exhibit maximum values for the pressure in the range from 3.5 MPa to 7 MPa.

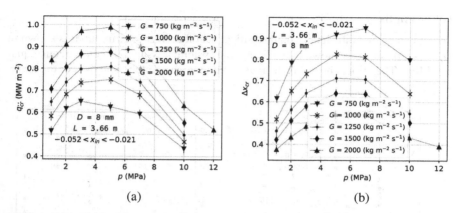

(a)                                                                          (b)

**Figure 9.2**  Critical heat flux $q''_{cr}$ (a) and critical quality increase $\Delta x_{cr}$ (b) as a function of the pressure $p$ and the mass flux $G$ in a round tube with inner diameter $D$ = 8 mm and the heated length $L$ = 3.66 m, cooled with water with inlet subcooling $x_{in} = -0.035$. Data from [208].

## Inlet Quality

The influence of the inlet quality on the critical heat flux in $D = 8$ mm and $L = 3.66$ m tube is shown in Fig. 9.3. For all cases the system pressure $p$ is constant and equal to 7 MPa. Only two values of the inlet quality are available, corresponding to the inlet

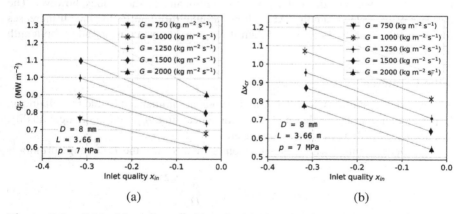

(a)                                                                          (b)

**Figure 9.3**  Critical heat flux $q''_{cr}$ (a) and critical quality increase $\Delta x_{cr}$ (b) as a function of the inlet quality $x_{in}$ and the mass flux $G$ in a round tube with an inner diameter $D = 8$ mm and heated length $L = 3.66$ m, cooled with water at the system pressure $p = 7$ MPa. Data from [208].

subcooling 10 K and 100 K. It can be seen that with a decreasing inlet quality, both $q''_{cr}$ and $\Delta x_{cr}$ decrease.

## Tube Length

The influence of a tube length on the critical heat flux in $D = 8$ mm tubes is shown in Fig. 9.4. For all cases shown in the figure, the system pressure $p$ is constant and equal to 7 MPa. The critical heat flux decreases with increasing tube length over the entire

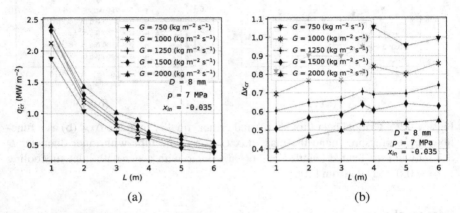

(a)                                                                      (b)

**Figure 9.4** Critical heat flux $q''_{cr}$ (a) and critical quality increase $\Delta x_{cr}$ (b) as a function of the length $L$ and the mass flux $G$ in a round tube with an inner diameter $D = 8$ mm and inlet mass flux $x_{in} = -0.035$, cooled with water at system pressure $p = 7$ MPa. Data from [208].

investigated range. The drop of $q''_{cr}$ is more significant for short tubes, however. The increase of the critical quality, $\Delta x_{cr}$, shows a quite different trend since it increases for short tubes (up to 3.66 m) and then it remains almost constant and tube-length independent.

## Tube Diameter

Boiling crisis conditions, both DNB and dryout, are influenced by the tube diameter. Experimental data shows that both the critical heat flux, $q''_{cr}$, and the critical quality, $x_{cr}$, decreases with an increasing diameter.

For DNB in tubes with a diameter other than 8 mm, the following formula is recommended [57],

$$q''_{cr}\,|_D = q''_{cr}\,|_{8\,\text{mm}} \left(\frac{8}{D}\right)^{1/2}, \tag{9.1}$$

where $q''_{cr}\,|_{8\,\text{mm}}$ is the critical heat flux in a tube with 8 mm inner diameter and $D$ is the tube's inner diameter in mm.

For dryout, the following formula is recommended,

$$x_{cr}\,|_D = x_{cr}\,|_{8\,\text{mm}} \left(\frac{8}{D}\right)^{0.15}, \tag{9.2}$$

where $x_{cr}\,|_{8\,\text{mm}}$ is the critical quality in a tube with 8 mm inner diameter.

**Channel Geometry**

The channel geometry, and in particular, its flow cross-section area, has a significant influence on the boiling crisis. In general, the highest values of the critical heat flux or the critical quality is obtained in round tubes. For channels with complex cross sections, such as fuel rod assemblies, a significant reduction of these critical parameters is observed. In addition, both the DNB and dryout are influenced by the design details of fuel rod assemblies and in particular by the presence of spacer grids.

### 9.1.3 SCALING OF BOILING CRISIS

The boiling crisis phenomena are primarily of interest for water-cooled reactors. Thus the models and correlations that are used for the prediction of the thermal safety margins in LWRs should be derived and valid mainly for water systems. However, since experiments with water require high pressures, temperatures, and heating powers, the "scaling fluid" approach has been frequently used.

One of the most important scaling constraints for boiling crisis simulation is the preservation of the density ratio $\rho_f/\rho_g$ the same as in the target fluid. To obtain the same density ratio as for water at 7 MPa, Refrigerant-12 has to be pressurized to only 1.04 MPa. In addition, this scaling fluid, when used for boiling crisis investigations, requires only about 6% of the heat input needed for water. Thus huge savings in experimentation costs are possible and the approach was tested in various test sections such as round tubes, annuli, and rod bundles. However, scaling invariably introduces additional uncertainties, which can be circumvented by utilizing prototypical fluids under prototypical operating conditions and conducting tests in the full-scale geometry of the fuel rod bundle.

## 9.2 POOL BOILING CHF CORRELATIONS

Thanks to well-defined conditions, pool boiling belongs to the best-investigated area of boiling heat transfer in general, and boiling crisis in particular. Already early theoretical investigations led to the following relation, valid for pool boiling critical heat flux,

$$\frac{q''_{cr}}{\rho_g i_{fg} \left[ \sigma g \left( \rho_f - \rho_g \right) / \rho_g^2 \right]^{1/4}} = \Phi_{cr}, \tag{9.3}$$

where $q''_{cr}$ is the critical heat flux and $\Phi_{cr}$ is a constant. This relationship can be derived both from a dimensional analysis and hydrodynamic instability considerations.

## 9.3 DNB CORRELATIONS

The *departure from nucleate boiling* (DNB) is a critical safety issue, especially in the context of pressurized water reactors (PWRs). DNB causes a local vapor layer on the fuel rod surface, leading to a dramatic reduction of heat transfer capability and a rapid cladding temperature increase. This phenomenon occurs in the subcooled or

**TABLE 9.1**

**Constant $\Phi_{cr}$ in Eq. (9.3) For Selected Pool Boiling CHF Correlations**

| Constant $\Phi_{cr}$ | Effect | Authors |
|---|---|---|
| 0.131 | Hydrodynamic instability | [243, 244] |
| 0.16 | Dimensional analysis | [137, 138] |
| $0.171\frac{(1+0.324\cdot10^{-3}\theta^2)^{1/4}}{(0.018\theta)^{1/2}}$ | Contact angle | [126] |
| $0.0044(\pi-\theta)^3 R_a^{0.125}$ | Contact angle and roughness | [183] |

low-quality region in the fuel rod assembly. The behavior of this type of boiling crisis depends on many flow conditions, such as pressure, mass flux, and local quality.

One of the key safety requirements of PWRs is that DNB will not occur during steady-state and transient normal operation and *anticipated operational occurrences* (AOOs). The *departure from nucleate boiling ratio* (DNBR) is a measure of the margin to boiling crisis. It is defined as the critical heat flux at a specific location and specific coolant parameters divided by the operating local heat flux at that location. The minimum value of DNBR, which is known as MDNBR, defines thermal margins. Fuel cladding integrity will be maintained if the MDNBR remains above the 95/95 DNBR limit (a 95% probability at a 95% confidence level).

The specific DNB correlation used for PWR safety analysis can vary depending on the specific design and safety requirements. In this section, we present only a few examples of DNB correlation. A more comprehensive description of DNB correlations and an extensive reference list can be found in [221].

### 9.3.1 SIMPLE CHANNELS

For DNB in vertical tubes with an inner diameter of 8 mm, uniformly heated and cooled with water flowing upward, the following correlation is proposed [144],

$$q''_{cr} = \left[10.3 - 7.8\frac{p}{98} + 1.6\left(\frac{p}{98}\right)^2\right]\left(\frac{G}{1000}\right)^{1.2\{[0.25(p-98)/98]-x_e\}} e^{-1.5x_e}, \quad (9.4)$$

where $p$ is the pressure (bar), $G$ is the mass flux (kg m$^{-2}$ s$^{-1}$), $x_e$ is the thermodynamic equilibrium quality, and $q''_{cr}$ is the critical heat flux (MW m$^{-2}$). The correlation is valid in the following ranges of parameters: $29.4 < p < 196$ bar, $750 < G < 5000$ kg m$^{-2}$ s$^{-1}$, and its root-mean-square error is 15%. This correlation can be applied to other tube diameters by using the formula given by Eq. (9.1).

Bowring proposed the following formula [21]:

$$q''_{cr} = \frac{C_1 + D_h G \Delta i_i/4}{C_2 + L}, \quad (9.5)$$

where,

$$C_1 = \frac{0.579 F_1 D_h G i_{fg}}{1 + 0.0143 F_2 D_h^{1/2} G} , \tag{9.6}$$

$$C_2 = \frac{0.077 F_3 D_h G}{1 + 0.347 F_4 (G/1356)^n} . \tag{9.7}$$

Here $D_h$ is the hydraulic diameter (m), $G$ is the mass flux (kg m$^{-2}$ s$^{-1}$), $\Delta i_i = i_f - i_i$ is the inlet subcooling (J kg$^{-1}$), $i_{fg}$ is the latent heat (J kg$^{-1}$), and $L$ is the tube length (m). The correlation parameters: $n$, $F_1$, $F_2$, $F_3$, $F_4$ are functions of pressure and are given as,

$$n = 2.0 - 0.5 p_R , \tag{9.8}$$

$$p_R = \frac{p}{6.895 \cdot 10^6} , \tag{9.9}$$

$$F_1 = \begin{cases} \dfrac{p_R^{18.942} \exp[20.8(1-p_R)] + 0.917}{1.917} & \text{for} \quad p_R \leq 1 \\[2mm] p_R^{-0.368} \exp[0.648(1-p_R)] & \text{for} \quad p_R > 1 \end{cases} , \tag{9.10}$$

$$\frac{F_1}{F_2} = \begin{cases} \dfrac{p_R^{1.316} \exp[2.444(1-p_R)] + 0.309}{1.309} & \text{for} \quad p_R \leq 1 \\[2mm] p_R^{-0.448} \exp[0.245(1-p_R)] & \text{for} \quad p_R > 1 \end{cases} , \tag{9.11}$$

$$F_3 = \begin{cases} \dfrac{p_R^{17.023} \exp[16.658(1-p_R)] + 0.667}{1.667} & \text{for} \quad p_R \leq 1 \\[2mm] p_R^{-0.219} & \text{for} \quad p_R > 1 \end{cases} , \tag{9.12}$$

$$\frac{F_4}{F_3} = p_R^{1.649} . \tag{9.13}$$

The correlation is based on a fit to 3792 experimental data points in the following parameter ranges: $136 < G < 18600$ kg m$^{-2}$ s$^{-1}$, $0.2 < p < 19$ MPa, $2 < D_h < 44.7$ mm, and $0.15 < L < 3.66$ m. The correlation is applicable for the prediction of boiling crisis in uniformly heated round tubes with vertical flow within the pressure range 0.7 to 17 MPa. When compared to all experimental data points, the correlation mean error $\varepsilon = (q_p'' - q_e'')/q_e''$ (where $q_p''$, $q_e''$ = predicted and experimental heat flux) was $-0.29\%$ and the RMS error, defined as $(\sum \varepsilon^2/N)^{1/2}$, was 7%.

## 9.3.2  SUBCHANNELS AND ROD BUNDLES

For uniformly heated fuel rod bundles, Tong developed the following formula, known as the W-3 DNB correlation:

$$\begin{aligned} q_{cr,U}'' = A \Big[ &2.022 - 0.0004302 p_R + (0.1722 - 0.0000984 p_R) \\ &\times e^{x(18.177 - 0.004129 p_R)} \Big] (1.157 - 0.869x) \\ &\times [(0.1484 - 1.596x + 0.1729x|x|) G_R + 1.037] \\ &\times \left( 0.2664 + 0.8357 e^{-3.151 D_R} \right) (0.8258 + 0.000794 \Delta i_R) F_s \end{aligned} \tag{9.14}$$

The parameters used in the correlations are as follows:

$q''_{cr,U}$ = critical heat flux in MW m$^{-2}$,
$A = 0.31544$,
$p_R = p/6.8947 \cdot 10^3$, $p$ - pressure in Pa,
$x$ = thermodynamic equilibrium quality,
$G_R = G/1.3562 \cdot 10^{-3}$, $G$ - mass flux in kg m$^{-2}$ s$^{-1}$,
$D_R = D_e/25.4$, $D_e$ - equivalent diameter,
$\Delta i_R = (i_f - i_{in})/2326$, $i_f$ - specific enthalpy of saturated liquid, $i_{in}$ - specific
enthalpy at the bundle inlet, both in J kg$^{-1}$ in mm,
$F_s$ = dimensionless grid or spacer factor.

For non-uniform power distribution in the channel, the following shape factor should
be used,

$$F_c \equiv \frac{q''_{cr,U}}{q''_{cr,NU}} = \frac{C}{q''(z_{NU})(1 - e^{-C \cdot z_U})} \int_0^{z_{NU}} q''(z) e^{-C(z_{NU}-z)} \mathrm{d}z \,, \tag{9.15}$$

where $z_U$ is the axial location at which DNB occurs with the uniform power distribution, $z_{NU}$ is the axial location at which DNB occurs with the nonuniform power
distribution, $q''(z_{NU})$ is the local heat flux at DNB location with nonuniform power
distribution and $q''(z)$ is the axial nonuniform heat flux distribution. The empirical
constant $C$ is given as,

$$C = 185.6 \frac{(1 - x_{cr})^{4.31}}{G_R^{0.478}} \,, \tag{9.16}$$

where $C$ is in m$^{-1}$ and $x_{cr}$ is the thermodynamic equilibrium quality at the DNB
location.

## 9.4 DRYOUT CORRELATIONS

Dryout correlations are used to predict the conditions under which boiling crisis
occurs in boiling channels with high thermodynamic equilibrium quality. For that
reason, dryout is a critical safety issue for boiling water reactors (BWRs). The occurrence of dryout depends on various flow conditions such as pressure, mass flux,
and heat flux distribution.

In a similar manner as for PWRs, one of the key safety requirements of BWRs is
that dryout will not occur during steady-state and transient normal operation and during anticipated operational occurrences. The *critical power ratio* (CPR) is a measure
of the margin to dryout. It is defined as the power of a fuel rod assembly at which
dryout occurs divided by the actual power of the fuel rod assembly. The minimum
value of CPR, which is known as MCPR, defines thermal margins. Fuel cladding
integrity will be maintained if the MCPR remains above the 95/95 CPR limit (a 95%
probability at a 95% confidence level).

The specific dryout correlation used for BWR safety analysis can vary depending
on the specific design of the fuel rod bundle and safety requirements. In this section,
we present only a few examples of dryout correlations. A more detailed description
of correlations applicable to BWRs and an extensive reference list can be found
in [139, 221].

### 9.4.1 SIMPLE CHANNELS

Dryout correlations have usually the following form:

$$x_{cr} = x_{cr}(G, p, D_h, L_B, \ldots), \tag{9.17}$$

where $x_{cr}$ is the critical quality, $G$ is the mass flux, $p$ is the pressure, $D_h$ is the hydraulic diameter and $L_B$ is the *boiling length*, which is the length of a channel over which a saturated boiling takes place. Using experimental data obtained in uniformly heated tubes with an inner diameter of 8 mm, Levitan and Lantsman derived the following formula for determining the critical quality [144]:

$$x_{cr}|_{8\,mm} = \left[0.39 + 1.57\frac{p}{98} - 2.04\left(\frac{p}{98}\right)^2 + 0.68\left(\frac{p}{98}\right)^3\right]\left(\frac{G}{1000}\right)^{-0.5}, \tag{9.18}$$

where $p$ is the pressure in bars and $G$ is the mass flux in kg m$^{-2}$ s$^{-1}$. The correlation can be used for tubes with other inner diameters by using Eq. (9.2). The correlation is based on experimental data with the system pressure in the range from 0.98 to 16.66 MPa and the mass flux in the range from 750 to 3000 kg m$^{-2}$ s$^{-1}$, and predicted $x_{cr}$ agrees with the data within $\pm 0.05$.

### 9.4.2 SUBCHANNELS AND ROD BUNDLES

CHF correlations applicable to rod bundles, in addition to the parametric influences observed in tubes, have to take into account the bundle geometry and the spatial power distribution details. One of the first expressions of this kind was the GE-CISE correlation that took the form of predicting critical quality as a function of the *boiling length* and a local peaking factor,

$$x_{cr} = \frac{C_1 L_B^*}{C_2 + L_B^*}\left(\frac{1.24}{R_f}\right). \tag{9.19}$$

Here $x_{cr}$ is the critical quality, $L_B^* = L_B/0.0254$, where $L_B$ is the boiling length (m), $R_f$ is the peaking factor (a ratio of maximum to average rod power), and $C_1$, $C_2$ are terms dependent on the mass flux and the pressure given as,

$$C_1 = 1.055 - 0.013\left(\frac{p_R - 600}{400}\right)^2 - 1.233G_R + 0.907G_R^2 - 0.285G_R^3, \tag{9.20}$$

$$C_2 = 17.98 + 78.873G_R - 35.464G_R^2, \tag{9.21}$$

with

$$G_R = G/1356.23 \quad p_R = p/6894.757, \tag{9.22}$$

where $G$ is the mass flux (kg m$^{-2}$ s$^{-1}$) and $p$ is the pressure (Pa). The correlation is primarily applicable to 7×7 bundles with possible extension to 8×8 bundles once replacing $B$ with $B/1.12$.

The EPRI correlation takes a form similar to Eq. (9.19) and is as follows [86],

$$x_{cr} = \frac{C_1 L_B^+}{C_2 + L_B^+} \left[ 2 - J_1 + \frac{0.19}{G_R} (J_1 - 1)^2 + J_3 \right] + C_3, \qquad (9.23)$$

where $x_{cr}$ is the critical quality, $L_B^+$ is a non-dimensional boiling length, $J_1$, $J_3$ are terms to accommodate the local peaking effects, and $C_1$, $C_2$, $C_3$ are terms that accommodate the effects of mass flux and pressure. The terms included in the EPRI correlation are given as follows,

$$L_B^+ = \frac{N \pi D_R L_B}{A}, \qquad (9.24)$$

$$C_1 = 0.5 G_R^{-0.43}, \qquad (9.25)$$

$$C_2 = 165 + 115 G_R^{2.3}, \qquad (9.26)$$

$$C_3 = 0.006 - 0.0157 \left( \frac{p_R - 800}{1000} \right) - 0.0714 \left( \frac{p_R - 800}{1000} \right)^2, \qquad (9.27)$$

$$J_1 = \begin{cases} \frac{1}{32} \left( 25 f_p + 3 \sum_{i=1}^2 f_i + f_j \right) & \text{for corner rods} \\ \frac{1}{32} \left( 22 f_p + 3 \sum_{i=1}^2 f_i + \sum_{j=1}^2 f_j + 2 f_k \right) & \text{for side rods} \\ \frac{1}{32} \left( 20 f_p + 2 \sum_{i=1}^4 f_i + \sum_{j=1}^4 f_j \right) & \text{for central rods} \end{cases} \qquad (9.28)$$

$$J_3 = \begin{cases} 0 & \text{for corner rods} \\ \frac{0.07}{G_R + 0.25} - 0.05 & \text{for side rods} \\ \frac{0.14}{G_R + 0.25} - 0.10 & \text{for central rods} \end{cases} \qquad (9.29)$$

where $N$ is the number of rods in a bundle, $D_R$ is the rod diameter (m), $L_B$ is the boiling length (m), $A$ is the bundle flow area (m$^2$), $p_R = p/6894.757$ and $p$ is the pressure (Pa), $G_R = G/1356.23$ and $G$ is the mass flux (kg m$^{-2}$ s$^{-1}$). The definitions of rod peaking factors for corner, side, and center rods are shown in Fig. 9.5.

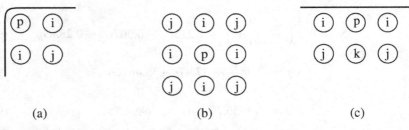

(a)                    (b)                    (c)

**Figure 9.5** Rod peaking factor definitions in the EPRI correlation for: (a) corner rod peaking, (b) center rod peaking, (c) side rod peaking [86].

The CHF data base for the EPRI correlation is primarily 16-rod, with heated length up to 3.66 m, with axially uniform and non-uniform heat flux, in a pressure range from 4.14 to 9.65 MPa, and a mass velocity range from 339 to 2034 kg m$^{-2}$ s$^{-1}$. The correlation predicts the critical power of the test data used to create the correlation with a standard deviation of 5.5% and a mean value of 0.9948. When all applicable BWR CHF data are compared to the correlation, it predicts the test bundle critical power with a standard deviation of 7.5%, with a mean value of 1.0295.

## 9.5  GENERALIZED CHF CORRELATIONS

Generalized critical heat flux (CHF) correlations attempt to predict the conditions under which a boiling crisis occurs across a broad spectrum of flow and heat flux conditions. In practical terms, these correlations are applicable to both PWRs and BWRs. The majority of these correlations are derived from experimental data obtained in simple channels, such as circular tubes. These correlations can take the form of an analytical function that represents the CHF in terms of operating conditions and channel size, or they can be presented as look-up tables, where numerical values of the CHF are tabulated as a function of specific numerical values that represent the operating conditions.

### 9.5.1  SIMPLE CHANNELS

The list of existing correlations for predicting CHF in simple channels is quite extensive. For those interested in further exploration, detailed information on some of these correlations, along with additional references, can be found in [28, 221]. As an example, we provide a simple analytical form of a generalized CHF correlation that was developed by Kim and Lee, who proposed the following formula [124],

$$q''_{cr} = C_1 C_2 \left( \frac{\sigma \rho_f}{G^2 L} \right)^{C_3} \left[ C_4 + \left( \frac{\Delta i_i}{i_{fg}} \right)^{C_5} \right]^{C_6} \exp\left[ C_7 (L/D)^{C_8} \right] \left( \frac{\rho_g}{\rho_f} \right)^{C_9} G i_{fg} , \quad (9.30)$$

where the variables used in the correlation are as follows: $q''_{cr}$–critical heat flux, $G$– mass flux, $L$–tube heated length, $D$–tube internal diameter, $p$–pressure, $\Delta i_i$–inlet subcooling enthalpy, $i_{fg}$–latent heat of evaporation, $\sigma$–surface tension, $\rho_g$–vapor density, and $\rho_f$–liquid density. The coefficients $C_1$ through $C_9$ are given as,

$$C_1 = \begin{cases} 6.471 \cdot 10^{-3} & \text{for} & K \leq 4.446 \cdot 10^{-7} \\ 9.411 \cdot 10^{-2} & \text{for} & 4.446 \cdot 10^{-7} < K \leq 2.848 \cdot 10^{-5} \\ 3.33 \cdot 10^{-2} & \text{for} & K > 2.848 \cdot 10^{-5} \end{cases} , \quad (9.31)$$

$$C_2 = \begin{cases} 0.962 & \text{for} & R \leq 4.705 \cdot 10^{-3} \\ 1.452 & \text{for} & 4.705 \cdot 10^{-3} < R \leq 3.849 \cdot 10^{-2} \\ 1.0 & \text{for} & 3.849 \cdot 10^{-2} < R \leq 1.262 \cdot 10^{-1} \\ 2.3 & \text{for} & R > 1.262 \cdot 10^{-1} \end{cases} , \quad (9.32)$$

$$C_3 = \begin{cases} 5.677 \cdot 10^{-2} & \text{for} & K \leq 4.446 \cdot 10^{-7} \\ 0.2398 & \text{for} & 4.446 \cdot 10^{-7} < K \leq 2.848 \cdot 10^{-5} \\ 0.141 & \text{for} & K > 2.848 \cdot 10^{-5} \end{cases}, \quad (9.33)$$

$$C_4 = 0.316, \quad C_5 = 1.1, \quad C_6 = 0.764, \quad C_7 = -0.121, \quad C_8 = 0.443, \quad (9.34)$$

$$C_9 = \begin{cases} 3.766 \cdot 10^{-2} & \text{for} & R \leq 4.705 \cdot 10^{-3} \\ 0.115 & \text{for} & 4.705 \cdot 10^{-3} < R \leq 3.849 \cdot 10^{-2} \\ 0.0 & \text{for} & 3.849 \cdot 10^{-2} < R \leq 1.262 \cdot 10^{-1} \\ 0.402 & \text{for} & R > 1.262 \cdot 10^{-1} \end{cases}, \quad (9.35)$$

where

$$K = \frac{\sigma \rho_f}{G^2 L}, \quad R = \frac{\rho_g}{\rho_f}. \quad (9.36)$$

The correlation has been derived from an experimental database with 12 879 data points and is valid in the following range of parameters:

$$11.67 \leq L/D \leq 855,$$

$$6.22 \cdot 10^{-4} \leq R \leq 0.35,$$

$$3.07 \cdot 10^{-9} \leq K \leq 1.98 \cdot 10^{-2},$$

$$0.1 \leq p \leq 20 \text{ MPa},$$

$$38 \leq G \leq 34\,200 \text{ kg m}^{-2} \text{ s}^{-1}.$$

In look-up tables, numerical values of the CHF are tabulated as a function of specific numerical values that represent the operating conditions. They are essentially normalized data banks for a vertical 8 mm water-cooled tube. The 2006 CHF look-up table is based on a database containing more than 30 000 data points. The table provides CHF values at 24 pressures, 20 mass fluxes, and 23 qualities, covering the full range of conditions of practical interest [81].

### 9.5.2 SUBCHANNELS AND ROD BUNDLES

EPRI used the COBRA-IIIC subchannel code to develop a generalized CHF correlation covering PWR and BWR operating conditions as well as postulated loss-of-coolant accident conditions. In addition to the influence of local flow and heat transfer conditions, the correlation includes the effects of cold walls, grid spacers, and axial heat flux distribution. The developed correlation is as follows [184],

$$q''_{R,cr} = \frac{C_1 F_a - x_{in}}{C_2 F_c F_g F_{nu} + \left[\frac{x - x_{in}}{q''_R}\right]}, \quad (9.37)$$

where

$$C_1 = P_1 p_R^{P_2} G_R^{(P_5 + p_R P_7)}, \quad (9.38)$$

$$C_2 = P_3 p_R^{P_4} G_R^{(P_6 + p_R P_8)}, \quad (9.39)$$

$q''_{R,cr} = q''_{cr}/3.1544$ and $q''_R = q''/3.1544$, where $q''_{cr}$ and $q''$ are critical and local heat fluxes (MW m$^{-2}$), $x_{in}$ and $x$ are the inlet and local qualities, $G_R = G/1356.23$, where $G$ is the mass flux (kg m$^{-2}$ s$^{-1}$), $p_R$ is the reduced pressure ($p/p_{cr}$), and $P_1$ through $P_8$ are constants,

$$P_1 = 0.5328, \ P_2 = 0.1212, \ P_3 = 1.6151, \ P_4 = 1.4066,$$

$$P_5 = -0.3040, \ P_6 = 0.4843, \ P_7 = -0.3285, P_8 = -2.0749 \ .$$

The CHF limits in subchannels with unheated walls are found to be different from CHF limits in internal subchannels, surrounded by four heated rods. This effect is captured by correction factors $F_a$ and $F_c$ which are equal to 1 for internal subchannels, and for subchannels with unheated wall, they are given by,

$$F_a = G_R^{0.1} \quad \text{and} \quad F_c = 1.183 G_R^{0.1} \ .$$

The correlation was extended to encompass all varieties of spacer grids through the incorporation of the following grid correction factor,

$$F_g = 1.3 - 0.3\xi_g \ ,$$

where $\xi_g$ is the grid's local loss coefficient.

The predictability of the correlation improved significantly when the upstream effects were taken into account. As a measure of the non-uniformity of the axial heat flux distribution, a parameter introduced by Bowring was used [22],

$$Y = \frac{1}{z} \int_0^z \frac{q''(z')dz'}{q''(z)} \ , \tag{9.40}$$

where $z$ is the distance from the inlet. Bowring examined experimental data and concluded that the upstream heat flux effect decreases as mass flux increases and suggested a correction factor in terms of the axial heat flux parameter $Y$ as,

$$F_{nu} = 1 + \frac{Y-1}{1+G_R} \ . \tag{9.41}$$

The correlation covers a wide range of fluid conditions: the pressure from 1.38 to 16.9 MPa, the mass flux from 271 to 5560 kg m$^{-2}$ s$^{-1}$, the local quality from -0.25 to 0.75. The correlation is based on 3607 CHF data points from 65 fuel assemblies (including $3\times3$, $4\times4$, and $5\times5$ bundles) simulating both PWR and BWR cores. These test sections had uniform as well as non-uniform radial heat flux distributions, and covered the following parameter ranges: inlet quality from $-1.10$ to 0.0, length from 0.76 to 4.27 m, hydraulic diameter from $8.89 \cdot 10^{-3}$ to $13.97 \cdot 10^{-3}$ m, and rod diameter from $9.65 \cdot 10^{-3}$ to $16 \cdot 10^{-3}$ m.

The accuracy of the correlation was determined by evaluating the following error statistics:

the averaged ratio:

$$R_{av} = \frac{1}{N} \sum_{i=1}^N R_i \ , \tag{9.42}$$

the root-mean-square error:

$$e_{RMS} = \left[ \frac{1}{N} \sum_{i=1}^{N} (R_i - 1)^2 \right]^{1/2}, \tag{9.43}$$

the standard deviation:

$$S = \left[ \frac{1}{N} \sum_{i=1}^{N} (R_i - R_{av})^2 \right]^{1/2}, \tag{9.44}$$

where $R_i$ is the ratio of predicted CHF to experimental CHF of the $i$th data point. The following accuracy indicators were reported by the correlation developers: $R_{av} = 0.995$, $e_{RMS} = 7.20\%$ and $S = 7.20\%$.

## 9.6   PREDICTION OF CHF IN A BOILING CHANNEL

In a heated channel, the local thermal-hydraulic conditions change along its length due to increasing enthalpy of the coolant and changing pressure due to friction and local pressure losses. The coolant enthalpy can be determined at any cross-section of the channel by solving the mass, momentum, and energy conservation equations, and applying proper boundary conditions. In this section, we perform such analysis for a tube with uniform heat flux distribution. Assuming steady-state conditions and neglecting pressure drop along the tube, the energy balance for a tube with length $L$ and diameter $D$ gives,

$$i_{ex} = i_{in} + \frac{4L}{GD} q'', \tag{9.45}$$

where $i_{ex}$ is the specific enthalpy at the exit from the tube, $i_{in}$ is the specific enthalpy at the tube's inlet, $q''$ is the heat flux, and $G$ is the coolant mass flux. We can express this equation in terms of the thermodynamic equilibrium quality $x \equiv (i - i_f)/i_{fg}$ as,

$$x_{ex} = x_{in} + \frac{4L}{GDi_{fg}} q''. \tag{9.46}$$

It should be noted that $q''$ present in Eq. (9.45) is an average heat flux along the tube length. Since we assumed a constant heat flux along the tube, the distinction between the average and the local heat flux is not necessary.

With a non-uniform heat flux distribution and an arbitrary cross-section of the channel, Eq. (9.45) becomes,

$$i_{ex} = i_{in} + \frac{P_H L}{GA} \cdot \frac{1}{L} \int_0^L q''(z) \mathrm{d}z = i_{in} + \frac{P_H L}{GA} \langle q'' \rangle_1, \tag{9.47}$$

where $P_H$ and $A$ are the heated perimeter and cross-section area of the channel, respectively. Notation $\langle \rangle_1$ is used to indicate one-dimensional averaging of the heat flux. We can express this equation in terms of the thermodynamic equilibrium quality as,

$$x_{ex} = x_{in} + \frac{4L}{GD_H i_{fg}} \langle q'' \rangle_1, \tag{9.48}$$

where we introduced the *heated diameter* defined as,

$$D_H \equiv \frac{4A}{P_H}. \tag{9.49}$$

We can note that for a non-uniform heat flux distribution and an arbitrary channel cross-section, the energy balance equation can be expressed in the same form as for the circular tube with the uniform heat flux distribution given by Eq. (9.46). However, in the equation the tube diameter has to be replaced with the heated diameter and the heat flux has to be averaged along the channel length up to the boiling crisis point.

A boiling crisis in channels is usually investigated in dedicated test sections, where variation of several global parameters is possible. When the geometry of a test section is fixed and a type of coolant is selected, the variable global parameters are limited to $G$, $p$, $x_{in}$, and $q''$. The typical procedure to approach a boiling crisis is to fix all above mentioned global parameters while increasing the heat flux until the boiling crisis is registered. However, it should be remembered that in real reactor core conditions there are multiple paths to boiling crisis. In addition to the most obvious one, when the core thermal power (either globally or locally) increases, boiling crises conditions can be achieved by reducing flow through a channel, changing inlet subcooling to a channel or changing a system pressure. In a most complex situation, all these changes combined can lead to an onset of boiling crises. It was shown experimentally that various paths to achieve boiling crises lead to the same critical conditions [12].

Assuming a uniformly heated round tube with diameter $D$ and length $L$ as a test section, we can expect that the onset of a boiling crisis is described with the following relationship,

$$F(G, p, D, L, x_{in}, q''_{cr}) = 0. \tag{9.50}$$

This function merely indicates that there is a unique combination of these six parameters at any boiling crisis condition. Assuming fixed values of any five arguments of the function $F$, the onset of boiling crisis will occur by continuously changing the value of the remaining free parameter. As already mentioned, one could consider a continuous change of any parameter, including the tube diameter and the length, but for practical reasons, usually the influence of the remaining four parameters is investigated for a given geometry of a test section. We assume that the implicit formulation given by Eq. (9.50) can be represented as follows,

$$q''_{cr} = F_q(G, p, D, L, x_{in}), \tag{9.51}$$

which has a clear physical interpretation: for a certain combination of arguments of function $F_q$, a boiling crisis will occur when the heat flux reaches a value given by the function $F_q$.

Since for uniformly heated tubes, the boiling crisis point is always located at the tube outlet, we can write the following energy conservation equation,

$$i_{cr} = i_{in} + \frac{4L}{DG} q''_{cr}, \tag{9.52}$$

where $i_{cr}$ is a "critical" specific enthalpy and $i_{in}$ is a specific enthalpy at the channel inlet. It is customary to express the energy equation in terms of the critical quality, $x_{cr} \equiv (i_{cr} - i_f)/i_{fg}$, rather than in terms of the critical enthalpy as follows,

$$x_{cr} = x_{in} + \frac{4L}{DGi_{fg}} q''_{cr}. \tag{9.53}$$

Combining Eqs. (9.51) and (9.53) gives the relationship in terms of the critical quality,

$$x_{cr} = x_{in} + \frac{4L}{DGi_{fg}} F_q(G, p, D, L, x_{in}) \equiv F_x(G, p, D, L, x_{in}). \tag{9.54}$$

The two functions $F_q$ and $F_x$, representing the critical heat flux (CHF) and the critical quality, respectively, have been extensively investigated both experimentally and analytically. However, as revealed by Eq. (9.54), these functions are not independent from each other for uniformly heated channels, and in particular, for uniformly heated circular tubes.

To predict the DNBR, we need to divide the critical heat flux by the actual heat flux at any location in a heated channel,

$$\text{DNBR}(z) = \frac{q''_{cr}(G, p, x_e^*(z), D, ...)}{q''_a(z)}, \tag{9.55}$$

where $x_e^*(z)$ is the thermodynamic equilibrium quality at location $z$ and $q''_a(z)$ is the actual heat flux at that location. Strictly speaking, the remaining independent variables of the correlation for the critical heat flux are location-dependent as well, but we assume that their variability is small and can be neglected.

A common misunderstanding about $x_e^*(z)$ is that this thermodynamic equilibrium quality can be found from the energy balance in the channel using the actual heat flux value. This type of approach is known in the literature as the *direct substitution method* (DSM) [80, 81]. However, such an approach is not correct, since the correlation for the critical heat flux is always developed for "critical" values of the parameters, including the critical local equilibrium quality. It means that $x_e^*(z)$ should be found from the energy balance in the channel using the critical heat flux value. This type of approach is known as the *heat balance method* (HBM). The applicability of both methods to CHF prediction has been extensively discussed in [30, 111, 218]. Using HBM, we have,

$$x_e^*(z) = x_{in} + \frac{4z}{DGi_{fg}} q''_{cr}. \tag{9.56}$$

It should be noted that iterations are needed to satisfy Eqs. (9.55) and (9.56).

Figure 9.6 shows a comparison of the critical heat flux predictions, obtained through various methods, with the experimental data derived from uniformly heated tubes.

The figure shows that application of the direct substitution method can lead to significant discrepancies between predictions and measurements. To elucidate the reason for these discrepancies, it is instructive to use the relationship between the

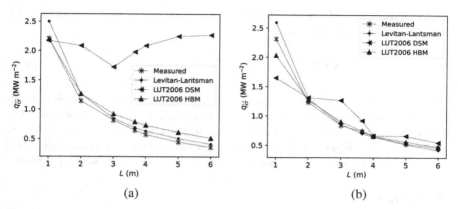

**Figure 9.6**   Critical heat flux $q''_{cr}$ as a function of the length $L$ in a round tube with inner diameter $D = 8$ mm, inlet subcooling $x_{in} = -0.035$, cooled with water at pressure (a) $p = 1$ MPa and (b) 7 MPa . Measured data from [208] and predictions based on the Levitan-Lantsman correlation [144] and look-up tables 2006 using the direct substitution method (DSM) and the heat balance method (HBM) [81].

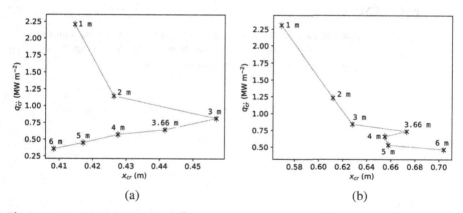

**Figure 9.7**   Critical heat flux, $q''_{cr}$, as a function of the critical quality, $x_{cr}$, measured in uniformly heated tubes with diameter $D = 8$ mm, variable length, $L$, with the inlet subcooling $x_{in} = -0.035$ and cooled with water at pressure (a) $p = 1$ MPa and (b) 7 MPa. Measured data from [208].

critical heat flux and the critical quality for various pipe lengths, as shown in Fig. 9.7. As can be seen, the relationship $q''_{cr}(x_{cr})$ is a multi-valued function making a critical heat flux correlations of this form questionable. At the same time it can be seen that the relationship $x_{cr}(q''_{cr})$ is a single-valued function and thus it is a proper form for the critical heat flux correlation.

To predict the CPR, we need to find a ratio of the critical power in a channel to the actual power of the channel,

$$\text{CPR} = \frac{q_{cr}}{q_a} = \frac{q_{cr}'' P_H L}{q_a'' P_H L} = \frac{q_{cr}''}{q_a''} = \frac{x_{cr} - x_{in}}{x_{ex} - x_{in}}, \tag{9.57}$$

where $x_{ex}$ is the thermodynamic equilibrium quality at the channel outlet with the actual heat flux applied. In this relationship the critical quality, $x_{cr}$, should be calculated in terms of other critical parameters. In particular, if $x_{cr}$ depends on the boiling length, the boiling length should be found for the value of the critical heat flux. Again iterations are needed to find the value of CPR, since the critical power is not known.

## PROBLEMS

### PROBLEM 9.1

Find the dryout power (W) in a uniformly heated tube with 10 mm inner diameter. The tube is cooled with water at 6.9 MPa pressure, inlet subcooling 10 K, and mass flux 1250 kg m$^{-2}$ s$^{-1}$. Neglect the pressure drop in the tube.

### PROBLEM 9.2

Use the Levitan and Lantsman correlation for dryout in a 8 mm tube with the length $L$ equal to 5 m. Apply the operating conditions as specified in Fig. 9.7(a). Compare your result with the values presented in the figure.

# Section III

Thermal Safety Margins

# 10 Thermal Performance of Fuel Elements

The analysis of the thermal performance of fuel elements is an integral part of nuclear reactors design. Such analysis provides distributions of temperature in a reactor core at normal and accidental conditions. In the first step, nominal temperature distributions are found. In simple geometries, such as plates, cylinders, and spheres, exact solutions of the heat conduction partial differential equation can be found. In more complex geometries the heat conduction equation can be solved numerically. However, such solutions do not account for such effects as uncertainties in theoretical analysis, fuel manufacturing tolerances, experimental errors on data used in the design, and uncertainties on adopted physical properties and correlations. These uncertainties are commonly expressed as dimensionless factors referred to as the hot channel or hot spot factors. The factors are greater than unity and, in the second step of analysis, the product of each of these factors by the corresponding nominal value of the temperature is used in the safety analysis. In this chapter we discuss the nominal solutions for temperature distributions in fuel elements. The methods to take into account the uncertainties are presented in chapter §11.

## 10.1 GOVERNING EQUATIONS

Temperature distributions in fuel elements can be obtained from solutions of the heat conduction partial differential equation (see §4.8.1) together with case-dependent initial and boundary conditions. The considered set of equations to be solved is as follows,

$$\rho c_p \frac{\partial T(\mathbf{r},t)}{\partial t} = \nabla [\lambda \nabla T(\mathbf{r},t)] + q'''(\mathbf{r},t), \tag{10.1}$$

where $T$ is the temperature, $\mathbf{r}$ is the position vector, $t$ is time, $\rho$ is the density, $c_p$ is the specific heat capacity, $\lambda$ is the thermal conductivity, and $q'''$ is the heat source rate per unit volume. In general, the physical properties $\rho$, $c_p$, and $\lambda$ depend on the position vector, temperature, and time. However, the time and space dependencies are rather weak and are usually neglected. For some materials, for example $UO_2$, the temperature influence on the physical properties is strong and has to be taken into account.

If we consider temperature distribution in an arbitrary volume $V$ with a boundary $S$, the following initial condition in $V$ is needed,

$$T(\mathbf{r},0) = T_0(\mathbf{r}) \quad \text{for} \quad \mathbf{r} \in V, \tag{10.2}$$

where $T_0(\mathbf{r})$ is a given function describing the initial temperature distribution in $V$.

In general four different conditions can be considered on the boundary surface $S$. For a boundary condition of the first kind, also known as the *Dirichlet boundary condition*, the temperature distribution is known on $S_1 \in S$ and is given as follows,

$$T(\mathbf{r},t)\,|_{\mathbf{r}\in S_1} = T_1(\mathbf{r}_1,t) \quad \text{for} \quad \mathbf{r}_1 \in S_1. \tag{10.3}$$

Here $T_1(\mathbf{r}_1,t)$ is a given function describing the temperature distribution on boundary surface $S_1$.

When heat flux is known at boundary $S_2 \in S$, the boundary condition of the second kind, also known as the *Neumann boundary condition*, is specified as follows,

$$-\lambda \nabla T(\mathbf{r},t)\,|_{\mathbf{r}\in S_2} \cdot \mathbf{n}_2 = q''(\mathbf{r}_2,t) \quad \text{for} \quad \mathbf{r}_2 \in S_2. \tag{10.4}$$

Here $\mathbf{n}_2$ is a unit vector normal to surface $S_2$ pointing outward from the volume $V$ and $q''(\mathbf{r}_2,t)$ is the known function representing heat flux distribution on surface $S_2$. This function is defined as positive when the heat flows out from the volume $V$.

For a body surface $S_3$ surrounded by a fluid with known far-field temperature $T_\infty$, it is convenient to specify a boundary condition of the third kind, also referred to as the *Robin boundary condition*, given as,

$$-\lambda \nabla T(\mathbf{r},t)\,|_{\mathbf{r}\in S_3} \cdot \mathbf{n}_3 = h(\mathbf{r}_3,t)\,[T(\mathbf{r}_3,t) - T_\infty] \quad \text{for} \quad \mathbf{r}_3 \in S_3. \tag{10.5}$$

Here $h$ is the heat transfer coefficient that in general depends on the location vector $\mathbf{r}_3$ and time, and $\mathbf{n}_3$ is a unit vector normal to surface $S_3$ pointing out from the volume $V$. It should be noted that the far-field fluid temperature $T_\infty$ is considered to be given and in general, varies both with the location vector and time.

Ideal direct contact boundary conditions describe the equality of temperatures and heat fluxes at surface $S_4$ that is common for two heat-conducting bodies $A$ and $B$. The conditions are given as,

$$T_A(\mathbf{r}_4,t) = T_B(\mathbf{r}_4,t) \quad \text{and} \quad \lambda_A \nabla T_A(\mathbf{r}_4,t) \cdot \mathbf{n}_A = -\lambda_B \nabla T_B(\mathbf{r}_4,t) \cdot \mathbf{n}_B. \tag{10.6}$$

It should be noted that unit vectors $\mathbf{n}_A$ and $\mathbf{n}_B$ are normal to surface $S_4$ and are pointing outward from bodies $A$ and $B$, respectively. Thus, they satisfy the following condition at any location $\mathbf{r}_4 \in S_4$: $\mathbf{n}_A = -\mathbf{n}_B$.

In reality there is a heat transfer resistance at the contact surface due to nonperfect alignment of the two bodies. As a result, the temperatures of bodies $A$ and $B$ are not equal at the contact surface $S_4$ and satisfy the following relationship,

$$T_A - T_B = R_{AB} q''_{AB} = \frac{q''_{AB}}{H_{AB}}. \tag{10.7}$$

Here $q''_{AB}$ is the heat flux normal to surface $S_4$, directed from body $A$ to body $B$, $R_{AB}$ ($m^2$ K $W^{-1}$) is the *thermal insulance*, and $H_{AB} \equiv 1/R_{AB}$ (W $m^{-2}$ $K^{-1}$) is the *thermal conductance* of the contact surface.

For steady-state conditions, the conductivity equation is as follows,

$$\nabla(\lambda \nabla T) = -q''', \tag{10.8}$$

where, for simplicity, we dropped notation to show the dependence of temperature on the spatial coordinates. This differential equation still requires boundary conditions to be solved.

## 10.2 CYLINDRICAL FUEL ELEMENTS

Cylindrical fuel elements, or fuel rods, are most common in nuclear power reactors. Their general structure is presented in §2.3. The temperature distribution in fuel rods is most conveniently considered in a cylindrical coordinate system, in which, the heat conduction equation takes the following form,

$$\rho c_p \frac{\partial T}{\partial t} = \frac{1}{r} \frac{\partial}{\partial r} \left( \lambda r \frac{\partial T}{\partial r} \right) + \frac{1}{r^2} \frac{\partial}{\partial \theta} \left( \lambda \frac{\partial T}{\partial \theta} \right) + \frac{\partial}{\partial z} \left( \lambda \frac{\partial T}{\partial z} \right) + q''', \tag{10.9}$$

where $(r, \theta, z)$ are coordinates in the radial, circumferential, and axial direction, respectively.

### 10.2.1 STEADY-STATE HEAT TRANSFER

When the heat source term in Eq. (10.9) and the boundary conditions are not changing with time, a steady-state heat transfer problem can be considered and the partial time derivative in Eq. (10.9) can be dropped. In this section we will consider several special cases of the heat conduction equation making additional simplifying assumptions.

### One-Dimensional Approximation

We will first consider steady-state heat conduction in an infinite cylindrical fuel element with outer radius $R_{Fo}$, constant heat source rate per unit volume $q_F'''$, and constant thermal conductivity $\lambda_F$. We assume that the temperature of the fuel element at the outer cylindrical surface is everywhere constant and equal to $T_{Fo}$. For such conditions, one-dimensional axisymmetric temperature distribution will exist, providing a symmetry condition at $r = 0$. Thus, the following set of equations can be written,

$$\frac{1}{r} \frac{d}{dr} \left( r \frac{dT_F}{dr} \right) = -\frac{q_F'''}{\lambda_F}, \tag{10.10}$$

$$\left. \frac{dT_F}{dr} \right|_{r=0} = 0, \tag{10.11}$$

$$T_F(r)|_{r=R_{Fo}} = T_{Fo}. \tag{10.12}$$

Double integration of Eq. (10.10) yields,

$$T_F(r) = -\frac{q_F''' r^2}{4\lambda_F} + C_1 \ln r + C_2, \tag{10.13}$$

where $C_1$ and $C_2$ are constants of integration that can be determined from boundary conditions. The solution symmetry condition at $r = 0$ requires that $C_1 = 0$, whereas the given temperature at the cylinder surface gives,

$$C_2 = T_{Fo} + \frac{q_F''' R_{Fo}^2}{4\lambda_F}. \tag{10.14}$$

Substituting constants $C_1$ and $C_2$ into Eq. (10.13) yields,

$$T_F(r) = \frac{q_F''' R_{Fo}^2}{4\lambda_F}\left[1 - \left(\frac{r}{R_{Fo}}\right)^2\right] + T_{Fo}. \tag{10.15}$$

We can now find the maximum temperature in the fuel element substituting $r = 0$ into the solution:

$$T_F(0) \equiv T_{Fc} = \frac{q_F''' R_{Fo}^2}{4\lambda_F} + T_{Fo}. \tag{10.16}$$

The temperature rise in the cylindrical fuel element, $\Delta T_F$, is thus given as,

$$\Delta T_F \equiv T_{Fc} - T_{Fo} = \frac{q_F''' R_{Fo}^2}{4\lambda_F}. \tag{10.17}$$

We will now consider a similar case, assuming a hollow cylindrical fuel element with inner radius $R_{Fi}$ and outer radius $R_{Fo}$. The governing differential equation and the boundary condition at the outer surface remains the same as in the case with solid fuel element, however, the boundary condition at the inner surface becomes as follows,

$$\left.\frac{dT_F}{dr}\right|_{r=R_{Fi}} = 0. \tag{10.18}$$

This condition requires that there is no net heat transfer between the fuel element's inner surface and the central empty space. The constants of integration are now as follows,

$$C_1 = \frac{q_F''' R_{Fi}^2}{2\lambda_F}, \tag{10.19}$$

$$C_2 = \frac{q_F''' R_{Fo}^2}{4\lambda_F} - \frac{q_F''' R_{Fi}^2}{2\lambda_F}\ln R_{Fo} + T_{Fo}. \tag{10.20}$$

Thus, the temperature distribution in the hollow cylindrical fuel element is as follows,

$$T_F(r) = \frac{q_F''' R_{Fo}^2}{4\lambda_F}\left[1 - \left(\frac{r}{R_{Fo}}\right)^2 - 2\left(\frac{R_{Fi}}{R_{Fo}}\right)^2\ln\left(\frac{r}{R_{Fo}}\right)\right] + T_{Fo}. \tag{10.21}$$

The corresponding temperature rise is found as,

$$\Delta T_F = \frac{q_F''' R_{Fo}^2}{4\lambda_F}\left[1 - \left(\frac{R_{Fi}}{R_{Fo}}\right)^2 - 2\left(\frac{R_{Fi}}{R_{Fo}}\right)^2\ln\left(\frac{R_{Fi}}{R_{Fo}}\right)\right]. \tag{10.22}$$

Comparing Eqs. (10.17) and (10.22) it can be seen that, keeping all other parameters equal, the temperature rise in the hollow fuel element is less than the corresponding temperature rise in the solid fuel element. Thus introduction of hollow fuel elements is an effective means of reducing the maximum fuel temperature.

Instead of expressing the fuel temperature rise in terms of the heat source rate per unit volume, it is frequently preferred to introduce linear fuel power $q'_F$. Taking a cylindrical fuel element with length $\Delta z$, a steady-state energy balance gives,

$$q'_F \Delta z = q'''_F \pi R_{Fo}^2 \Delta z. \tag{10.23}$$

Thus

$$q'''_F = \frac{q'_F}{\pi R_{Fo}^2}. \tag{10.24}$$

For hollow fuel element, the following expression is obtained,

$$q'''_F = \frac{q'_F}{\pi \left( R_{Fo}^2 - R_{Fi}^2 \right)}. \tag{10.25}$$

Thus, in terms of a linear power, the temperature rise in the solid and hollow fuel element are given as, respectively,

$$\Delta T_F = \frac{q'_F}{4\pi \lambda_F}, \tag{10.26}$$

$$\Delta T_F = \frac{q'_F R_{Fo}^2}{4\pi \lambda_F \left( R_{Fo}^2 - R_{Fi}^2 \right)} \left[ 1 - \left( \frac{R_{Fi}}{R_{Fo}} \right)^2 - 2 \left( \frac{R_{Fi}}{R_{Fo}} \right)^2 \ln \left( \frac{R_{Fi}}{R_{Fo}} \right) \right]. \tag{10.27}$$

We can note that the fuel temperature rise for the solid cylindrical fuel element depends only on the linear power density and the thermal conductivity. In the case of the hollow fuel element, the additional parameter on which the fuel temperature rise depends is the ratio of the inner radius to the outer radius of the fuel element.

Fuel rods of nuclear reactors have a cylindrical composite structure, with the central part containing fissile material, surrounded by a cladding tube. The fissile material and the inner surface of the cladding tube are separated from each other with a gas gap. Thus a fuel rod can be considered as a cylinder with concentric layers having different physical properties.

The heat transfer in a gas gap is very complex due to gas motion and the presence of various heat transfer modes, including natural convection, conduction, and radiation heat transfer. The heat transfer capability of the gas gap is thus frequently expressed in terms of gas conductance, which combines many effects including the gas thermal conductivity and the gas gap thickness. For the purpose of the present analysis, we assume that gas gap can be treated as a solid cylinder with known inner and outer radius equal to $R_{Gi}$ and $R_{Go}$, respectively, and with a known thermal conductivity $\lambda_G$. With these assumptions, the temperature distribution in an infinite cylindrical gas gap is described with the following equation,

$$\frac{1}{r} \frac{d}{dr} \left( r \lambda_G \frac{dT_G}{dr} \right) = 0. \tag{10.28}$$

After integration we get,

$$T_G(r) = C_1 \ln r + C_2. \tag{10.29}$$

We now apply ideal boundary conditions of the fourth kind at $r = R_{Gi} = R_{Fo}$ as follows,

$$T_G(r)|_{r=R_{Gi}} = T_F(r)|_{r=R_{Fo}}, \quad \lambda_G \frac{dT_G(r)}{dr}\bigg|_{r=R_{Gi}} = \lambda_F \frac{dT_F(r)}{dr}\bigg|_{r=R_{Fo}}. \quad (10.30)$$

Substituting the solutions given by Eqs. (10.13) and (10.29) into the boundary conditions, the integration constants $C_1$ and $C_2$ can be determined and the temperature distribution in the gas gap is obtained as,

$$T_G(r) = -\frac{q_F''' R_{Fo}^2}{2\lambda_G} \ln\left(\frac{r}{R_{Fo}}\right) + T_{Fo} \quad \text{for} \quad R_{Fo} \leq r \leq R_{Go}. \quad (10.31)$$

The temperature rise in the gas gap is obtained as,

$$\Delta T_G \equiv T_{Fo} - T_{Go} = \frac{q_F''' R_{Fo}^2}{2\lambda_G} \ln\left(\frac{R_{Go}}{R_{Fo}}\right). \quad (10.32)$$

In a similar manner we obtain the temperature distribution in the cladding tube, $T_C(r)$, as follows,

$$T_C(r) = -\frac{q_F''' R_{Fo}^2}{2\lambda_C} \ln\left(\frac{r}{R_{Go}}\right) + T_{Go} \quad \text{for} \quad R_{Go} \leq r \leq R_{Co}, \quad (10.33)$$

where $R_{Co}$ is the cladding tube outer radius. The corresponding temperature rise in the cladding tube is obtained as,

$$\Delta T_C \equiv T_{Go} - T_{Co} = \frac{q_F''' R_{Fo}^2}{2\lambda_C} \ln\left(\frac{R_{Co}}{R_{Go}}\right). \quad (10.34)$$

The total temperature rise in the fuel rod, $\Delta T_R$, can be found as,

$$\Delta T_R \equiv T_{Fc} - T_{Co} = \frac{q_F''' R_{Fo}^2}{4\lambda_F} \left[1 + \frac{2\lambda_F}{\lambda_G} \ln\left(\frac{R_{Go}}{R_{Fo}}\right) + \frac{2\lambda_F}{\lambda_C} \ln\left(\frac{R_{Co}}{R_{Go}}\right)\right]. \quad (10.35)$$

In terms of the linear power, the total temperature rise in the rod is as follows,

$$\Delta T_R = \frac{q_F'}{4\pi\lambda_F} \left[1 + \frac{2\lambda_F}{\lambda_G} \ln\left(\frac{R_{Go}}{R_{Fo}}\right) + \frac{2\lambda_F}{\lambda_C} \ln\left(\frac{R_{Co}}{R_{Go}}\right)\right]. \quad (10.36)$$

The derived expression for the temperature rise in a composite fuel rod is valid for steady state heat transfer in an infinite rod with uniform radial and axial power distribution. For such conditions the temperature varies only in the radial direction and heat is conducted radially from fuel through the gas gap and the cladding tube.

## Two-Dimensional Temperature Distributions

Fuel rods have non-uniform temperature distributions in both radial and axial directions. The radial temperature distribution is mainly governed by heat conduction from the fuel pellet region, through the gas gap and cladding tube, to the coolant. Also fission power density in a fuel rod cross-section has non-uniform distribution due to various effects, such as the non-uniform neutron flux distribution and non-uniform distribution of fissile material in the pellet. In the axial direction, fuel rods span the whole height of the core and thus the axial power distributions in fuel rods correspond to the axial power distribution in the entire core.

Using a one-group diffusion equation, the axial neutron flux, and thus the axial linear power density has a cosine distribution as follows,

$$q'_F(z) = q'_{F0} \cos\left(\frac{\pi z}{\hat{H}}\right), \tag{10.37}$$

where $\hat{H} = H + 2d$ is the extrapolated core height, $H$ is the physical core height, and $d$ is the extrapolation length for the core. Here $q'_F(z)$ is a linear power density at the axial coordinate $z$ and $q'_{F0} \equiv q'_F(0)$ is the linear power density at $z = 0$.

We will now consider a cooling channel attached to the fuel rod with an inlet at $z = -H/2$ and an outlet at $z = H/2$. A steady state energy balance for the cooling channel segment from the inlet to an axial location $z$ gives,

$$W\left(i_b(z) - i_{in}\right) = \int_{-H/2}^{z} q'_F(z')dz', \tag{10.38}$$

where $W$ (kg s$^{-1}$) is the coolant mass flow rate, $i_b(z)$ is the bulk specific enthalpy of coolant at axial location $z$, and $i_{in}$ is the coolant specific enthalpy at the channel inlet. After integration, we obtain the following expression for the bulk specific enthalpy in the channel,

$$i_b(z) = \frac{q'_{F0}}{W} \frac{\hat{H}}{\pi} \left[\sin\left(\frac{\pi z}{\hat{H}}\right) + \sin\left(\frac{\pi H}{2\hat{H}}\right)\right] + i_{in}. \tag{10.39}$$

Introducing an equivalent specific heat capacity of coolant defined as,

$$\bar{c}_p(z) \equiv \frac{i_b(z) - i_{in}}{T_b(z) - T_{in}}, \tag{10.40}$$

the coolant bulk temperature at axial location $z$ in the channel is given as,

$$T_b(z) = \frac{q'_{F0}}{W\bar{c}_p(z)} \frac{\hat{H}}{\pi} \left[\sin\left(\frac{\pi z}{\hat{H}}\right) + \sin\left(\frac{\pi H}{2\hat{H}}\right)\right] + T_{in}, \tag{10.41}$$

where $T_{in}$ is the coolant temperature at the inlet. The equivalent specific heat capacity is a function of the axial distance, however, to simplify further analysis, we will assume that its value is constant in the entire channel: $\bar{c}_p(z) = \bar{c}_p = const.$

Using the Newton equation of cooling, the temperature of the cladding outer surface at location $z$ can be found as,

$$T_{Co}(z) = T_b(z) + \frac{q''_{Co}(z)}{h}, \tag{10.42}$$

where $q_{Co}''(z) = q_F'(z)/(2\pi R_{Co})$ is the heat flux at the outer surface of the cladding tube and $h$ is the heat transfer coefficient. Combining Eqs. (10.41) and (10.42) yields,

$$T_{Co}(z) = \frac{q_{F0}'}{W\bar{c}_p(z)} \frac{\hat{H}}{\pi} \left[ \sin\left(\frac{\pi z}{\hat{H}}\right) + \sin\left(\frac{\pi H}{2\hat{H}}\right) \right] + \frac{q_{F0}'}{2\pi R_{Co}h} \cos\left(\frac{\pi z}{\hat{H}}\right) + T_{in}. \quad (10.43)$$

We note that this equation has the following general form,

$$T_{Co}(z) = A + B\sin\left(\frac{\pi z}{\hat{H}}\right) + C_{Co}\cos\left(\frac{\pi z}{\hat{H}}\right), \quad (10.44)$$

where constants $A$, $B$, and $C_{Co}$ are defined as,

$$B = \frac{q_{F0}'}{W\bar{c}_p(z)} \frac{\hat{H}}{\pi}, \quad A = B\sin\left(\frac{\pi H}{2\hat{H}}\right) + T_{in}, \quad C_{Co} = \frac{q_{F0}'}{2\pi R_{Co}h}. \quad (10.45)$$

To find the maximum temperature of the cladding outer surface, we take the first derivative of the function given by Eq. (10.44) and set it to zero:

$$\frac{dT_{Co}(z)}{dz} = B\cos\left(\frac{\pi z}{\hat{H}}\right) - C_{Co}\sin\left(\frac{\pi z}{\hat{H}}\right) = 0. \quad (10.46)$$

Thus the maximum temperature of the cladding outer surface is located at

$$z_{Co,max} = \frac{\hat{H}}{\pi} \arctan\left(\frac{B}{C_{Co}}\right). \quad (10.47)$$

Substituting this value into Eq. (10.44), the maximum temperature of the cladding outer surface is obtained as,

$$T_{Co,max} = A + \sqrt{B^2 + C_{Co}^2}. \quad (10.48)$$

Similarly, the cladding inner surface temperature can be obtained as,

$$T_{Ci}(z) = T_{Co}(z) + \Delta T_C = A + B\sin\left(\frac{\pi z}{\hat{H}}\right) + C_{Ci}\cos\left(\frac{\pi z}{\hat{H}}\right), \quad (10.49)$$

where $\Delta T_C$ is given by Eq. (10.34) and,

$$C_{Ci} = \frac{q_{F0}'}{2\pi} \left[ \frac{1}{\lambda_C} \ln\left(\frac{R_{Co}}{R_{Ci}}\right) + \frac{1}{R_{Co}h} \right]. \quad (10.50)$$

Thus, in analogy with the previous derivation, the location and the value of the maximum temperature of the cladding inner surface is found as, respectively,

$$z_{Ci,max} = \frac{\hat{H}}{\pi} \arctan\left(\frac{B}{C_{Ci}}\right), \quad (10.51)$$

$$T_{Ci,max} = A + \sqrt{B^2 + C_{Ci}^2}. \quad (10.52)$$

The maximum fuel temperature is found at the rod centerline at the axial location given as,

$$z_{Fc,max} = \frac{\hat{H}}{\pi} \arctan\left(\frac{B}{C_{Fc}}\right), \tag{10.53}$$

where,

$$C_{Fc} = \frac{q'_{F0}}{2\pi}\left[\frac{1}{2\lambda_F} + \frac{1}{\lambda_G}\ln\left(\frac{R_{Go}}{R_{Gi}}\right) + \frac{1}{\lambda_C}\ln\left(\frac{R_{Co}}{R_{Ci}}\right) + \frac{1}{R_{Co}h}\right], \tag{10.54}$$

and the corresponding maximum fuel temperature is given by,

$$T_{Fc,max} = A + \sqrt{B^2 + C_{Fc}^2}. \tag{10.55}$$

## Variable Thermal Conductivity of Fuel

As shown in §3.2, the thermal conductivity of fuel is a strong function of temperature. This variability has to be taken into account once predicting temperature distribution in fuel pellets. For steady-state heat conduction in an infinite cylinder, the governing equation is as follows,

$$\frac{1}{r}\frac{d}{dr}\left(r\lambda_F\frac{dT_F}{dr}\right) = -q'''_F = -\frac{q'_F}{\pi R_{Fo}^2}, \tag{10.56}$$

with the following boundary conditions:

$$\left.\frac{dT_F}{dr}\right|_{r=0} = 0, \tag{10.57}$$

$$T_F(r)|_{r=R_{Fo}} = T_{Fo}. \tag{10.58}$$

Single integration of Eq. (10.56) yields,

$$r\lambda_F\frac{dT_F}{dr} + \frac{q'_F}{\pi R_{Fo}^2}\frac{r^2}{2} = C, \tag{10.59}$$

where $C$ is the integration constant. This constant has to be zero, as can be seen from the symmetry boundary condition at $r = 0$. Thus, the differential equation becomes,

$$\lambda_F dT_F = -\frac{q'_F}{2\pi R_{Fo}^2} r dr. \tag{10.60}$$

Integration of both sides of the equation from $r = 0$ to $r = R_{Fo}$ yields,

$$\int_{T_{Fc}}^{T_{Fo}} \lambda_F dT_F = -\frac{q'_F}{4\pi}. \tag{10.61}$$

The equation can be written in the following form,

$$q'_F = 4\pi \int_{T_{Fo}}^{T_{Fc}} \lambda_F dT_F. \tag{10.62}$$

We obtained an important equation that relates the linear heat generation rate with an integral of the thermal conductivity in the temperature range from the pellet outer surface to its center. This integral is a special case of the *conductivity integral* defined as,

$$I_C(T) = \int_{T_{ref}}^{T} \lambda(\tau)d\tau, \qquad (10.63)$$

where $T_{ref}$ is a reference temperature and $\lambda$ is a temperature-dependent thermal conductivity. Using the conductivity integral, Eq. (10.62) can be expressed as,

$$q_F' = 4\pi \left( \int_{T_{Fo}}^{T_{ref}} \lambda_F dT_F + \int_{T_{ref}}^{T_{Fc}} \lambda_F dT_F \right) = 4\pi \left[ I_C(T_{Fc}) - I_C(T_{Fo}) \right] . \qquad (10.64)$$

Thus, for known $q_F'$ and $T_{Fo}$, we can find the conductivity integral for the fuel temperature at the centerline,

$$I_C(T_{Fc}) = I_C(T_{Fo}) + \frac{q_F'}{4\pi}. \qquad (10.65)$$

Knowing the value of the conductivity integral at the centerline, the fuel temperature at that location can be found from its inverse function as $T_{Fc} = I_C^{-1}$.

## Fuel Restructuring

In the previous section we assumed that the composition of the fuel and the geometry of the fuel rod do not change. In reality high-temperature irradiation of the uranium-dioxide or MOX fuels results in local restructuring of the fuel. When the fuel temperature in the central zones exceeds 1623–1673 K, an equiaxed-grain structure is formed. For higher temperatures, around 1873–1973 K, restructuring into columnar grain forms is taking place. These processes are accompanied with increase of local fuel density and creation of the void in the center of the pellet.

Extensive fuel restructuring have been first observed in fuel rods after irradiation in Fast Breeder Reactors (FBRs). During commercial operation of Light Water Reactors (LWRs) the high-temperature fuel restructuring is considered to occur less likely thanks to a relatively low *linear heat generation rate* (LHGR), which is limited by a maximum permitted power, usually referred to as *thermo-mechanical operational limit* (TMOL). However, intensive power ramps in LWR fuel rods were shown to result in a discernible high-temperature restructuring and central void formation.

We will consider a radial fuel pellet restructuring which results in creation of the following cylindrical regions:

1. Cylindrical central-void region (region 0) with outer radius $R_0$.
2. Annular columnar-grain region (region 1) with radius in a range from $R_0$ to $R_1$.
3. Annular equiaxed-grain region (region 2) with radius in a range from $R_1$ to $R_2$.
4. Annular as-fabricated region (region 3) with radius in a range from $R_2$ to $R_{Fo}$.

Due to the fuel density change in regions 1 and 2, the fission power density in these regions changes as well. We first note that the fuel restructuring is not affecting the linear heat generation rate $q_F'$, since the fissile material moves only radially. It is also reasonable to assume that the power density in regions 1 and 2 changes proportionally to the density changes. Thus, we have,

$$q_1''' = q_F''' \frac{\rho_1}{\rho_F}, \quad q_2''' = q_F''' \frac{\rho_2}{\rho_F}, \quad q_3''' = q_F''', \tag{10.66}$$

where $q_i'''$ and $\rho_i$ are the power density and fuel mass density in region $i$, respectively, and $q_F'''$ and $\rho_F$ are the power density and the fuel mass density in the as-fabricated fuel, respectively.

From fuel mass conservation in a cross section we have,

$$\rho_1 \left( R_1^2 - R_0^2 \right) + \rho_2 \left( R_2^2 - R_1^2 \right) = \rho_F R_2^2. \tag{10.67}$$

From this equation we can determine the central void radius in terms of known densities and sizes of regions 1 and 2:

$$R_0 = \sqrt{\frac{\rho_1 - \rho_2}{\rho_1} R_1^2 + \frac{\rho_2 - \rho_F}{\rho_1} R_2^2}. \tag{10.68}$$

With known fission power density in each region, we solve the heat conductivity equation and the following expressions for the conductivity integrals are found,

$$\int_{T_{Fo}}^{T_2} \lambda_{F3} dT = \frac{q_F'}{4\pi} \left[ 1 - \left( \frac{R_2}{R_{Fo}} \right)^2 \right], \tag{10.69}$$

$$\int_{T_2}^{T_1} \lambda_{F2} dT = \frac{q_F'}{4\pi} \frac{\rho_2}{\rho_3} \left( \frac{R_2}{R_{Fo}} \right)^2 \left[ 1 - \left( \frac{R_1}{R_2} \right)^2 - 2 \left( 1 - \frac{\rho_3}{\rho_2} \right) \ln \left( \frac{R_2}{R_1} \right) \right], \tag{10.70}$$

$$\int_{T_1}^{T_0} \lambda_{F1} dT = \frac{q_F'}{4\pi} \frac{\rho_1}{\rho_3} \left( \frac{R_1}{R_{Fo}} \right)^2 \left[ 1 - \left( \frac{R_0}{R_1} \right)^2 - 2 \left( \frac{R_0}{R_1} \right)^2 \ln \left( \frac{R_1}{R_0} \right) \right]. \tag{10.71}$$

Here $T_0$, $T_1$, and $T_2$ are temperatures at $R_0$, $R_1$, and $R_2$, respectively, and $\lambda_{F1}$, $\lambda_{F2}$, and $\lambda_{F3}$ are thermal conductivities in regions 1–3.

## Variable Cladding Thermal Conductivity

For cladding tube materials, such as Zircaloy 2 and Zircaloy 4, the thermal conductivity can be assumed to be a linear function of temperature:

$$\lambda_C = a + b T_C, \tag{10.72}$$

where $a$ and $b$ are constants. The heat conductivity equation in the cladding with temperature-dependent thermal conductivity is as follows,

$$\frac{1}{r} \frac{d}{dr} \left( r \lambda_C \frac{dT_C}{dr} \right) = 0, \tag{10.73}$$

with the following boundary condition:

$$-\lambda_C \frac{dT_C}{dr}\bigg|_{r=R_{Co}} = q''_{Co} = \frac{q'_F}{2\pi R_{Co}}, \tag{10.74}$$

Single integration of Eq. (10.73) yields,

$$r\lambda_C \frac{dT_C}{dr} = C, \tag{10.75}$$

and application of the boundary condition yields,

$$r\lambda_C \frac{dT_C}{dr} + \frac{q'_F}{2\pi} = 0. \tag{10.76}$$

After integration from $R_{Ci}$ to $R_{Co}$ we get the following equation for temperature $T_{Ci}$,

$$a(T_{Co} - T_{Ci}) + \frac{b}{2}\left(T_{Co}^2 - T_{Ci}^2\right) + \frac{q'_F}{2\pi}\ln\left(\frac{R_{Co}}{R_{Ci}}\right) = 0. \tag{10.77}$$

Solution of this second-order algebraic equation gives,

$$T_{Ci} = \frac{1}{b}\left[\sqrt{(a+bT_{Co})^2 + \frac{bq'_F}{\pi}\ln\left(\frac{R_{Co}}{R_{Ci}}\right)} - a\right]. \tag{10.78}$$

### Pellet-Cladding Gap Behavior

Fresh fuel rods contain pellets separated from cladding tube with a small gap. During fabrication, the gap is filled with pressurized helium gas.[1] Helium is inert and has a relatively high thermal conductivity so that the temperature rise across the gap is minimized during fuel initial operation. During irradiation in the core, small quantities of gaseous fission products such as krypton and xenon are released and mixed with the helium gas in the gap. Since these gases have fairly low thermal conductivity, the resulting mixture thermal conductivity becomes lower as well. In addition to the gas composition change, the gas gap undergoes several other transformations during reactor operation. The main additional parameters that affect thermal properties of the gap include gas pressure, temperature, and average distance between fuel and cladding surfaces. A variation of all these parameters during fuel irradiation have to be followed in order to properly predict the temperature distribution across the gap and in the whole fuel rod.

Gap thermal conductance (frequently referred to as *gap conductance*) normally is defined as,[2]

$$H_G \equiv \frac{q''_{Gi}}{\Delta T_G}, \tag{10.79}$$

---

[1] The initial rod internal pressure ranges from 0.3 to 3.45 MPa for light water reactors and increases with irradiation.

[2] This definition is not always followed, however. For example, in [139] the heat flux at outer gap surface is used, whereas in [219] a mean-in-gap heat flux is adopted.

where $q''_{Gi}$ is a heat flux at the gap inner surface and $\Delta T_G$ is the temperature rise across the gas gap. Since several heat transfer mechanisms co-exist in the gap, the gap conductance is calculated as a superposition of these mechanisms as follows [222],

$$H_G = H_g + H_s + H_r. \qquad (10.80)$$

Here $H_g$, $H_s$, and $H_r$ are contributions due to fill gas conductance, solid contact conductance, and direct thermal radiation, respectively.

The fill gas conductance depends on the heat transfer kinetics between gas molecules, which depends on the value of the Knudsen number defined as,

$$Kn = \frac{\mathscr{L}}{D}, \qquad (10.81)$$

where $\mathscr{L}$ is a molecular mean free path and $D$ is a characteristic distance. Four heat transfer regimes are usually considered: continuum, slip, transition, and free-molecular.

In the continuum flow regime ($Kn < 10^{-3}$) we assume that the Fourier law of conductivity is valid and the fill gas conductance can be derived from Eq. (10.32) as follows,

$$H_g = \frac{\lambda_g}{R_{Gi} \ln \left( \frac{R_{Go}}{R_{Gi}} \right)}, \qquad (10.82)$$

where $\lambda_g$ is the fill gas thermal conductivity.

In the *slip flow* regime, when the distance between cladding and fuel surfaces is decreasing, the energy transfer between gas molecules and the surfaces is affected. The concept of the *temperature jump distance* has been developed to account for a temperature discontinuity at the surfaces. The models for the temperature jump distance can be summarized in the following generalized form [75],

$$g = \frac{C \lambda_g \sqrt{T_g}}{p_g} F(a, f, M), \qquad (10.83)$$

where $g$ (m) is the temperature jump distance, $C$ is a constant, $\lambda_g$ (W m$^{-1}$ K$^{-1}$) is thermal conductivity of the gas mixture, $p_g$ (Pa) is gas pressure, $T_g$ (K) is mean gas temperature, and $F$ is a function of thermal accommodation coefficient $a$, mole fraction $f$, and molecular weight of gas species $M$ (kg mol$^{-1}$).

Taking into account the temperature discontinuities at gas-solid surfaces, the fill gas conductance in the slip flow regime is given as,

$$H_g = \frac{\lambda_g}{R_{Gi} \left[ \ln \left( \frac{R_{Go}}{R_{Gi}} \right) + \frac{g_{Gi}}{R_{Gi}} + \frac{g_{Go}}{R_{Go}} \right]}, \qquad (10.84)$$

where $g_{Gi}$ is the temperature jump distance for the fuel-gap surface and $g_{Go}$ is the temperature jump distance for the gap-cladding surface.

In the transition flow regime, when the mean free path length of a molecule is of the same order as the distance between surfaces, a coupling will exist between the incident and reflected energy of the molecule between surfaces. Under such conditions, both the spatially distributed molecular structure, as well as the kinetic energy exchange between solid surface atoms and molecules will determine the gas conductance between the solid surfaces.

In the *free-molecular flow* regime (Kn $\gg$ 1), inter-molecular collisions are negligible and the fill gas conductance is given as,

$$H_g = H_{FM} = (T_{Gi} - T_{Go}) \frac{1 + \kappa}{2} \frac{c_v p}{(2\pi R_u T)^{1/2}} \frac{a_{Gi} a_{Go}}{a_{Go} + a_{Gi}(1 - a_{Go}) R_{Gi}/R_{Go}}, \quad (10.85)$$

where $c_v$ is the gas specific heat at constant volume, $\kappa = c_p/c_v$ is the specific heat ratio, $c_p$ is the gas specific heat at constant pressure, $a_{Gi}$ and $a_{Go}$ are thermal accommodation coefficients for the gap inner and outer surface, respectively, and $p$ and $T$ are pressure and temperature of a Maxwellian gas having the same density as the gas in the gap [121].

The solid contact conductance is an additional heat transfer term resulting from partial or complete contact between the fuel and cladding. Taking into account the effect of surface roughness on the contact area, the following expression for the contact conductance was proposed [189],

$$H_s = \frac{\lambda_m}{a_0 R^{1/2}} \frac{p_c}{H}, \quad (10.86)$$

where $H$ (Pa) is the Mayer hardness of softer solid, $p_c$ (Pa) is the contact pressure between solids, and $a_0 = 0.05$ m$^{-1/2}$ is a constant derived from experimental data obtained for the contact pressure in a range from 4.9 to 53.9 MPa. The parameter $R$ takes into account the arithmetic mean roughness of the fuel outer surface $R_{a,Fo}$ and the cladding inner surface $R_{a,Ci}$, and is defined as,

$$R = \left( \frac{R_{a,Fo}^2 + R_{a,Ci}^2}{2} \right)^{1/2}. \quad (10.87)$$

The harmonic mean of the thermal conductivities $\lambda_F$ and $\lambda_C$ of the contacting surfaces of fuel and cladding is found as,

$$\lambda_m = \frac{2\lambda_F \lambda_C}{\lambda_F + \lambda_C}. \quad (10.88)$$

The radiation component of the gap conductance can be found by considering the heat transfer due to thermal radiation between two concentric cylinders. Assuming that fuel and cladding surfaces behave like gray bodies with emissivities $e_{Fo}$ and $e_{Ci}$, respectively, the gap radiation conductance becomes,

$$H_s = \frac{\sigma_{SB}(T_{Fo}^2 + T_{Ci}^2)(T_{Fo} + T_{Ci})}{\frac{1}{e_{Fo}} + \left( \frac{1}{e_{Ci}} - 1 \right) \frac{R_{Fo}}{R_{Ci}}}, \quad (10.89)$$

where $\sigma_{SB}$ is the Stefan-Boltzmann constant ($5.67 \cdot 10^{-8}$ W m$^{-2}$ K$^{-4}$).

## Crud and Oxidation Layers

During operation, nuclear fuel rods are immersed in the primary water, causing waterside corrosion of the cladding and $crud^3$ buildup on its surface. The corrosion reaction of zirconium metal in water is written as,

$$Zr + 2H_2O \longrightarrow ZrO_2 + H_2. \tag{10.90}$$

The terms on the right-hand side of the equation are the products of the reaction: the formation of an oxide layer and the generation of hydrogen, some of which gets picked up by the metal. After oxide layer formation, direct contact between the water and the metal no longer exists, thereby preventing the corrosion reaction from occurring directly. Thus, the zirconium oxide layer is mostly protective. As a result, the corrosion rate is a nonlinear function of time well described by an empirical law of the form,

$$w = At^n, \tag{10.91}$$

where $w$ is the weight gained in mass per unit area, $t$ is the exposure time, and $A$ and $n$ are constants derived from experimental data. The value of $n$ in particular is characteristic of each zirconium alloy [167].

Crud is a form of fouling in LWRs caused by the deposition of iron and nickel particles as a result of corrosion in the primary system. For PWRs, the primary issues caused by crud deposition are crud-induced localized corrosion and crud-induced power shifts, also known as the *axial offset anomaly*, caused by boron buildup in crud. Since crud tends to grow on upper spans of PWR fuel rods, where nucleate subcooled boiling usually occurs, a downward axial shift in the power distribution takes place.

Crud thermal conductivity changes with its composition and temperature. Since it is a porous material with complex morphology, the effective thermal conductivity is calculated based on known thermal conductivities of the solid components (mainly Ni, NiO, $Fe_3O_4$, and $ZrO_2$) and water [201].

The oxide and crud layers create additional sources of temperature rise across nuclear fuel elements. Assuming for simplicity a constant thermal conductivity in both layers, the temperature distribution can be found as,

$$T_{OX}(r) = -\frac{q_F''' R_{Fo}^2}{2\lambda_{OX}} \ln\left(\frac{r}{R_{OXo}}\right) + T_{Co} \quad \text{for} \quad R_{Co} \leq r \leq R_{OXo}, \tag{10.92}$$

$$T_{CR}(r) = -\frac{q_F''' R_{Fo}^2}{2\lambda_{CR}} \ln\left(\frac{r}{R_{CRo}}\right) + T_{OXo} \quad \text{for} \quad R_{OXo} \leq r \leq R_{CRo}. \tag{10.93}$$

Here $T_{OX}(r)$ and $T_{CR}(r)$ are the temperature distributions in the oxide and crud layers, respectively.

---

[3]Phenomenon known as Chalk River Unidentified Deposit, after the location of its discovery, but recently also an acronym for the Corrosion-Related Unidentified Deposit.

## PROBLEMS

### PROBLEM 10.1

Calculate temperature rises in the fuel pellet, the gas gap, the cladding, and the thermal boundary layer using the following data: $d_{Fo} = 8.25$ mm, $d_{Go} = 8.43$ mm, $d_{Co} = 9.7$ mm; average thermal conductivity for cladding 11 W/(m K), for gas gap 0.6 W/(m K), for fuel 2.5 W/(m K); heat transfer coefficient 45 kW/(m$^2$K), and linear power density 41 kW/m.

### PROBLEM 10.2

For the conditions as in Example 10.1, calculate the maximum allowed linear power density for the fuel not to exceed 3073 K temperature.

### PROBLEM 10.3

Knowing the pellet surface temperature equal to 1100 K and the fuel melting temperature equal to 2900 K, calculate the maximum allowed linear power before the fuel starts melting. Assume MOX fuel (80%U+20%Pu) with 95% theoretical density and O/M = 2.0.

# 11 Uncertainty Treatment

Uncertainties result from imperfect or unknown information. In nuclear reactor applications, uncertainties are present both in measurements and calculations of reactor parameters. Variability in the results of repeated measurements arises because variables that can affect the measurement result are impossible to hold constant. In calculations, uncertainties result from unknown exact values of input parameters used in the calculations and also can result from imperfections in the computational models and procedures.

The true value of any physical parameter cannot be absolutely determined. Thus in analyses the true values are replaced with certain reference values that are different from the true values. The difference between the true value and the reference value is called an error. In experiments, error is a difference between a measurement and the true value of the quantity being measured. Clearly, since the true values are not known, the errors are not known either. However, the errors can be controlled and characterized.

The total error usually results from a combination of systematic and random errors. The systematic error tends to shift all measurements in a systematic way so that the mean value is displaced or varies in a predictable way. Systematic errors in experiments can be corrected by proper instrument calibration. On the contrary, the random error varies in an unpredictable way and cannot be controlled.

In this chapter we introduce some basic aspects of uncertainty treatment in nuclear power safety analyses. After describing various types of uncertainties in §11.1, we continue with a discussion of main sources of uncertainties in §11.2, and presentation of useful statistical methods for uncertainty analysis in §11.3.

## 11.1 TYPES OF UNCERTAINTIES

Uncertainty is the component of the reference value that characterizes the range within which the true value is asserted to lie with a certain level of confidence. A reliable assessment of uncertainties should include separation of aleatory and epistemic sources of uncertainties. *Aleatory uncertainty* is uncertainty inherent in a phenomenon and is of relevance for events or phenomena that occur in a random manner. Thus, this type of uncertainty is closely related to inherently random processes, such as the turbulence-influenced heat transfer at the cladding surface or the variation in fuel pellet diameter during the manufacturing process.

*Epistemic uncertainty* is uncertainty attributable to incomplete knowledge about a phenomenon and that can be reduced when the knowledge about the phenomenon is increased. For example, this type of uncertainty can, result from imperfect modeling of heat transfer on the cladding surface, or from neglect of fuel deformation during reactor operation.

DOI: 10.1201/9781003255000-11

The different sources of uncertainty require different treatments when performing the uncertainty analysis. A comparison of code predictions with experimental data is the preferred means to quantify the epistemic uncertainties. Other approaches include a combination of sensitivity studies, code-to-code comparisons, and expert judgments. The preferred means for assessing aleatory uncertainties is the collection of data from nuclear power plants or initial and boundary conditions that are relevant to the events being considered [109].

## 11.2 SOURCES OF UNCERTAINTIES

Safety analyses of nuclear reactors rely on a very large experimental database containing nuclear and thermo-mechanical property data. In addition, to perform modern safety analyses, it is required to use a chain of sophisticated and specialized computational tools. To perform calculations, reliable information is needed about the initial and boundary conditions that describe the plant state at the beginning and during the progression of the analyzed scenario. Needless to say that all these uncertainty-bearing components of the safety analysis require special attention to provide meaningful and reliable computational results.

Consideration of data uncertainty in calculations is an active area of research in data and software development in general, and in nuclear power safety development in particular. One of the important first steps in the nuclear reactor safety analysis is identification of the sources of uncertainties. In general, the following major categories of uncertainties can be distinguished:

1. Input data uncertainties ("Using the right background information").
2. Initial and boundary condition data uncertainties ("Formulating the right problem to be solved").
3. Model formulation uncertainties ("Solving the right equations").
4. Numerical method uncertainties ("Solving the equations right").
5. Results interpretation uncertainties ("Right understanding of the results").

All these uncertainty categories are broadly described in the literature (e.g., [102, 186, 234]) and are shortly summarized in the following sections.

### 11.2.1 INPUT DATA

Input data needed for reactor safety analysis consists of nuclear data, such as cross sections for various reactions and nuclides, fission product yield data, and fission product decay data. Uncertainties associated with these data affect uncertainties in predictions of power distributions, spent fuel isotopic concentrations, and source terms.

Main non-nuclear input parameters necessary for thermal-hydraulic calculations include gap conductance, fuel and cladding thermal conductivity and specific heat capacity, peaking factors, pellet power distribution, and fuel-to-coolant heat transfer coefficient. Uncertainties in these data have significant influence on uncertainties in prediction of the fuel and cladding maximum temperatures.

## 11.2.2 INITIAL AND BOUNDARY CONDITIONS

Any safety analysis requires proper formulation of initial and boundary conditions that correspond to the problem to be solved. Parameters such as initial power distribution, mass flow rate distribution, water free-surface levels, and system pressure have to be specified. Boundary conditions should reflect the behavior of system boundaries during the considered scenario and provide values of inflows and outflows of coolant, inflow and outflow of heat, and changes in the system topology or geometry. Uncertainties in these data affect uncertainties in prediction of global system parameters such as power level, coolant flow distribution, and system pressure.

## 11.2.3 MODEL FORMULATION

The basis of nuclear power safety analysis is to solve the conservation equations for mass, momentum, and energy, provided that the system geometry and its initial state and boundary conditions are known. Since the analyzed physical phenomena include very complex interactions on various spatial and temporal scales, additional relationships are required to close the system of equations. Clearly, many simplifying assumptions have to be adopted to perform the analysis with a reasonable computational effort. The main paths of introduction of modeling uncertainties are as follows:

- averaging governing equations in time and space,
- assuming constant or "effective" physical properties of fluids and solids,
- employing closure relationships derived at conditions different from reactor conditions.

The averaging of governing equations is practically always necessary to reduce the computational effort. Fluid flow in long channels with constant cross section can be treated as one-dimensional problem. Similarly, a subchannel analysis model can be applied to predict the axial flow in fuel rod assemblies. Time averaging of governing equations is used to eliminate the necessity to resolve turbulent fluctuations of the velocity, pressure, and temperature fields.

Assumption of constant properties of fluids is very common, even though sometimes it can lead to different types of fluid flow behaviors. For example, assuming constant fluid density leads to incompressible flow, in which pressure perturbations travel with an infinite speed. In some applications such effects may lead to wrong results. In two phase flow applications, mixture models can be used, in which physical properties are replaced with "effective" properties. Such models can be adequate when two phase flow under consideration has a homogeneous structure. However, the models can lead to wrong results for two phase flows in which strong dynamic or thermodynamic nonequilibrium prevails.

Closure relationships are needed for parameters that cannot be obtained from solutions of first-principle conservation equations. The most important parameters are heat transfer coefficients, friction and shape pressure loss factors, and interfacial mass, momentum, and energy transfer terms in two-phase flows. Since these closure relationships are predominantly obtained from experimental data, it is important that the experimental conditions accurately represent the conditions used in the safety analysis. In particular, scaling analysis of fluid flow and heat transfer phenomena has to be properly performed.

A model or code verification is usually used to check its suitability to solve a particular problem. During verification, the results of predictions are compared to predictions obtained from another model or code.

## 11.2.4   NUMERICAL METHODS

Any nontrivial computation task involves application of numerical methods to solve a set of the governing equations. Thus, it is necessary to demonstrate that the equations are solved correctly, with a certain order of accuracy, and always in a consistent manner.

The error in a discretization is the difference between the solution of the original problem and the solution of the discrete problem. To render this quantitative we need to select a norm with which to measure the error. The choice of norm is important and must reflect the goal of the computation. For example, in some applications, a large error at a single point of the domain could be a disaster, while in others only the average error over the domain is significant.

Consistency of a discretization refers to a quantitative measure of the extent to which the exact solution satisfies the discrete problem. The discrepancy between the exact solution and the solution of the discrete problem is called the consistency error or sometimes also the truncation error.

Stability of a discretization refers to a quantitative measure of the well-posedness of the discrete problem. If a problem in differential equations is well-posed, then, by definition, the solution depends continuously on the right-hand-side terms.

A discretization of a differential equation always entails a certain amount of error. If the error is not small enough, one generally refines the discretization and a whole sequence of solutions is obtained. The scheme is called convergent if the error between the exact solution and the discrete solutions tends to zero as the discretization parameter size tends to zero. Clearly, convergence is a highly desirable property. It means that we can achieve any level of accuracy we need, provided that we use sufficiently fine computational grid. According to one of the fundamental theorems of numerical analysis, a discretization scheme that is consistent and stable is convergent.

## 11.2.5   INTERPRETATION OF RESULTS

The interpretation of results should involve a qualitative and general validity check to determine whether important phenomena are present in the obtained results. This can include global conservation of mass and energy in the whole system, physically motivated bounds for computed variable values, and expected trends for variable distributions in time and space. Here comparison of computed target parameters with simple estimates can be very useful. At the same time it is important to check whether all other predicted parameters, even if they are not important from the safety point of

view, have reasonable values. The purpose of these checks is to eliminate the possibility to obtain "correct" values of the target parameter (for example peak cladding
temperature that agrees well with experimental data) once the overall quality of the
solution is unacceptable.

## 11.3   STATISTICAL METHODS

Uncertainties, both aleatory and epistemic, are present in experimental observations
and in model predictions. Probabilistic models are one way to quantify the effects of
uncertainty on experimental and computational results. Since uncertainty contains a
random component, it can be represented by *cumulative distribution function*,

$$F(x) = \mathscr{P}(X \leq x), \quad -\infty < x < \infty, \tag{11.1}$$

and the *probability density function*,

$$f(x) = \frac{\mathrm{d}}{\mathrm{d}x} F(x), \tag{11.2}$$

where $X$ is a random variable and $\mathscr{P}$ denotes the probability. In applications, we
often only know some statistics of the random variable, such as the absolute moment,

$$E[X^p] = \int_{-\infty}^{\infty} x^p f(x) \mathrm{d}x, \ p \geq 1, \tag{11.3}$$

and the central moment,

$$E[(X - \mu)^p] = \int_{-\infty}^{\infty} (x - \mu)^p f(x) \mathrm{d}x, \ p \geq 1. \tag{11.4}$$

Taking $p = 1$ we obtain the mean value of $X$ denoted by $\mu$ and taking $p = 2$ gives
the variance of $X$ denoted by $\sigma^2$. It should be noted that $X$ is only partially defined
by its moments. A complete definition requires either a probability density function
or a cumulative distribution function.

### 11.3.1   PROBABILITY DISTRIBUTIONS

Random variables that are present in the analysis of the thermal reliability of nuclear reactors may have different probabilistic distribution functions, such as uniform
probability distribution, normal distribution, and chi-square distribution. In addition,
some discrete distributions, such as the binomial and Poisson distributions, are of
interest.

### The Continuous Uniform Distribution

The simplest of all continuous probability distributions is the continuous uniform
distribution, with a constant probability density function. A physical quantity that is
measured with an instrument that has a limited resolution and the measured value is

rounded can be modeled with the continuous uniform distribution. If $X$ is a continuous uniform random variable, its probability density function, mean, and variance are as follows,

$$f(x) = \begin{cases} \frac{1}{b-a} & \text{for} \quad a \leq x \leq b \\ 0 & \text{for} \quad x < a \quad \text{or} \quad x > b \end{cases}, \tag{11.5}$$

$$\mu = E(X) = \int_a^b \frac{x}{b-a}\,\mathrm{d}x = \left.\frac{0.5x^2}{b-a}\right|_a^b = \frac{a+b}{2}, \tag{11.6}$$

$$\sigma^2 = V(X) = \int_a^b \frac{\left(x - \frac{a+b}{2}\right)^2}{b-a}\,\mathrm{d}x = \left.\frac{\left(x - \frac{a+b}{2}\right)^3}{3(b-a)}\right|_a^b = \frac{(b-a)^2}{12}. \tag{11.7}$$

## The Normal Distribution

The *normal distribution*, also referred to as the *Gaussian distribution*, is the most widely used distribution for modeling random variables. A random variable $X$ with probability density function

$$f(x; \mu, \sigma) = \frac{1}{\sqrt{2\pi}\sigma} e^{\frac{-(x-\mu)^2}{2\sigma^2}} \quad \text{for} \quad -\infty < x < \infty, \tag{11.8}$$

has a normal distribution with parameters

$$E(X) = \mu \quad \text{and} \quad V(X) = \sigma^2. \tag{11.9}$$

The random variable

$$Z = \frac{X - \mu}{\sigma} \tag{11.10}$$

is called a *standard normal random variable* with mean value $E(Z) = \mu = 0$ and $V(Z) = \sigma^2 = 1$.

The cumulative distribution function of a standard normal random variable is denoted as

$$\Phi(z) = \mathscr{P}(Z \leq z) = 0.5 + \frac{1}{\sqrt{2\pi}} \int_0^z e^{-\zeta^2/2}\,\mathrm{d}\zeta = \frac{1 + \mathrm{erf}\left(\frac{z}{\sqrt{2}}\right)}{2}, \tag{11.11}$$

where

$$\mathrm{erf}\,x = \frac{2}{\sqrt{\pi}} \int_0^x e^{-\xi^2}\,\mathrm{d}\xi \tag{11.12}$$

is known as the error function. The values of $1 - \Phi(z)$ function are provided in Appendix D including the region with the low probability values, which is of special interest in nuclear power safety applications.

### Example 11.1: Probability of dryout during measurements

During dryout experiments in a uniformly heated tube, a single thermocouple is used to measure tube wall temperature in the vicinity of the exit from the tube.

During normal heat transfer, the maximum wall temperature is 463 K. Dryout is assumed to occur whenever the wall temperature exceeds that value. Calculate what is the probability that the dryout occurred when the measured temperature was 461 K and the measurement uncertainty for the thermocouple has a normal distribution with a standard deviation 2.2 K.

* * *

*Solution*: We find the standard normal random variable $z = (463 - 461)/2.2 = 0.909$. Thus the probability that the dryout did not occur (that is, the true wall temperature is less than 463 K) is: $p_{ndo} = [1 + \text{erf}(0.909/\sqrt{2})]/2 \cong 0.818$. Thus, the probability that the dryout occurred is $p_{do} \cong 1 - 0.818 = 0.182$.

## The Chi-square Distribution

The *chi-square distribution* is one of the most useful sampling distributions. It is a continuous, nonsymmetrical distribution used for the analysis of the variance of samples in a population and to determine the goodness of fit of a distribution for a particular application. A random variable $X = Z_1^2 + Z_2^2 + ... + Z_k^2$, where $Z_1, Z_2, ..., Z_k$ are normally and independently distributed random variables with mean $\mu = 0$ and variance $\sigma^2 = 1$, has the probability density function

$$f(x) = \frac{1}{2^{k/2}\Gamma\left(\frac{k}{2}\right)} x^{(k/2)-1} e^{-x/2}, \quad \text{for } x > 0, \tag{11.13}$$

and is said to follow the chi-square distribution with $k$ degrees of freedom, abbreviated $\chi_k^2$. The mean and variance of this distribution are $\mu = k$ and $\sigma^2 = 2k$.

## The $t$ Distribution

Let $Z$ have a normal distribution with $\mu = 0$ and $\sigma^2 = 1$ and $V$ be a chi-square random variable with $k$ degrees of freedom. If $Z$ and $V$ are independent, then the random variable $T = Z/\sqrt{V/k}$ has the probability density function

$$f(x) = \frac{\Gamma[(k+1)/2]}{\sqrt{\pi k}\Gamma(k/2)} \cdot \frac{1}{[(x^2/k) + 1]^{(k+1)/2}} \quad -\infty < x < \infty \tag{11.14}$$

and is said to follow the $t$ distribution with $k$ degrees of freedom, abbreviated $t_k$. The $t$ distribution is similar to the standard normal distribution, but it has heavier tails than the normal. As the number of degrees of freedom $k \to \infty$, the limiting form of $t$ distribution is the standard normal distribution.

## 11.3.2  TOLERANCE INTERVALS

A *tolerance interval* is constructed from a random sample so that a specified proportion of the population is contained within the interval. Statistical tolerance intervals have a probabilistic interpretation whereas engineering tolerances are specified outer limits of acceptability usually prescribed by a design engineer. Tolerance intervals can be one-sided if, for example, we want to know what interval guarantees that a specified proportion of the population will not exceed an upper limit. A two-sided tolerance interval is used to determine what interval contains a specified proportion of the population.

The two-sided tolerance interval is defined by two limits, lower ($L_L$) and upper ($L_U$), which are computed as,

$$L_L = \bar{x} - ks, \quad L_U = \bar{x} + ks, \tag{11.15}$$

where $\bar{x}$ is the sample mean, $s$ is the sample standard deviation, and $k$ factor is determined so that the interval covers at least a certain required proportion of the population with confidence $\alpha$. The proportion of population distribution that is between the limits is called coverage.

For example, for the normal distribution with known mean $\mu$ and known variance $\sigma^2$, the population limits are defined by,

$$L_L = \mu - z_{\alpha/2}\sigma, \quad L_U = \mu + z_{\alpha/2}\sigma, \tag{11.16}$$

whereas the coverage is the area under the standard normal distribution between these limits. A bound that exactly covers 95% of the population is given by $\mu \pm 1.96\sigma$. This interval is the tolerance interval, and the points $\mu - 1.96\sigma$ and $\mu + 1.96\sigma$ are the lower and the upper tolerance limits, respectively. Because $\mu$ and $\sigma^2$ are known, the coverage of 95% provided by this interval is exact.

In most practical situations, $\mu$ and $\sigma^2$ are not known, and they must be estimated from a random sample. In this case, the tolerance limits are random variables because $\mu$ and $\sigma$ are replaced by $\bar{x}$ and $s$, and the proportion of the population covered by the interval is not exact. Consequently, $\bar{x} \pm ks$ will not cover a specified proportion of the population all the time, and we determine $k$ so that we can state with confidence $100(1 - \alpha)$ percent that the limits contain at least the specified proportion of the population.

### Example 11.2: Tolerance Interval

Cladding tubes for PWR fuel rods are machined in a fuel factory with the tube outer diameter normally distributed with unknown mean and variance. A random sample of $n = 10$ tubes produced by a single machine is selected, and the diameters are found to be 10.24, 10.25, 10.27, 10.26, 10.23, 10.25, 10.22, 10.27, 10.24, and 10.23 mm. Find a 95% tolerance interval that contains at least 95% of the tube diameters produced by this machine.

\* \* \*

*Solution*: For given tube diameter data we find $\bar{x} = 10.246$ mm and $s = 0.017$ mm. From Table D.2 in Appendix D with $p = 0.95$, $1 - \alpha = 0.95$, and $n = 10$ we find that $k = 3.379$. Thus, the desired tolerance interval is $10.246 \pm 3.379 \times 0.017$ mm, or from 10.189 to 10.303 mm. We are 95% confident that at least 95% of the cladding tubes will have diameter between 10.189 mm and 10.303 mm.

**NOTE ON TOLERANCE INTERVAL**

It is easy to get confused about the difference between tolerance intervals and confidence intervals, however, they are fundamentally different. Confidence intervals are used to estimate a parameter of a population, whereas tolerance intervals are used to define the limits where we can expect to find a proportion of a population. In particular, the difference is visible when $n$ approaches infinity. As a result, the length of a confidence interval approaches zero, while the length of a tolerance interval approaches a finite non-zero value. In particular, for $p = 0.95$, $k$ in Eq. (11.15) approaches 1.96, as can be seen in Table D.2 in Appendix D.

### 11.3.3  FAILURE PROBABILITY ESTIMATION

Probabilistic methods are required for the estimation of the failure probability due to the presence of uncertainties in the actual load value and the value of the system strength limit. As an illustration, we will first consider a case when the strength limit value is known exactly, whereas the actual load value is a stochastic variable with a known probability density function. This situation is presented in Fig. 11.1(a), where we assumed that the actual load has a normal distribution. The probability of failure

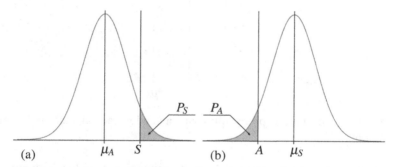

**Figure 11.1**   The failure probability: (a) a random actual load with mean value $\mu_A$ and a fixed strength limit $S$, (b) a fixed actual load $A$ and a random strength limit with mean value $\mu_S$.

can be found as,

$$P_S = \mathscr{P}(Z \geq z_S) = \int_{z_S}^{\infty} \frac{1}{\sqrt{2\pi}} e^{-\frac{z^2}{2}} \, dz , \qquad (11.17)$$

where $z_S = (S - \mu_A)/\sigma_A$ is the standardized value of the strength limit, $S$ is the known strength limit value, $\mu_A$ is the mean value of the actual load, and $\sigma_A$ is the standard deviation of the actual load. This probability is shown in the figure as the shaded area under the normal probability density function.

In a similar manner we can calculate the probability of failure when the strength limit value is a stochastic variable with a known probability density function and the actual load value is known exactly. This situation is shown in Fig. 11.1(b). Assuming that the strength limit value has the normal distribution, the probability of failure can be found as,

$$P_A = \mathscr{P}(Z \leq z_A) = \int_{-\infty}^{z_A} \frac{1}{\sqrt{2\pi}} e^{-\frac{z^2}{2}} dz \,, \tag{11.18}$$

where $z_A = (A - \mu_S)/\sigma_S$ is the standardized value of the actual load, $A$ is the known value of the actual load, $\mu_S$ is the mean value of the strength limit, and $\sigma_S$ is the standard deviation of the strength limit value. This probability is shown in the figure as the shaded area under the normal probability density function.

In reality, both the actual load value and the strength limit value are stochastic variables. Let us assume that $X_A$, $\mu_A$, and $\sigma_A$ denote the actual load, its mean value, and its standard deviation, respectively. The corresponding parameters for the strength limit value are $X_S$, $\mu_S$, and $\sigma_S$. Fig. 11.2 shows the two probabilistic distribution functions. As can be seen, the functions cross each other at a certain point $C$

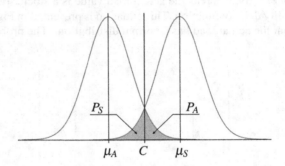

**Figure 11.2** The failure probability with a random actual load and a random strength limit.

and create a common area which is located simultaneously below both curves. We can notice that this area consists of two parts, $P_A$ and $P_S$, that can be found as follows,

$$P_A = \mathscr{P}(Z \geq z_C) = \int_{z_C}^{\infty} \frac{1}{\sqrt{2\pi}} e^{-\frac{z^2}{2}} dz \,, \tag{11.19}$$

where $z = (X_A - \mu_A)/\sigma_A$ and $z_C = (C - \mu_A)/\sigma_A$,

$$P_S = \mathscr{P}(Z \leq z_C) = \int_{-\infty}^{z_C} \frac{1}{\sqrt{2\pi}} e^{-\frac{z^2}{2}} dz \,, \tag{11.20}$$

where $z = (X_S - \mu_S)/\sigma_S$ and $z_C = (C - \mu_S)/\sigma_S$.

The areas $P_A$ and $P_S$ have a clear physical interpretation. $P_A$ is equal to the probability that the actual load will exceed $C$, and $P_S$ is equal to the probability that the strength limit value will be less than $C$. The product of these two probabilities corresponds to such stochastic events for which $X_S \leq C \leq X_A$, in which case the failure will occur, since $X_S \leq X_A$. However, this product is not equal to the probability of failure since such stochastic events as $X_S \leq X_A \leq C$ or $C \leq X_S \leq X_A$, in which failures will occur since $X_S \leq X_A$, are not included. Thus, this product can be treated as a lower estimate of the failure probability, thus,

$$P_{fail} > P_A P_S . \tag{11.21}$$

The failure will not occur for all stochastic events when $X_A < C < X_S$. The probability of all such events is equal to the following product: $(1 - P_A)(1 - P_S)$. However, this is not equal to the probability of non-failure, since such stochastic events as $X_A < X_S < C$ or $C < X_A < X_S$ are not included. Thus, this product can be treated as a lower estimate of the non-failure probability, thus,

$$P_{no-fail} > (1 - P_A)(1 - P_S) = 1 - P_S - P_A + P_A P_S . \tag{11.22}$$

Since

$$P_{no-fail} + P_{fail} = 1 , \tag{11.23}$$

we have,

$$P_A P_S < P_{fail} < P_S + P_A - P_S P_A . \tag{11.24}$$

These inequalities indicate that the failure probability is greater than the product of probabilities $P_A$ and $P_S$, and less than the sum of probabilities, represented by the shaded area in the figure, minus the product of probabilities. This probability estimate is not of great use, however. For example, if we take $P_A = P_S = 10^{-3}$, the probability of failure will be between $10^{-6}$ and $1.999 \cdot 10^{-3}$, which is a quite wide interval and a rather poor probability estimate.

A much better estimate can be obtained by introducing a new stochastic variable, $X_F = X_S - X_A$, that has a normal distribution with the mean value $\mu_F = \mu_S - \mu_A$ and the standard deviation $\sigma_F = \sqrt{\sigma_S^2 + \sigma_A^2}$, as shown in Fig. 11.3. Thus the failure is

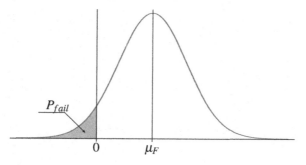

**Figure 11.3**  The failure probability estimation based on a difference between a random load and a random strength limit.

represented by all events when $X_F \leq 0$ and the probability of failure is given as,

$$P_{fail} = \mathscr{P}(Z \leq z_0) = \int_{-\infty}^{z_0} \frac{1}{\sqrt{2\pi}} e^{-\frac{z^2}{2}} dz , \qquad (11.25)$$

where $z = (X_F - \mu_F)/\sigma_F$ and $z_0 = -\mu_F/\sigma_F$. This probability is shown in the figure as the shaded area under the probability density function.

## PROBLEMS

### PROBLEM 11.1

Explain the difference between the aleatory and epistemic uncertainties. Give examples of each of them in the context of a nuclear reactor core.

### PROBLEM 11.2

Assume two random variables $X_A$ and $X_S$, with standard normal distributions as shown in Fig. 11.2, such that $P_S = P_A = 0.05$. Find the following probability: $P = \mathscr{P}(X_S \leq X_A)$.

### PROBLEM 11.3

The critical heat flux $q''_{cr}$ and the actual heat flux $q''_a$ in a reactor core are random variables with normal distributions. Derive the exact expression for the distribution of the random variable that is equal to a ratio of the two: DNBR $= q''_{cr}/q''_a$. Does it have the normal distribution as well?

# 12 Thermal Safety Margins

Maintaining safety margins in both the design and operation of nuclear reactors is a primary objective for reactor designers and operators. Among the various safety margins imposed on nuclear reactors, thermal safety margins are crucial to prevent fuel melting or cladding damage. Up to this point, we have discussed several methods for predicting the nominal values of fuel and cladding temperatures. To prevent fuel melting, the core must be designed such that the linear power density never exceeds a specific maximum value. For the protection of cladding in light water reactors, a constraint is imposed that requires the surface heat flux or the total power of the channel to always stay below its critical limit. In the case of liquid metal-cooled and gas-cooled reactors, the coolant temperature must be maintained below its maximum permissible value.

Thus, quantifying the thermal safety margin of an operating reactor involves calculating a specific safety parameter within the reactor and comparing it with a predetermined limiting value. The safety margin is then defined as the difference between this limiting value and the actual value of the parameter. The presence of safety margins ensures safe operation of the plant under conditions assumed in the calculations. Typically, both the limiting value and the actual value are not known with absolute precision, necessitating the estimation of uncertainties.

In the design of early pressurized water reactors, hot spot and hot channel factors were introduced to quantify the degree to which actual safety parameters might depart from their nominal values. The heat flux hot spot factor was defined as a ratio of the highest possible heat flux anywhere in the core, to the average heat flux across the entire core. The channel in which the hot spot occurs, or along which the maximum increase in coolant enthalpy takes place, is called a hot channel. Similarly, a hot channel factor was defined as a ratio of the highest assembly power output to the average assembly power output in the core. A more comprehensive discussion of the traditional view on hot channel factors can be found in, e.g., [60, 220].

In this chapter we discuss various methods to determine safety margins. Commonly used definitions of safety margins and limits are presented in §12.1. In §12.2 we discuss some selected aspects of hot channel factor analysis. Application of deterministic safety analyses to quantify margins is presented in §12.3.

## 12.1 DEFINITIONS OF MARGINS AND LIMITS

We will extend the concept of safety margins introduced in §1.4 to take into account various aspects of nuclear power plant safety, including design, operation, and licensing. In each of these areas, specific definitions of limits and margins are used, as schematically shown in Fig. 12.1. The definitions presented in this section should be treated as indicative, however, since there is no consensus in that matter. There are many definitions of a safety margin, depending on the application of this

DOI: 10.1201/9781003255000-12

important concept. Specific definitions are used by regulators, operators, licensees, and designers. Some of these definitions are provided below.

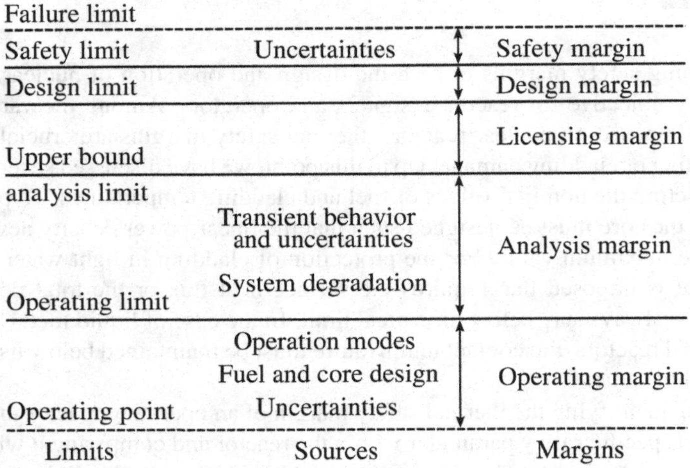

| Failure limit | | |
| --- | --- | --- |
| Safety limit | Uncertainties | Safety margin |
| Design limit | | Design margin |
| | | Licensing margin |
| Upper bound | | |
| analysis limit | | |
| | Transient behavior and uncertainties | Analysis margin |
| | System degradation | |
| Operating limit | | |
| | Operation modes | |
| | Fuel and core design | Operating margin |
| Operating point | Uncertainties | |
| Limits | Sources | Margins |

**Figure 12.1**  Limits and safety margins.

## 12.1.1  FAILURE POINT LIMIT

An important reference value of any safety variable is its *failure point* or *ultimate capacity point*. Examples of such limiting values include the fuel melting temperature or the cladding temperature at which it breaks. The failure point can be most often established experimentally, but in some cases it is determined based on best-estimate calculations. It is important to note that the exact failure point is not known due to uncertainties in experimental and analytical methods.

## 12.1.2  SAFETY LIMIT

The concept of a *safety limit*, also referred to as a *licensing acceptance limit*, is directly related to the prevention of unacceptable releases of radioactive materials from a nuclear power plant. This can be achieved through the application of limits on the temperatures of fuel and fuel cladding, coolant pressure, and other safety-related parameters. The safety limits should usually be stated as the maximum acceptable values which ensure the integrity of the barriers for radioactive releases, taking into account uncertainties. Thus the safety limit is more restrictive than the failure point limit. Regulators decide about safety limit values for various safety parameters.

## 12.1.3  DESIGN LIMIT

A *design limit* can be proposed by the designer and is based on engineering calculations. Usually the design limit is more restrictive than the safety limit, but sometimes, for example for design basis accident analyses, the design limit is taken identical to the safety limit.

### 12.1.4  OPERATING LIMIT

An *operating limit*, also referred to as a *safety analysis limit*, is the limit value of a safety parameter corresponding to the operating point that should not be reached during normal operations and anticipated operational occurrences. Limits for normal operation is set at any level between the range of steady-state operation and the actuation setting for the safety system. The monitored safety parameters may exceed the steady state range as a result of load changes or imbalance of the control system, for example. Operating limit provides enough flexibility to bring the plant back to normal steady state operation without actuating the safety systems.

### 12.1.5  APPARENT MARGIN

An *apparent margin*, also referred to as a *margin to failure*, is defined as a distance between the operating point and the failure point. This margin can be further divided into a *margin available to the licensee* and a *margin controlled by the regulator*. This distinction clarifies the role of both the regulator and the licensee in the overall margin management. The safety limit is the boundary between the two margins.

### 12.1.6  SAFETY MARGIN

The distance between the safety limit and the failure limit is called a *safety margin*. This margin is entirely managed by the regulator and consists of uncertainties on the failure limit and possibly additional provisions for unknown factors.

### 12.1.7  DESIGN MARGIN

To ensure that the safety limit is met in all applicable plant states, the designer can introduce a *design margin*, which is the distance between the design limit and the safety limit. This margin is usually greater than zero for conditions prevailing during normal operation and during anticipated operational occurrences, meaning that the design limit is more restrictive than the safety limit. However, during design basis accidents, the two limits are usually considered identical.

### 12.1.8  ANALYSIS AND LICENSING MARGINS

A difference between the operating limit and the design limit corresponds to two additional margins. The first one is a *licensing margin* that accounts for the safety analysis results, such as the cladding peak temperature or critical heat flux, that exceed the design limit values. The second one is an *analysis margin* that takes into account provisions for system degradation and other unknown factors, transient behavior, and uncertainties in transient calculations.

### 12.1.9 OPERATING MARGIN

To ensure a flexible and reliable plant operation, there is a certain distance between the operating point and the operating limit called an *operating margin*. It includes provisions for all operational modes and transients, provisions for fuel and core design, and uncertainties in operating point.

## 12.2 HOT SPOTS AND HOT CHANNEL FACTORS

An analysis of hot channel factors has been an integral part of nuclear reactor design since the very early days of civilian nuclear power development. Results of nominal design calculations do not account for uncertainties in theoretical and experimental data, manufacturing tolerances, and other deviations of parameters from their nominal values. These deviations, in turn, lead to deviations of nuclear reactors from their nominal operation regime. To ensure safe, economical, and reliable operation, it is important to estimate these deviations. In the application of this task, the concept of hot channel factors has been used to quantitatively estimate the thermal safety margins in nuclear reactors. These factors are commonly expressed as dimensionless factors greater than unity that are used to multiply the nominal or calculated values of the considered quantities. Such modified quantities are used in the design and safety analyses.

The hot spot factors consist of systematic and random components that originate from various sources. The systematic factors are due to uncertainties on such parameters as reactor thermal power, power distribution, coolant flow rate, coolant flow distribution, and inlet coolant temperature. The random factors are mainly due to manufacturing tolerances and uncertainties in physical properties.

The impact of uncertainties on temperature predictions is typically accounted for through the assessment of hot channel factors, which consider the increase in reactor temperatures due to specific uncertainties. By reducing the values of the hot channel factor, we can increase the nominal peak cladding temperature. This increase would lead to enhanced reactor power and, consequently, economic benefits.

Hot channel factors are employed to accommodate the effects of uncertainties, stemming from theoretical and experimental analyses, on the design of the reactor. The thermal-hydraulic design of a reactor must conform to a set of design criteria. Many of these relate to fuel, cladding, and coolant exit temperature. For example, a typical design criterion related to fuel is that no fuel melting occurs at some specific overpower, that is reactor power greater by some fraction than the nominal reactor power. In order to ensure the integrity of the fuel cladding, its temperature must be kept below a certain limit. Additionally, the highest permissible coolant temperature must guarantee the structural integrity of the area above the core.

For any safety parameter, the hot channel factor is a ratio of the maximum value of that parameter to its nominal value. The factors are based on a combination of experimental data and validated analytical methods. However, their exact definition and values will evolve with new reactor designs and with the development of new computational techniques.

There exist multiple methodologies for conducting hot channel factor analysis. The typical procedure commences with the analysis of coolant inlet flow and temperature. Subsequently, we proceed axially through the core up to a specific level, followed by a radial progression through the fuel pin structure. At each step, a hot channel factor multiplies a parameter representing a temperature difference, rendering the outcome independent of the temperature units utilized.

Two types of uncertainties might influence a particular safety parameter: *aleatory uncertainties* and *epistemic uncertainties*, or biases. Usually we assume that aleatory uncertainties follow a normal distribution which can be characterized by the variance, $\sigma^2$, and the standard deviation, $\sigma$. The two types of uncertainties propagate to yield the overall uncertainty of the safety parameter. However, as described below, the method of propagation differs for the two types of uncertainties.

If for a random safety variable the requirement is that its value should lie below a specific limiting value with a certain probability, the hot channel factor for that variable would be unity plus standard deviation multiplied by a constant. For example, if the required probability is 99.9%, the nominal value of the safety parameter must be $3\sigma$ less than the design limit, and the corresponding hot channel factor is $1 + 3\sigma$. For some parameters 97.5% is considered satisfactory and the hot channel factor is then $1 + 2\sigma$.

## 12.2.1 DETERMINISTIC METHOD

The development of the hot spot approach originally began with the application of the deterministic method, a technique utilized during the initial phases of light water reactor development. However, this method was found to be overly conservative, leading to unnecessarily large margins between the design limits and the actual operating conditions of the plant. To address this issue, statistical and semi-statistical methods were introduced.

Every method of hot spot analysis provides a means to determine the aggregate effect on a core variable, resulting from the simultaneous existence of various uncertainties. These methods can be applied to any variable that can be expressed in the following analytical form,

$$y = \sum_{i=1}^{m} F_i(x_1, ..., x_n) , \qquad (12.1)$$

where $x_j$ is a variable whose value is uncertain, $n$ is the number of uncertain variables, $F_i$ is a function of the uncertain variables, and $m$ is the number of terms in $y$.

A sub-factor which expresses the effect of a single uncertainty on a single term in Eq. (12.1) is given as,

$$f_{ij} = \frac{F_i(\overline{x}_1, ..., \overline{x}_j + \Delta x_j, ..., \overline{x}_n)}{F_i(\overline{x}_1, ..., \overline{x}_n)}, \qquad (12.2)$$

where $\overline{x}_j$ is the nominal value of $x_j$ and $\Delta x_j$ is an error in $x_j$ relative to $\overline{x}_j$. If the uncertainty in $x_j$ is a random variable then $\Delta x_j$ is defined as one standard deviation. If the uncertainty is deterministic then $\Delta x_j$ is the bias error in $x_j$.

The deterministic method assumes all uncertainties occur with their most unfavorable values at the same location and at the same time. The location is that of peak nominal conditions. Then from Eqs. (12.1) and (12.2) we get the approximation of the hot spot value as,

$$y = \sum_{i=1}^{m} \prod_{j=1}^{n} f_{ij} F_i(\bar{x}_1, ..., \bar{x}_n) . \tag{12.3}$$

This method is appropriate for representing errors whose magnitudes remain constant in time and space and can be estimated a priori.

## 12.2.2  STATISTICAL METHOD

The likelihood of all uncertainties simultaneously reaching their most unfavorable values at the same location and time is extremely low. This realization led to the creation of the statistical method. In this approach, each uncertainty is considered a normal random variable with a known standard deviation and a mean value of zero. The uncertainties are assumed to be independent. If the uncertainties are of small magnitude, then the two-sigma bound on the hot spot can be determined using the following formula,

$$y^{2\sigma} = \sum_{i=1}^{m} F_i(\bar{x}_1, ..., \bar{x}_n) + 2 \left[ \sum_{j=1}^{n} \sum_{i=1}^{m} (f_{ij} - 1) F_i(\bar{x}_1, ..., \bar{x}_n)^2 \right]^{1/2} , \tag{12.4}$$

where $f_{ij}$ is a sub-factor given by Eq. (12.2) that is based on one standard deviation of uncertainty. Uncertainties that are well represented through the statistical method include manufacturing variability of fuel pellet fissile material content, dimensional tolerances, and experimental uncertainties in correlations. However, this method is not suitable for handling biases and uncertainties that are dependent.

## 12.2.3  SEMI-STATISTICAL METHOD

The semi-statistical method is a combined approach that merges the deterministic and statistical methods. This provides the designer with greater flexibility in determining whether an uncertainty occurs as a constant bias or assumes a statistical distribution of values. However, to meaningfully apply this method, one must provide a valid justification for the chosen approach for each uncertainty, which may not always be straightforward. Similar to the statistical method, the semi-statistical method is typically used to establish a two-sigma bound.

## 12.2.4  TREATMENT OF STATISTICAL FACTORS

The treatment of statistical factors depends on the form of the relationship between a safety variable and the input random parameters. If a safety variable, $y$, can be expressed as a linear function of independent random variables, $x_i$,

$$y = c_0 + c_1 x_1 + c_2 x_2 + ... + c_n x_n, \tag{12.5}$$

then the variance of $y$ is equal to the sum of the variances of the terms $c_i x_i$,

$$\sigma_y^2 = \sum_{i=1}^{n} c_i^2 \sigma_i^2. \tag{12.6}$$

The safety variable, $y$, will have a normal distribution if the independent variables have normal distributions. As the number of independent variables becomes large, the variable $y$ will tend to approach a normal distribution even if the independent variables, $x_i$, do not have normal distributions.

Let us now assume that the safety variable, $y$, is an arbitrary function of the independent variables, $x_i$,

$$y = y(x_1, x_2, ..., x_n). \tag{12.7}$$

When each independent variable is at its mean value, $\bar{x}_i$, the safety variable takes the value $\bar{y}$. For independent variables departing from their mean values by some arbitrary amount $\delta x_i$, the value of the safety variable can be approximated by a Taylor series expansion about $y = \bar{y}$ in which only linear terms are retained,

$$y \simeq \bar{y} + \sum_{i=1}^{n} \overline{\frac{\partial y}{\partial x_i}} \delta x_i, \tag{12.8}$$

where partial derivatives are evaluated at mean values of independent variables. Thus, using the expression for the variance of a linear function we have,

$$\sigma_y^2 \simeq \sum_{i=1}^{n} \left( \overline{\frac{\partial y}{\partial x_i}} \right)^2 \sigma_i^2. \tag{12.9}$$

With this variance and a normal distribution of $y$, the probability that $\bar{y} + 2\sigma_y$ will not be exceeded is 97.73% and the probability that $\bar{y} + 3\sigma_y$ will not be exceeded is 99.87%.

For random variables with specified bounds, such as dimensions, it is reasonable to assume that the mean value has a uniform probability distribution. For such distribution, the bounds correspond to $\pm\sqrt{3}\sigma$.

For unknown variances of independent variables, the population variances can be estimated from a variance of a random sample as,

$$s_i^2 = \frac{\sum_{j=1}^{n_i} (x_{ij} - \bar{x}_i)^2}{n_i - 1}. \tag{12.10}$$

If the distribution of $y$ is normal, and the number of observations is small, tolerance limits on the possible values of $y$ can still be given. For the case where $y$ is a linear function of $x_i$'s, $i = 1, 2, ..., n$, the upper tolerance limit is given by [27],

$$\bar{y} + t_{\alpha, v_e} s_e, \tag{12.11}$$

where,

$$s_e = \left[ \sum_{i=1}^{n} \frac{(c_i^2 s_i^2)(n_i + 1)}{n_i} \right]^{1/2}, \tag{12.12}$$

$$V_e = \frac{s_e^4}{\sum_{i=1}^{n} \frac{c_i^4 s_i^4 (n_i+1)^2}{v_i n_i^2}}, \tag{12.13}$$

$$\bar{x}_i = \frac{1}{n_i} \sum_{j=1}^{n_i} x_{ij}, \tag{12.14}$$

$s_i^2$ is found from Eq. (12.10), and $t_{\alpha, v_e}$ is the upper $\alpha$ percent point of the $t$-distribution with $v_e$ degrees of freedom.

### 12.2.5 FUEL TEMPERATURE ANALYSIS

Hot spot factors are used to account for various uncertainties and to ensure that the specified maximum fuel temperature in the core does not exceed the fuel design limit at any time and any location during the normal operating conditions.

Fuel temperature at an arbitrary evaluation point is obtained by adding an inlet coolant temperature to the temperature difference at each position in the axial and radial position of the fuel rod structure. Thus the fuel temperature is evaluated by the following expression,

$$T_F = T_{in} + \sum_{i=1}^{5} F_i \cdot \overline{\Delta T_i}, \tag{12.15}$$

where $T_{in}$ is the core inlet coolant temperature, $F_i$ is the overall hot spot factor for component $i$, and $\overline{\Delta T_i}$ is the nominal temperature rise. The components $i = 1$–5 are as follows,

1. Coolant temperature rise from the core inlet to a specific axial location,
2. Film temperature rise in the coolant thermal boundary layer,
3. Temperature rise in the cladding,
4. Temperature rise in the gas gap,
5. Temperature rise in the fuel.

In order to determine $F_i$, the systematic and random factors are obtained individually based on the functional relationship between the uncertainty and the temperature rise. Te total systematic factor, $F_i^s$, is given by combining the systematic factors $f_{ij}^s$,

$$F_i^s = \prod_{j=1}^{n_s} f_{ij}^s, \tag{12.16}$$

where $n_s$ is the number of systematic factors. The random factors, $f_{ij}^r$, are combined to give the total random factor as follows,

$$F_i^r = 1 + \left[ \sum_{k=1}^{n_r} (f_{ik}^r - 1)^2 \right]^{1/2}, \tag{12.17}$$

where $n_r$ is the number of random factors. The overall hot spot factor for component $i$ can be now found as,

$$F_i = F_i^s \cdot F_i^r, \tag{12.18}$$

and the fuel temperature becomes,

$$T_F = T_{in} + \sum_{i=1}^{5} \left\{ \prod_{j=1}^{n_s} f_{ij}^s \cdot \left[ 1 + \left( \sum_{k=1}^{n_r} (f_{ik}^r - 1)^2 \right)^{1/2} \right] \cdot \overline{\Delta T_i} \right\}. \tag{12.19}$$

Typically the systematic and random factors for each component $i$ are determined by sensitivity analysis.

## 12.3  DETERMINISTIC SAFETY ANALYSIS

The deterministic safety analysis can be used to demonstrate that the safety parameters of a nuclear power plant are kept below acceptable limits with adequate safety margins. Such safety analyses are performed for normal operation, anticipated operational occurrences, and accident conditions. The safety analyses have evolved from the initial conservative approaches to best estimate predictions with uncertainty quantification.

Fully conservative approaches were introduced to cover uncertainties due to the limited capability for modeling physical phenomena. However, such approaches are no longer recommended for safety analyses, since the obtained results may be misleading and the level of conservatism is in general unknown [102].

A more realistic, still frequently used approach in safety analyses is based on a best estimate computer code combined with conservative input data. In this approach, the system availability, initial conditions, and boundary conditions supplied to the code are conservative. The use of this approach is quite straightforward and, in some cases, just one calculation is sufficient to demonstrate safety. However, this approach provides only a rough estimate of uncertainties.

A much better estimate of uncertainties can be obtained with a best estimate computer code combined with best estimate input data with uncertainties.[1] The system availability can still be treated conservatively if more realistic data are not available.

Deterministic safety analysis and the consideration of uncertainties are addressed in IAEA Safety Standard Series [96, 97]. Moreover, the use of best estimate codes is generally recommended for deterministic safety analysis with the following two options to demonstrate sufficient safety margins:

1. Use a best estimate code with a reasonably conservative selection of input data and a sufficient evaluation of the uncertainties of the results through code-to-code comparisons, code-to-data comparisons, expert judgement, and sensitivity studies.
2. Use a best estimate code with realistic assumptions on initial and boundary conditions employing statistically combined uncertainties for plant conditions and code models.

An uncertainty analysis consists of identification and characterization of relevant input parameters and their uncertainty as well as of the methodology to quantify

---

[1] Such approach is frequently referred to as the "best estimate plus uncertainty" (BEPU).

the global influence of the combination of these uncertainties on selected output parameters and their uncertainties. Several methods have been developed for input uncertainty propagation or output uncertainty extrapolation, as described in [102].

Uncertainties in the outputs of a computer code occur when there are uncertainties in the values of parameters used in the code, either as random inputs (aleatory), or as uncertain constituents of the code itself (epistemic). In the case of the aleatory uncertainties, they represent a random variation in the input. The epistemic uncertainties represents shortcomings in the code such as, e.g., a simplification of the momentum equation in which the axial stress tensor is neglected. Owing to this simplification, random differences may be observed when compared to test data. There are various methods to deal with uncertainties in input parameters as described below.

## 12.3.1  MONTE CARLO METHODS

In Monte Carlo analysis, probability-based sampling is used to develop a mapping from input parameters to computed output parameters. The mapping provides a basis for the sensitivity analysis and uncertainty quantification. Several sampling procedures exists, including simple random sampling, stratified sampling, and Latin hypercube sampling.

A simple random sample from a population is a sample in which each member of the population has the same probability of being selected. This approach requires a very large sampling frame, which results in extensive sampling calculations and excessive costs. Advantage is that the method is free of classification errors and it requires minimum advance knowledge of the population.

When sub-populations vary considerably, it is advantageous to sample each of it independently. Then random sampling is applied within each sub-population.

In a Latin hypercube sampling the range of each uncertain input parameter is divided into $N$ equiprobable sections, where $N$ is the sample size. For each section of each input parameter, one value is selected at random. As a result, each element in the population has a known and equal probability of selection. This makes systematic sampling functionally similar to simple random sampling, however, it is much more efficient and less expensive to apply.

## 12.3.2  RESPONSE SURFACE METHODS

A response surface can be established by regression analysis to fit the calculated safety variable, such as the cladding picking temperature, in terms of the important input parameters. The purpose of the response function is to replace the code in prediction of the probabilistic density function of the output safety variable. In order to produce a reasonable estimate of the probability density function from an response function we must sample the surface in a statistically acceptable way. With algebraic surfaces a crude Monte Carlo sampler can be used. The number of uncertain input parameters is limited because of the required number of code calculations. The highest degree of polynomial that may be fitted to the predictions from a $p$-level factorial is $p - 1$.

## 12.3.3  TOLERANCE LIMIT METHODS

Let us consider the one-sided case in which we are interested in bracketing a value of a random variable $y_\alpha$ and assume that the probability of any output falling below $y_\alpha$ is $\alpha$. Thus, the probability of all $N$ outputs falling below $y_\alpha$ is $\alpha^N$ and the probability that $y_N$ will fall above $y_\alpha$ is

$$\beta = 1 - \alpha^N. \tag{12.20}$$

The probability that exactly one output will fall above $y_\alpha$ is $N(1-\alpha)\alpha^{N-1}$. Therefore the probability that two outputs will fall above $\alpha^N$ is,

$$\beta = 1 - \alpha^N - N(1-\alpha)\alpha^{N-1}. \tag{12.21}$$

Similarly it can be shown that for $m$ values in the range the probability is,

$$\beta = 1 - \sum_{i=N-m+1}^{N} \frac{N!}{i!(N-i)!}\alpha^i(1-\alpha)^{N-i}. \tag{12.22}$$

This equation can be used to determine the minimum number of code runs to obtain a value of a safety variable that is the largest in the sample and has probability $\beta$ of exceeding the $\alpha$th quantile of all possible safety variable values, which can be compared to a certain limiting allowed value of that safety variable. For example, if $\alpha = 0.95$, $\beta > 0.95$, $m = 1$ requires 59 runs of the code, $m = 2$ requires 93 runs, and $m = 10, 311$.

## 12.4  CORE THERMAL LIMITS

During reactor operation there are no direct indications of the fuel temperature and the limit to the critical heat flux. The in-core instrumentation system is used to provide flux maps of the core necessary to calculate *peaking factors*, which express a core power distribution in terms of peak-to-average ratios. The two peaking factors that are usually measured are the heat flux hot channel factor, $F_Q(z)$, and the nuclear enthalpy rise hot channel factor, $F_{\Delta H}^N$. The limiting values of these factors for normal operation conditions are specified in the technical specifications.

---

### PEAKING FACTORS VERSUS UNCERTAINTY FACTORS

A hot channel or hot spot factor is used to determine an expected maximum value of a safety variable during reactor operation. Frequently a mean-in-core value is used as a reference. We can partition such hot spot or hot channel factors into two major sub-factors:

1. A peaking sub-factor, which is equal to the ratio of the local nominal maximum value of the safety variable in the core to its core average value.
2. An uncertainty sub-factor, which takes into account a combined effect of all uncertainties on the peaking factor.

Thus the peaking sub-factor is resulting from a known nominal non-uniformity of the safety variable distribution in the core. A typical example of such a factor is the heat flux channel factor which represents a non-uniformity of the spatial distribution of heat flux in the core. The uncertainty sub-factor includes the core-wide, assembly, and pin-level uncertainties resulting from biases and random effects.

Control of the power distribution with respect to these two factors ensures that the local conditions in fuel rods and cooling channels do not challenge fuel pin integrity at any location in the core during normal operation or during postulated accident conditions. The following fuel design limits are valid:

1. The 95/95 criterion for the hottest fuel rod in the core that it does not experience CHF condition, during both normal operation and a loss of flow accident.
2. During large-break loss of coolant accident, the peak fuel clad temperature will not exceed the $1204°C$ limit.
3. During an ejected rod accident, the energy deposition to the fuel must not exceed 280 cal/g.
4. The control rods must be capable of shutting down the reactor with a minimum required shutdown margin with the highest worth control rod stuck fully withdrawn.

### 12.4.1  HEAT FLUX HOT CHANNEL FACTOR

The heat flux channel factor is defined as,

$$F_Q(z) = \frac{q'''(x_p, y_p, z)}{\frac{1}{V} \int_V q'''(x, y, z) \, dx dy dz}. \tag{12.23}$$

Here $q'''(x, y, z)$ is the thermal power density at location $x, y, z$ in the core and $V$ is the core volume. The fuel rods are oriented in the $z$-direction, which is the core axial direction, whereas coordinates $x, y$ describe any point in the core lateral direction. In particular, coordinates $x_p, y_p, z$ correspond to a point of maximum power density in the lateral plane located at $z$-coordinate.

The heat flux channel factor is traditionally limited according to the following relationship,

$$F_Q(z) \leq \begin{cases} \frac{F_{Q,lim}}{\phi_Q} K(z) & \text{for} \quad \phi_Q > 0.5 \\ \frac{F_{Q,lim}}{0.5} K(z) & \text{for} \quad \phi_Q \leq 0.5 \end{cases}, \tag{12.24}$$

where $F_{Q,lim}$ is a core-design dependent safety limit for the heat flux channel factor that includes the peaking sub-factor and all uncertainty sub-factors, $K(z)$ is a correction function governed by the time history of core uncovery and reflood during loss of coolant accident, and $\phi_Q$ is a ratio of the core actual thermal power to the core rated thermal power.

## 12.4.2 NUCLEAR ENTHALPY RISE HOT CHANNEL FACTOR

Nuclear enthalpy rise hot channel factor is defined as the ratio of the integral of the linear power along the rod with the highest total power to the average rod power. This factor is used in the critical heat flux calculations. It should be noted that the operator has a direct influence on heat flux hot channel factor, $F_Q$, through movements of control rods. However, it is not possible to directly control the nuclear enthalpy rise hot channel factor.

## 12.5 MULTISCALE AND MULTIPHISICS REACTOR CORE ANALYSIS

The sensitivity analysis and uncertainty quantification are integrated into nuclear reactor design and safety analyses. In such analyses, it is customary to subdivide complex reactor systems into smaller tasks with varying scales such as a pin cell, an assembly, and a core. Each of the tasks contributes to the total uncertainty of the final coupled system and proper uncertainty propagation through all steps is required.

The sources of input uncertainties in computer code simulations can be divided into the following three groups:

1. Data uncertainties, such as nuclear data, geometry, and materials.
2. Models and numerical methods, such as approximations in the numerical solutions, nodalization, and homogenization approaches.
3. Imperfect knowledge of boundary and initial conditions.

The two primary categories for quantifying uncertainty encompass deterministic methods and stochastic methods.

The deterministic method calculates the sensitivity of the system response $y$ with respect to uncertain input parameter $\delta x$ using perturbation theory and computes an estimate for the covariance matrix $\text{Cov}[y]$ by linearizing the response $y \approx S\delta x$. Here $S$ is the sensitivity matrix of the response vector. With the linearization, the covariance matrix can be calculated by folding sensitivities with the variance and covariance matrix $\text{Cov}[\delta x]$,

$$\text{Cov}[y] \approx \text{Cov}[S\delta x] = S\text{Cov}[\delta x]S^T. \tag{12.25}$$

The stochastic method relies on the sampling of the uncertain input parameters provided in the variance and covariance matrix and statistically analyzing the calculated output responses. The variance is computed as,

$$\text{Var}[y] = \frac{\sum_{i=1}^{N}(y_i - \bar{y})^2}{N - 1}, \tag{12.26}$$

where $N$ is the number of samples and $\bar{y}$ is the sample mean of the responses.

## 12.5.1 REACTOR PHYSICS CALCULATIONS

Multiscale modeling and simulation of neutronics consists of three steps, including cell physics, lattice physics and core physics. The main goal of the *cell physics analysis* is to produce multigroup microscopic cross-section libraries. At that stage it is convenient to use the perturbation method to compute the sensitivity coefficients of output variables with respect to nuclear data. Such studies performed for light water reactors show that some reactions dominate the contribution to the uncertainty of the multigroup cross sections and $k_\infty$. These output parameters are particularly sensitive to the neutron capture in $^{238}$U, neutron capture in $^{235}$U, and average number of neutrons per fission in $^{235}$U, $\bar{\nu}$ [92].

The lattice physics model is a 2D fuel assembly model containing all fuel rods, instrumentation tubes, guide tubes, and other fuel assembly components. During lattice calculations, the uncertainties are propagated to the target output uncertainties on few-group constants such as homogenized cross-sections and other nodal parameters. At the same time, the lattice physics calculations contribute with other sources of uncertainty due to modeling approximations embedded in the lattice codes.

The core effective multiplication factor, $k_{\mathrm{eff}}$, and the core power distribution are determined during core physics calculations. In the fully-deterministic approach, the perturbation calculation is performed at both lattice and core levels, and the variance and covariance matrix of the few-group homogenized constants generated in the lattice calculation is used to evaluate the uncertainty of core responses. Alternatively, one-step stochastic approach can be applied in which the stochastic sampling is performed on both lattice and core levels. In this way there is direct connection between lattice calculations to generate the few group cross-section library and the core calculation that relies on this library. In these calculations a standard reactor simulation procedure is followed in which homogenized constants are first generated, the core geometry is simplified, and a lower-order solver, such as the nodal diffusion method, is used. The major modeling variation is due to the choice of the spatial homogenization, which can be either on the pin cell level or on the assembly level.

### 12.5.2 FUEL BEHAVIOR CALCULATIONS

Similarly as the reactor physics analysis, fuel behavior calculations are performed at various levels of approximation. Detailed models are implemented in fuel performance codes whereas rather simplified lumped models are implemented in core simulation codes which combine neutronics, thermal-hydraulics, and fuel rod behavior. Propagation of uncertainties between these two approaches and between other core simulation tools need to be analyzed.

High-fidelity fuel performance codes are used to model a single fuel rod with detailed descriptions of various parameters, such as, e.g., geometry, fuel enrichment, and fuel burnup. Such codes also employ mechanistic models for fission gas release, cladding corrosion, swelling, and other performance-related parameters. The main parameter from the fuel feedback point of view is the so-called Doppler temperature needed in full-core coupled multiphysics calculations to calculate Doppler feedback. Other three important parameters that are obtained include fuel thermal conductivity, cladding thermal conductivity, and gap conductance. Usually the gap conductance is expressed in terms of temperature and reactor power, the fuel conductance in terms of temperature and burnup, and cladding conduction in terms of temperature. The mean values and uncertainties of these parameters are computed and parameterized for each type of fuel rod to be subsequently used in simplified fuel models in core analysis codes. The data are represented as lookup tables for steady-state analyses while for transient and accident analyses various analytical expressions are preferred.

### 12.5.3 THERMAL-HYDRAULIC CALCULATIONS

Thermal-hydraulic calculations provide values of parameters that have influence on core behavior through various feedback effects. The main parameters of interest are the coolant temperature, the coolant void fraction, and the cladding surface temperature. These parameters are highly dependent on the flow regime prevailing in the fuel assembly: single-phase turbulent flow, two-phase nucleate boiling flow, or film boiling flow with post-CHF heat transfer conditions.

As for neutronics and for fuel behavior analyses, the thermal-hydraulic calculations can be performed with varying levels of complexity. Computational fluid dynamics (CFD) approach provides solutions of Reynolds-averaged Navier-Stokes equations and takes into account geometry details of fuel assemblies with spacers. Due to high computational efforts, this approach is currently limited to few subchannels or to a single fuel assembly. With increasing computer power, larger core parts will be analyzed with CFD.

A subchannel code can be used as an efficient tool to propagate uncertainties on the thermal-hydraulic output parameters. The code runs are repeated many times with input samples from the specified input distributions of uncertainties on boundary conditions, geometry, and modeling. Through statistical analysis of the results, conclusions can be drawn about the behavior of the quantities of interest. CFD calculations can be used to assess uncertainties due to modeling and geometry approximations adopted in the subchannel analysis.

## 12.6 UNCERTAINTY PROPAGATION

A thorough safety margin prediction involves calculations of mean values and uncertainties on safety parameters of a core. These calculations combine three major areas of reactor core analysis: neutronics, thermal-hydraulics, and fuel performance analysis. In addition, the analyses can be performed on various core scales, starting from a single pin, through a fuel assembly to whole core analysis. Since all these types of physics and scales are treated separately, a careful uncertainty propagation is required to predict safety margins. In particular, it should be noted that uncertainties and variables are correlated. For example, nuclear data uncertainties introduce uncertainties on the predicted fuel composition, which in turn has impact on the neutronics parameters such as few-group cross sections, on the heat flux, and on the fuel modeling parameters, such as the gap conductance and the fuel thermal conductivity.

## PROBLEMS

### PROBLEM 12.1

Explain the reason why the operating margin in a nuclear power plant is needed and who decides about this margin.

## PROBLEM 12.2

Estimate the required minimum number of code runs for two-sided tolerance limit for one safety variable with the desired probability content $\alpha = 0.95$ and the confidence level $\beta = 0.95$.

## PROBLEM 12.3

In a hot channel factor analysis of a liquid-metal fuel assembly, the coolant temperature rise component has the following systematic sub-factors: the power level 1.03, the inlet flow maldistribution 1.02, the flow simulation bias 1.03, and the following random sub-factors: the inlet flow maldistribution 1.06, the simulation uncertainty 1.06, the subchannel flow area uncertainty, 1.02. Calculate the overall hot spot factor for this component.

## PROBLEM 12.4

Find an expression for the heat flux channel factor in a cylindrical core with radius $R$ and height $H$ at $r_p = 0$, assuming that the thermal power density has an axisymmetric distribution given as $q''' = q_0''' J_0(v_1 r/R) \cos(\pi z/H)$, where $v_1 \approx 2.405$ is the smallest root of $J_0(v) = 0$.

# A Notation

## A.1 NUMBER NOTATION

In the scientific notation very small and very large numbers are written as values between 1 and less than 10 multiplied by a power of 10. In the engineering notation large and small numbers are converted into a value between 1 and less than 1000, multiplied by a power of 10 in increments of three, such as $10^{-6}$, $10^{-3}$, $10^3$, $10^6$, etc. Thus, $3.142 \cdot 10^{11}$ in the scientific notation would be $314.2 \cdot 10^9$ in the engineering notation. In this book both systems are employed. For microscopic systems scientific notation is used, whereas for macroscopic and global systems (such as an energy system of a whole nuclear power plant) engineering notation is preferred.

## A.2 NOMENCLATURE AND SYMBOLS

For the reader's convenience, the meanings of variables used in equations are explained immediately after the equations appear. When a series of equations occurs and the same set of variables is used, any new variable is explained at the first appearance. The notation used in the book is also provided in the following lists.

### LIST OF ROMAN SYMBOLS

| Symbol | Unit | Description and Location where the Symbol First Appears |
|---|---|---|
| $a$ | $m^2\,s^{-1}$ | Thermal diffusivity |
| $\mathbf{a}$ or $a$ | $m\,s^{-2}$ | Acceleration, Eq. (4.3) |
| $a_i'''$ | $m^{-1}$ | Solid-liquid interfacial area density in a porous medium, Eq. (4.148) |
| $A$ | $m^2$ | Area, Eq. (4.155) |
| $\mathbf{b}$ | $N\,m^{-3}$ | Body force vector per unit volume, Eq. (4.39) |
| $c_p$ | $J\,kg^{-1}\,K^{-1}$ | Specific heat at constant pressure |
| $c_v$ | $J\,kg^{-1}\,K^{-1}$ | Specific heat at constant volume |
| $C_D$ | - | Drag coefficient |
| $C_f$ | - | Fanning friction factor |
| $d$ | $m$ | Particle diameter |
| đ | - | Path-dependent differential, Eq. (4.43) |
| $D$ | $m$ | Pipe diameter |
| $\mathbf{D}$ | $s^{-1}$ | Deformation tensor, Eq. (4.14) |
| $D_h$ | $m$ | Hydraulic diameter |
| $D_H$ | $m$ | Heated diameter |
| $e_I$ | $J\,kg^{-1}$ | Specific internal energy, Eq. (4.44) |
| $e_K$ | $J\,kg^{-1}$ | Specific kinetic energy |
| $e_{IK} = e_I + e_K$ | $J\,kg^{-1}$ | Stagnation specific internal energy: a sum of specific internal and kinetic energy, Eq. (4.100) |

DOI: 10.1201/9781003255000-A

## LIST OF ROMAN SYMBOLS (CONT.)

| Symbol | Unit | Description and Location where the Symbol First Appears |
|---|---|---|
| $e_P$ | $J\,kg^{-1}$ | Specific potential energy |
| $e_T = e_I + e_K + e_P$ | $J\,kg^{-1}$ | Specific total energy |
| $E_I$ | J | Internal energy, Eq. (4.44) |
| $E_K$ | J | Kinetic energy, Eq. (4.45) |
| $E_P$ | J | Potential energy, Eq. (4.44) |
| $E_T$ | J | Total energy, Eq. (4.43) |
| $\mathbf{e}_x, \mathbf{e}_y, \mathbf{e}_z$ | - | Unit vectors in the Cartesian coordinates, Eq. (4.1) |
| $\mathbf{F}$ | N | Force vector, Eq. (4.35) |
| $\mathbf{F}_B$ | N | Body force vector, Eq. (4.39) |
| $\mathbf{F}_S$ | N | Surface force vector, Eq. (4.38) |
| $\mathbf{g}$ or $g$ | $m\,s^{-2}$ | Acceleration due to gravity, Eq. (4.40) |
| $G$ | $kg\,m^{-2}\,s^{-1}$ | Mass flux |
| $H$ | $W\,m^{-2}\,K^{-1}$ | Gap conductance, Eq. (10.79) |
| $i$ | $J\,kg^{-1}$ | Specific enthalpy, Eq. (4.28) |
| $\mathbf{I}$ | - | Unit tensor, Eq. (4.20) |
| $\mathbf{J}$ | Varies | General diffusive flux of $\Psi$ per unit area and unit time, Eq. (4.60) |
| $m$ | kg | Mass, Eq. (4.30) |
| $\mathbf{n}$ | - | Outward-pointing unit vector normal to a surface, Eq. (4.32) |
| $\mathbf{n}_w$ | - | Unit vector normal to a wall, Eq. (4.163) |
| $p$ | Pa | Pressure, Eq. (4.20) |
| $\mathbf{p}$ | $kg\,m\,s^{-1}$ | Linear momentum vector, Eq. (4.35) |
| $P_H$ | m | Heated perimeter |
| $P_w$ | m | Wetted perimeter |
| $q$ | W | Thermal power |
| $q'$ | $W\,m^{-1}$ | Thermal linear power |
| $\mathbf{q}''$ or $q''$ | $W\,m^{-2}$ | Heat flux, Eq. (4.47) |
| $q'''$ | $W\,m^{-3}$ | Thermal power density, Eq. (4.48) |
| $Q$ | J | Heat, Eq. (4.43) |
| $Q_B$ | J | Bulk heat, Eq. (4.48) |
| $Q_S$ | J | Surface heat, Eq. (4.47) |
| $r$ | m | Radial coordinate |
| $\mathbf{r}$ | m | Position vector, Eq. (4.1) |
| $\mathbf{r}_C$ | m | Position vector of a centroid, Eq. (4.105) |
| $\mathbf{r}_G$ | m | Position vector of a center of mass, Eq. (4.113) |
| $R$ | m | Radius |
| $R_{sp}$ | $J\,kg^{-1}\,K^{-1}$ | Specific gas constant, Table 3.3 |
| $R_u$ | $J\,mol^{-1}\,K^{-1}$ | Universal gas constant, Eq. (3.52) |
| $s$ | $J\,kg^{-1}\,K^{-1}$ | Specific entropy, Table 4.1 |
| $\mathbf{s}$ | m | Position vector |
| $S$ | $m^2$ | Surface, Eq. (4.59) |
| $S_e$ | $m^2$ | External surface, Eq. (4.147) |
| $S_{ef}$ | $m^2$ | External fluid surface, Eq. (4.147) |
| $S_i$ | $m^2$ | Internal fluid-solid surface, Eq. (4.147) |
| $S_m$ | $m^2$ | Material surface, Eq. (4.32) |
| $t$ | s | Time, Eq. (4.1) |
| $T$ | K | Temperature |
| $\mathbf{T}$ | $N\,m^{-2}$ | Total stress tensor, Eq. (4.20) |
| $u, v, w$ | $m\,s^{-1}$ | Velocity vector components in Cartesian coordinates, Eq. (4.1) |
| $u$ | $J\,kg^{-1}$ | Specific thermal energy, Eq. (4.27) |
| $\mathbf{v}$ | $m\,s^{-1}$ | Velocity vector of fluid, Eq. (4.1) |

## LIST OF ROMAN SYMBOLS (CONT.)

| Symbol | Unit | Description and Location where the Symbol First Appears |
|---|---|---|
| $v_n = \mathbf{v} \cdot \mathbf{n}$ | m s$^{-1}$ | Velocity of fluid normal to surface $S(t)$, Eq. (4.147) |
| $\mathbf{v}_r = \mathbf{v} - \mathbf{v}_s$ | m s$^{-1}$ | Relative velocity vector of fluid at surface $S(t)$, Eq. (4.65) |
| $\mathbf{v}_s$ | m s$^{-1}$ | Velocity vector of surface $S(t)$, Eq. (4.65) |
| $V$ | m$^3$ | Volume, Eq. (4.53) |
| $V_m$ | m$^3$ | Material volume, Eq. (4.31) |
| $W$ | J | Work, Eq. (4.43) |
| $x, y, z$ | m | Cartesian coordinates, Eq. (4.1) |

## LIST OF GREEK SYMBOLS

| Symbol | Unit | Description and Location where the Symbol First Appears |
|---|---|---|
| $\alpha$ | - | Void fraction |
| $\varepsilon_3$ | - | Volume porosity, Eq. (4.138) |
| $\varepsilon_2$ | - | Area porosity, Eq. (4.147) |
| $\boldsymbol{\varepsilon}_2$ | - | Area porosity tensor, Eq. (4.153) |
| $\varepsilon_n$ | - | Porosity of $n$-dimensional space, Eq. (4.140) |
| $\theta$ | ° | Contact angle, Eq. (8.42) |
| $\kappa$ | - | Specific heat ratio |
| $\lambda$ | W m$^{-1}$ K$^{-1}$ | Thermal conductivity, Eq. (4.24) |
| $\mu$ | Pa s | Dynamic viscosity, Eq. (4.10) |
| $\mu'$ | Pa s | Bulk coefficient of viscosity, Eq. (4.19) |
| $\nu$ | m$^2$ s$^{-1}$ | Kinematic viscosity, Eq. (4.10) |
| $\rho$ | kg m$^{-3}$ | Mass density, Eq. (4.31) |
| $\sigma$ | N m$^{-1}$ | Surface tension |
| $\tau$ | N m$^{-2}$ | Shear stress tensor, Eq. (4.13) |
| $\upsilon$ | m$^3$ kg$^{-1}$ | Specific volume |
| $\upsilon_M$ | m$^3$ mol$^{-1}$ | Molar volume |
| $\phi$ | Varies | Volumetric source/sink of $\Psi$ per unit mass and time, Eq. (4.57) |
| $\psi$ | Varies | General extensive property $\Psi$ per unit mass, Eq. (4.55) |
| $\Gamma$ | Varies | Sources or sinks of property $\Psi$ per unit time, Eq. (4.54) |
| $\Delta_s$ | J m$^{-3}$ K$^{-1}$ s$^{-1}$ | Entropy source per unit volume and time, Table 4.1 |
| $\Gamma_s$ | Varies | Surface part of $\Gamma$, Eq. (4.56) |
| $\Gamma_{sc}$ | Varies | Convective part of $\Gamma_s$, Eq. (4.58) |
| $\Gamma_{sd}$ | Varies | Diffusive part of $\Gamma_s$, Eq. (4.58) |
| $\Gamma_v$ | Varies | Volumetric part of $\Gamma$, Eq. (4.56) |
| $\Psi$ | Varies | General extensive property, Eq. (4.54) |

## OVERLINES

| Symbol | Description and Location where the Symbol First Appears |
|---|---|
| $\overline{f}$ | Time average of function $f$, Eq. (4.74) |
| $\widetilde{f}$ | Mass-weighted time average of function $f$, Eq. (4.88) |

## BRACKETS

| Symbol | Description and Location where the Symbol First Appears |
|---|---|
| $\langle f \rangle_1$ | Line average of function $f$, Eq. (4.108) |
| $\langle f \rangle_2$ | Area average of function $f$, Eq. (4.107) |
| $\langle f \rangle_3$ | Volume average of function $f$, Eq. (4.105) |
| $\langle f \rangle_n$ | Average of function $f$ in $n$-dimensional space, Eq. (4.116) |
| $\langle f \rangle_{3\rho}$ | Mass-weighted, volume average of function $f$, Eq. (4.112) |
| $\langle f \rangle_{n\rho}$ | Mass-weighted average of function $f$ in $n$-dimensional space, Eq. (4.117) |
| $\langle f \rangle_{3S}$ | Volume superficial average of function $f$, Eq. (4.139) |
| $\langle f \rangle_{nS}$ | Superficial average of function $f$ in $n$-dimensional space, Eq. (4.140) |
| $\langle f \rangle_{nI}$ | Intrinsic average of function $f$ in $n$-dimensional space, Eq. (4.140) |
| $\langle f \rangle_{nI\rho}$ | Mass-weighted intrinsic average of $f$ in $n$-dimensional space, Eq. (4.142) |
| $\langle f \rangle_{2Ie}$ | External area-averaged function $f$, Eq. (4.148) |
| $\langle f \rangle_{2Ii}$ | Internal area-averaged function $f$, Eq. (4.148) |

## SUBSCRIPTS

| Symbol | Description and Location where the Symbol First Appears |
|---|---|
| $f$ | Saturated liquid phase |
| $g$ | Saturated vapor phase, gas phase |
| $h$ | Hydraulic |
| $H$ | Heated |
| $l$ | Liquid phase |
| $v$ | Vapor phase |
| $w$ | Wetted, wall |

## SUPERSCRIPTS

| Symbol | Description and Location where the Symbol First Appears |
|---|---|
| $f'$ | Fluctuation of function $f$ from $\overline{f}$, Eq. (4.75) |
| $f''$ | Fluctuation of function $f$ from $\widehat{f}$, Eq. (4.90) |
| $f^{\star 3}$ | Deviation of function $f$ from $\langle f \rangle_3$, Eq. (4.109) |
| $f^{\star n}$ | Deviation of function $f$ from $\langle f \rangle_n$, Eq. (4.116) |
| $f^{\star 3\rho}$ | Deviation of function $f$ from $\langle f \rangle_{3\rho}$, Eq. (4.114) |
| $f^{\star n\rho}$ | Deviation of function $f$ from $\langle f \rangle_{n\rho}$, Eq. (4.117) |
| $f^{\star nI\rho}$ | Deviation of function $f$ from $\langle f \rangle_{nI\rho}$, Eq. (4.144) |
| $\tau^T$ | Transpose of a tensor, Table 4.1 |

## OTHER SYMBOLS

| Symbol | Description and Location where the Symbol First Appears |
|---|---|
| $\equiv$ | Defined as |
| $\approx, \cong$ | Approximately equal to |
| $\nabla\cdot$ | Divergence operator, Eq. (4.33) |
| $\frac{\partial}{\partial t}$ | Time derivative in the Eulerian frame of reference, Eq. (4.5) |
| $\frac{D}{Dt}$ | Time derivative in the Lagrangian frame of reference, Eq. (4.6) |

# B Useful Mathematical Formulas

This appendix provides a concise overview of integral theorems, averaging rules, and different forms of balance equations. While the content here is designed to be somewhat self-contained, it does not delve into detailed explanations or rigorous proofs. For a deeper understanding, readers are encouraged to refer to specialized texts available on the subject.

## B.1 INTEGRAL THEOREMS

Integral theorems are used in the transformation of conservation equations. They are presented in the following subsections.

### DIVERGENCE THEOREM

The **divergence theorem**, also referred to as the **Gauss integral theorem**, states an equivalence of the volume integral over the volume $V$ and the surface integral over the boundary $S$ of the volume $V$. For any vector or tensor quantity $\Psi$ the divergence theorem is as follows,

$$\iiint_{V(t)} \nabla \cdot \Psi dV = \iint_{S(t)} \Psi \cdot \mathbf{n} dS, \tag{B.1}$$

where $\mathbf{n}$ is a unit vector directed normally outward from $S$. Closely related **theorem of the rotational** is as follows,

$$\iiint_{V(t)} \nabla \times \Psi dV = \iint_{S(t)} \mathbf{n} \times \Psi dS. \tag{B.2}$$

For any scalar function $\Phi$, the following **theorem of the gradient** is valid,

$$\iiint_{V(t)} \nabla \Phi dV = \iint_{S(t)} \Phi \mathbf{n} dS. \tag{B.3}$$

### LEIBNIZ'S RULES

**Leibniz's rules** are useful to find a derivative with respect to time of a time-dependent quantity integrated over a time dependent region. For one-dimensional case, assuming that the region is a segment of $x$-axis, the rule is as follows,

$$\frac{d}{dt} \left[ \int_{a(t)}^{b(t)} f(x,t) dx \right] = \int_{a(t)}^{b(t)} \frac{\partial f(x,t)}{\partial t} dx + f(b,t) \frac{db(t)}{dt} - f(a,t) \frac{da(t)}{dt}. \tag{B.4}$$

DOI: 10.1201/9781003255000-B

The physical interpretation of the rule will be more clear when it is written as follows,

$$\frac{d}{dt}\left[\int_{a(t)}^{b(t)} f(x,t)dx\right] = \int_{a(t)}^{b(t)} \frac{\partial f(x,t)}{\partial t}dx + f(b,t)\mathbf{v}_b \cdot \mathbf{n}_b + f(a,t)\mathbf{v}_a \cdot \mathbf{n}_a, \qquad \text{(B.5)}$$

where

$$\mathbf{v}_a = \frac{da(t)}{dt}, \ \mathbf{v}_b = \frac{db(t)}{dt}, \ \mathbf{n}_a = -\mathbf{e}_x, \ \mathbf{n}_b = \mathbf{e}_x. \qquad \text{(B.6)}$$

Here $\mathbf{e}_x$ is a unit vector in the $x$-axis direction. It can be seen that the two last terms in Eq. (B.5) contain the velocity of displacement of integration boundaries and they represent the effect of the boundary movement on the overall value of the time derivative.

For a two-dimensional case, assuming that the region is an area $A(t)$ located on the $x$-$y$ plane and surrounded with a contour $C(t)$, the corresponding formulation of Leibniz's rule is as follows,

$$\frac{d}{dt}\left[\iint_{A(t)} f(x,y,t)dA\right] = \iint_{A(t)} \frac{\partial f(x,y,t)}{\partial t}dA + \int_{C(t)} f(x,y,t)\mathbf{v} \cdot \mathbf{n}dC. \qquad \text{(B.7)}$$

For three-dimensional case, when integration is over a volume $V(t)$ surrounded with a surface $S(t)$, Leibniz' rule becomes,

$$\frac{d}{dt}\left[\iiint_{V(t)} f(x,y,z,t)dV\right] = \\ \iiint_{V(t)} \frac{\partial f(x,y,z,t)}{\partial t}dV + \iint_{S(t)} f(x,y,z,t)\mathbf{v} \cdot \mathbf{n}dS \qquad . \quad \text{(B.8)}$$

Vector $\mathbf{n}$ in Eqs. (B.7) and (B.8), similarly as for the one-dimensional case, is a unit normal vector pointing outwards from the region of integration. It should be remembered here that vector $\mathbf{v}$ on the right-hand-side of Eq. (B.8) represents the velocity of boundary $S(t)$.

The Leibniz rule is also applicable when integration is taken over a time interval $[t_1; t_2]$ as follows,

$$\frac{d}{dt}\left[\int_{t_1(t)}^{t_2(t)} f(x,\tau)d\tau\right] = \int_{t_1(t)}^{t_2(t)} \frac{\partial f(x,\tau)}{\partial \tau}d\tau + f(t_2,t)\frac{dt_2(t)}{dt} - f(t_1,t)\frac{dt_1(t)}{dt}. \qquad \text{(B.9)}$$

## REYNOLDS' TRANSPORT THEOREM

**Reynolds' transport theorem** is given as follows

$$\frac{D}{Dt}\left(\iiint_{V_m(t)} f(\mathbf{r},t)dV\right) = \iiint_{V_m(t)} \frac{\partial f(\mathbf{r},t)}{\partial t}dV + \iint_{S_m(t)} f(\mathbf{r},t)\mathbf{v} \cdot \mathbf{n}dS, \qquad \text{(B.10)}$$

where $\mathbf{n}$ is the unit normal vector pointing out of volume $V_m(t)$ at the surface $S_m(t)$ and $\mathbf{v}$ is the velocity vector. This equation is useful to transform the derivatives of integrals from material-based coordinates (Lagrangian frame of reference) on the left-hand side into spatial coordinates (Eulerian frame of reference) on the right-hand side. Note that the integration is performed over a material volume $V_m(t)$ bounded by a material surface $S_m(t)$.

## LEIBNIZ RULE FOR TIME AVERAGING

The **phase characteristic function**, $X_k(\mathbf{r},t)$, is defined as follows,

$$X_k(\mathbf{r},t) = \begin{cases} 1 & \text{if phase } k \text{ present at point } \mathbf{r} \text{ at time } t \\ 0 & \text{otherwise} \end{cases}. \tag{B.11}$$

An instant multiphase flow configuration is shown in Fig. B.1. We use three-phase flow of phases $A$, $B$, and $C$ as an example. Path $s$–$s$ in the figure represents a collec-

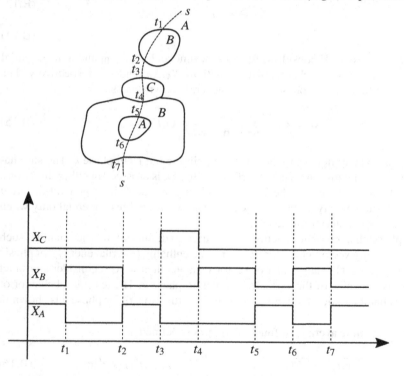

**Figure B.1** Phase characteristic functions for three-phase flow.

tion of traces in the three-phase flow when passing through a certain point $P(x,y,z)$ in the space. Let us denote the interface between phases $A$ and $B$ as $A \wr B$. At time instant $t_1$ the interface $A \wr B$ is crossing point $P$. At this time instant the characteristic function for phase $A$ jumps from 1 to 0, whereas the characteristic function for phase $B$ jumps from 0 to 1. This means that for $t < t_1$ point $P$ is occupied by phase $A$ and for time $t > t_1$ it is occupied by phase $B$. This is only possible if $\mathbf{n}_{AB} \cdot \mathbf{v}_i < 0$ at point $P$ and time $t_1$. Here $\mathbf{n}_{AB}$ is a vector normal to the interface $A \wr B$ pointing from phase $A$ to phase $B$ and $\mathbf{v}_i$ is the interface velocity vector while crossing point $P$. If this condition is not satisfied, the interface will not cross point $P$ and the point will remain within phase $A$. In a similar manner, at time $t_2$ the condition that the interface crosses point $P$ is that $\mathbf{n}_{AB} \cdot \mathbf{v}_i > 0$.

It can be noted that the characteristic functions for phases $A, B$, and $C$ can be represented as,

$$X_A = X_{A0} - \Theta(t - t_1) + \Theta(t - t_2) - \Theta(t - t_3) + \Theta(t - t_5) \atop - \Theta(t - t_6) + \Theta(t - t_7) \tag{B.12}$$

$$X_B = X_{B0} + \Theta(t - t_1) - \Theta(t - t_2) + \Theta(t - t_4) - \Theta(t - t_5) \atop + \Theta(t - t_6) - \Theta(t - t_7) \tag{B.13}$$

$$X_C = X_{C0} + \Theta(t - t_3) - \Theta(t - t_4). \tag{B.14}$$

Here $\Theta(t)$ is the step (Heaviside's) function of time and $X_{A0}, X_{B0}$, and $X_{C0}$ are initial values (at time $t = 0$) of characteristic functions $X_A, X_B$, and $X_C$, respectively. The characteristic function for phase $k$ can, in general, be written as,

$$X_k = X_{k0} + \sum_j \left( \frac{\mathbf{n}_{ki} \cdot \mathbf{v}_i}{|\mathbf{n}_{ki} \cdot \mathbf{v}_i|} \right)_j \Theta(t - t_j), \tag{B.15}$$

where $\mathbf{n}_{ki}$ is a vector normal to the interface pointing out from phase $k$. The summation is over all time instants when interface to phase $k$ is crossed (in either direction). It should be noted here that the interface velocity vector $\mathbf{v}_i$ is in general different from the phasic velocity at the interface $\mathbf{v}_{ki}$. The two velocities are equal only when the interface is a material surface.

The phase characteristic function can be used to define any phasic quantity, such as phasic velocity vector, phasic pressure, phasic enthalpy (internal energy), or phasic density. If $f(x, y, z, t)$ is any scalar or vector variable, then we can consider a product $f_k = X_k f$ as the value of the variable pertinent to phase $k$. The temporal behavior of the phase characteristic function that selects field function $f_k$ for phase $k$ is shown in Fig. B.2.

The phasic time average of function $f_k(\mathbf{r}, t)$ is defined as,

$$\overline{f_k(\mathbf{r}, t)}^T = \overline{X_k(\mathbf{r}, t) f(\mathbf{r}, t)}^T \equiv \frac{1}{T} \int_{t-T}^{t} X_k(\mathbf{r}, t') f(\mathbf{r}, t') \mathrm{d}t'. \tag{B.16}$$

Based on the $f_k$ function representation shown in Fig. B.2, the time-averaging integration can be expressed as the following sum,

$$\int_{t-T}^{t} X_k f \, \mathrm{d}t' = \int_{t-T}^{t_2} f \, \mathrm{d}t' + \int_{t_3}^{t_4} f \, \mathrm{d}t' + \ldots + \int_{t_{2n-1}}^{t} f \, \mathrm{d}t'. \tag{B.17}$$

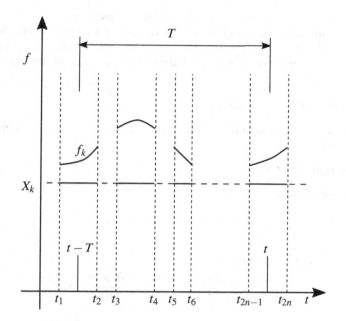

**Figure B.2** Temporal behavior of the phase characteristic function that selects field function $f$ for phase $k$.

Combining Eqs. (B.16) and (B.17) yields the following temporal partial derivative of time-averaged function,

$$\frac{\partial \left[\overline{X_k f(t)}^T\right]}{\partial t} = \frac{1}{T}\left[f(t) - f(t - T)\right], \tag{B.18}$$

which is consistent with a definition of a partial derivative of a function. In addition, we have,

$$\overline{X_k \frac{\partial f}{\partial t}}^T = \frac{1}{T}\left(\int_{t-T}^{t_2} \frac{\partial f}{\partial t'}dt' + \int_{t_3}^{t_4} \frac{\partial f}{\partial t'}dt' + \ldots + \int_{t_{2n-1}}^{t} \frac{\partial f}{\partial t'}dt'\right). \tag{B.19}$$

Integrating the functions on the right hand side yields,

$$\overline{X_k \frac{\partial f}{\partial t}}^T = \frac{1}{T}\left[f(t) - f(t_{2n-1}^+) + \ldots + f(t_4^-) - f(t_3^+) + f(t_2^-) - f(t - T)\right]. \tag{B.20}$$

Combining Eqs. (B.20) and (B.18) gives,

$$\frac{\partial \left[\overline{X_k f(t)}^T\right]}{\partial t} = \overline{X_k \frac{\partial f}{\partial t}}^T - \frac{1}{T}\left[-f(t_{2n-1}^+) + \ldots + f(t_4^-) - f(t_3^+) + f(t_2^-)\right]. \tag{B.21}$$

As can be seen, the time derivative of the time average of a discontinuous function $f_k$ is equal to the time average of the time derivative of a function minus an additional term resulting from the function discontinuities during the averaging period. The plus or minus signs at time instants $t_i$ indicate that the function value should be taken when phase $k$ is present.

## B.2 BALANCE EQUATIONS

Differential balance equations of mass, linear momentum, and total energy are given in this section.

### MASS BALANCE EQUATION

*In the general form:*

$$\frac{\partial \rho}{\partial t} + \nabla \cdot (\rho \mathbf{v}) = 0 . \tag{B.22}$$

*In the Cartesian coordinates $(x, y, z)$:*

$$\frac{\partial \rho}{\partial t} + \frac{\partial (\rho u)}{\partial x} + \frac{\partial (\rho v)}{\partial y} + \frac{\partial (\rho w)}{\partial z} = 0 . \tag{B.23}$$

*In the cylindrical coordinates $(r, \theta, z)$:*

$$\frac{\partial \rho}{\partial t} + \frac{1}{r}\frac{\partial (\rho r u)}{\partial r} + \frac{1}{r}\frac{\partial (\rho v)}{\partial \theta} + \frac{\partial (\rho w)}{\partial z} = 0 . \tag{B.24}$$

*In the spherical coordinates $(r, \theta, \phi)$:*

$$\frac{\partial \rho}{\partial t} + \frac{1}{r^2}\frac{\partial}{\partial r}\left(\rho r^2 u\right) + \frac{1}{r\sin\theta}\frac{\partial}{\partial \theta}\left(\rho v \sin\theta\right) + \frac{1}{r\sin\theta}\frac{\partial}{\partial \phi}\left(\rho w\right) = 0 . \tag{B.25}$$

### MOMENTUM BALANCE EQUATION

In terms of the total shear stress tensor $\mathbf{T}$:

$$\rho\left(\frac{\partial \mathbf{v}}{\partial t} + \mathbf{v} \cdot \nabla \mathbf{v}\right) = \nabla \cdot \mathbf{T} + \rho \mathbf{g}. \tag{B.26}$$

In terms of the pressure gradient $\nabla p$ and the viscous shear stress $\boldsymbol{\tau}$:

$$\rho\left(\frac{\partial \mathbf{v}}{\partial t} + \mathbf{v} \cdot \nabla \mathbf{v}\right) = -\nabla p + \nabla \cdot \boldsymbol{\tau} + \rho \mathbf{g}. \tag{B.27}$$

### In terms of the components of the shear stress tensor:

*In the Cartesian coordinates $(x, y, z)$:*

*x*-component:

$$\rho\left(\frac{\partial u}{\partial t} + u\frac{\partial u}{\partial x} + v\frac{\partial u}{\partial y} + w\frac{\partial u}{\partial z}\right) = -\frac{\partial p}{\partial x} + \frac{\partial \tau_{xx}}{\partial x} + \frac{\partial \tau_{yx}}{\partial y} + \frac{\partial \tau_{zx}}{\partial z} + \rho g_x, \tag{B.28}$$

$y$-component:

$$\rho \left( \frac{\partial v}{\partial t} + u\frac{\partial v}{\partial x} + v\frac{\partial v}{\partial y} + w\frac{\partial v}{\partial z} \right) = -\frac{\partial p}{\partial y} + \frac{\partial \tau_{xy}}{\partial x} + \frac{\partial \tau_{yy}}{\partial y} + \frac{\partial \tau_{zy}}{\partial z} + \rho g_y, \quad \text{(B.29)}$$

$z$-component:

$$\rho \left( \frac{\partial w}{\partial t} + u\frac{\partial w}{\partial x} + v\frac{\partial w}{\partial y} + w\frac{\partial w}{\partial z} \right) = -\frac{\partial p}{\partial z} + \frac{\partial \tau_{xz}}{\partial x} + \frac{\partial \tau_{yz}}{\partial y} + \frac{\partial \tau_{zz}}{\partial z} + \rho g_z. \quad \text{(B.30)}$$

The components of the shear stress tensor:

$$\tau_{xx} = 2\mu \frac{\partial u}{\partial x} - \left( \frac{2}{3}\mu - \mu' \right) \left( \frac{\partial u}{\partial x} + \frac{\partial v}{\partial y} + \frac{\partial w}{\partial z} \right), \quad \text{(B.31)}$$

$$\tau_{yy} = 2\mu \frac{\partial v}{\partial y} - \left( \frac{2}{3}\mu - \mu' \right) \left( \frac{\partial u}{\partial x} + \frac{\partial v}{\partial y} + \frac{\partial w}{\partial z} \right), \quad \text{(B.32)}$$

$$\tau_{zz} = 2\mu \frac{\partial w}{\partial z} - \left( \frac{2}{3}\mu - \mu' \right) \left( \frac{\partial u}{\partial x} + \frac{\partial v}{\partial y} + \frac{\partial w}{\partial z} \right), \quad \text{(B.33)}$$

$$\tau_{xy} = \tau_{yx} = \mu \left( \frac{\partial v}{\partial x} + \frac{\partial u}{\partial y} \right), \quad \text{(B.34)}$$

$$\tau_{yz} = \tau_{zy} = \mu \left( \frac{\partial w}{\partial y} + \frac{\partial v}{\partial z} \right), \quad \text{(B.35)}$$

$$\tau_{zx} = \tau_{xz} = \mu \left( \frac{\partial u}{\partial z} + \frac{\partial w}{\partial x} \right), \quad \text{(B.36)}$$

Based on the Stokes hypothesis, it can be assume that $\mu' = 0$.

*In the cylindrical coordinates* $(r, \theta, z)$:

$r$-component:

$$\rho \left( \frac{\partial u}{\partial t} + u\frac{\partial u}{\partial r} + \frac{v}{r}\frac{\partial u}{\partial \theta} + w\frac{\partial u}{\partial z} - \frac{v^2}{r} \right) = -\frac{\partial p}{\partial r} + $$
$$\frac{1}{r}\frac{\partial(r\tau_{rr})}{\partial r} + \frac{1}{r}\frac{\partial \tau_{\theta r}}{\partial \theta} + \frac{\partial \tau_{zr}}{\partial z} - \frac{\tau_{\theta\theta}}{r} + \rho g_r \quad \text{(B.37)}$$

$\theta$-component:

$$\rho \left( \frac{\partial v}{\partial t} + u\frac{\partial v}{\partial r} + \frac{v}{r}\frac{\partial v}{\partial \theta} + w\frac{\partial v}{\partial z} + \frac{uv}{r} \right) = -\frac{1}{r}\frac{\partial p}{\partial \theta} + $$
$$\frac{1}{r^2}\frac{\partial(r^2\tau_{r\theta})}{\partial r} + \frac{1}{r}\frac{\partial \tau_{\theta\theta}}{\partial \theta} + \frac{\partial \tau_{z\theta}}{\partial z} + \rho g_\theta \quad \text{(B.38)}$$

$z$-component:

$$\rho\left(\frac{\partial w}{\partial t}+u\frac{\partial w}{\partial r}+\frac{v}{r}\frac{\partial w}{\partial \theta}+w\frac{\partial w}{\partial z}\right)=-\frac{\partial p}{\partial z}+$$
$$\frac{1}{r}\frac{\partial(r\tau_{rz})}{\partial r}+\frac{1}{r}\frac{\partial \tau_{\theta z}}{\partial \theta}+\frac{\partial \tau_{zz}}{\partial z}+\rho g_z \qquad \text{(B.39)}$$

The components of the shear stress tensor:

$$\tau_{rr}=2\mu\frac{\partial u}{\partial r}-\left(\frac{2}{3}\mu-\mu'\right)\left(\frac{1}{r}\frac{\partial}{\partial r}(ru)+\frac{1}{r}\frac{\partial v}{\partial \theta}+\frac{\partial w}{\partial z}\right), \qquad \text{(B.40)}$$

$$\tau_{\theta\theta}=2\mu\left(\frac{1}{r}\frac{\partial v}{\partial \theta}+\frac{u}{r}\right)-\left(\frac{2}{3}\mu-\mu'\right)\left(\frac{1}{r}\frac{\partial}{\partial r}(ru)+\frac{1}{r}\frac{\partial v}{\partial \theta}+\frac{\partial w}{\partial z}\right), \qquad \text{(B.41)}$$

$$\tau_{zz}=2\mu\frac{\partial w}{\partial z}-\left(\frac{2}{3}\mu-\mu'\right)\left(\frac{1}{r}\frac{\partial}{\partial r}(ru)+\frac{1}{r}\frac{\partial v}{\partial \theta}+\frac{\partial w}{\partial z}\right), \qquad \text{(B.42)}$$

$$\tau_{r\theta}=\tau_{\theta r}=\mu\left[r\frac{\partial}{\partial r}\left(\frac{v}{r}\right)+\frac{1}{r}\frac{\partial u}{\partial \theta}\right], \qquad \text{(B.43)}$$

$$\tau_{\theta z}=\tau_{z\theta}=\mu\left(\frac{1}{r}\frac{\partial w}{\partial \theta}+\frac{\partial v}{\partial z}\right), \qquad \text{(B.44)}$$

$$\tau_{zr}=\tau_{rz}=\mu\left(\frac{\partial u}{\partial z}+\frac{\partial w}{\partial r}\right). \qquad \text{(B.45)}$$

## For Newtonian Fluids with Constant Density and Viscosity

*In the general form:*

$$\rho\left(\frac{\partial \mathbf{v}}{\partial t}+\mathbf{v}\cdot\nabla\mathbf{v}\right)=-\nabla p+\mu\nabla^2\mathbf{v}+\rho\mathbf{g}. \qquad \text{(B.46)}$$

*In the Cartesian coordinates $(x,y,z)$:*

$x$-component:

$$\rho\left(\frac{\partial u}{\partial t}+u\frac{\partial u}{\partial x}+v\frac{\partial u}{\partial y}+w\frac{\partial u}{\partial z}\right)=-\frac{\partial p}{\partial x}+\mu\left[\frac{\partial^2 u}{\partial x^2}+\frac{\partial^2 u}{\partial y^2}+\frac{\partial^2 u}{\partial z^2}\right]+\rho g_x, \qquad \text{(B.47)}$$

$y$-component:

$$\rho\left(\frac{\partial v}{\partial t}+u\frac{\partial v}{\partial x}+v\frac{\partial v}{\partial y}+w\frac{\partial v}{\partial z}\right)=-\frac{\partial p}{\partial y}+\mu\left[\frac{\partial^2 v}{\partial x^2}+\frac{\partial^2 v}{\partial y^2}+\frac{\partial^2 v}{\partial z^2}\right]+\rho g_y, \qquad \text{(B.48)}$$

$z$-component:

$$\rho\left(\frac{\partial w}{\partial t}+u\frac{\partial w}{\partial x}+v\frac{\partial w}{\partial y}+w\frac{\partial w}{\partial z}\right)=-\frac{\partial p}{\partial z}+\mu\left[\frac{\partial^2 w}{\partial x^2}+\frac{\partial^2 w}{\partial y^2}+\frac{\partial^2 w}{\partial z^2}\right]+\rho g_z. \qquad \text{(B.49)}$$

*In the cylindrical coordinates $(r, \theta, z)$:*

$r$-component:

$$\rho \left( \frac{\partial u}{\partial t} + u \frac{\partial u}{\partial r} + \frac{v}{r} \frac{\partial u}{\partial \theta} + w \frac{\partial u}{\partial z} - \frac{v^2}{r} \right) = -\frac{\partial p}{\partial r} +$$
$$\mu \left[ \frac{\partial}{\partial r} \left( \frac{1}{r} \frac{\partial (ru)}{\partial r} \right) + \frac{1}{r^2} \frac{\partial^2 u}{\partial \theta^2} + \frac{\partial^2 u}{\partial z^2} - \frac{2}{r^2} \frac{\partial v}{\partial \theta} \right] + \rho g_r \qquad \text{(B.50)}$$

$\theta$-component:

$$\rho \left( \frac{\partial v}{\partial t} + u \frac{\partial v}{\partial r} + \frac{v}{r} \frac{\partial v}{\partial \theta} + w \frac{\partial v}{\partial z} + \frac{uv}{r} \right) = -\frac{1}{r} \frac{\partial p}{\partial \theta} +$$
$$\mu \left[ \frac{\partial}{\partial r} \left( \frac{1}{r} \frac{\partial (rv)}{\partial r} \right) + \frac{1}{r^2} \frac{\partial^2 v}{\partial \theta^2} + \frac{\partial^2 v}{\partial z^2} + \frac{2}{r^2} \frac{\partial u}{\partial \theta} \right] + \rho g_\theta \qquad \text{(B.51)}$$

$z$-component:

$$\rho \left( \frac{\partial w}{\partial t} + u \frac{\partial w}{\partial r} + \frac{v}{r} \frac{\partial w}{\partial \theta} + w \frac{\partial w}{\partial z} \right) = -\frac{\partial p}{\partial z} +$$
$$\mu \left[ \frac{1}{r} \frac{\partial}{\partial r} \left( r \frac{\partial w}{\partial r} \right) + \frac{1}{r^2} \frac{\partial^2 w}{\partial \theta^2} + \frac{\partial^2 w}{\partial z^2} \right] + \rho g_z \qquad \text{(B.52)}$$

## ENERGY BALANCE EQUATION

### In terms of the rate of change of the kinetic energy

$$\rho \left( \frac{\partial e_K}{\partial t} + \mathbf{v} \cdot \nabla e_K \right) = \nabla \cdot [(\boldsymbol{\tau} - p\mathbf{I}) \cdot \mathbf{v}] + p \nabla \cdot \mathbf{v} + \mathbf{v} \cdot \rho \mathbf{g} - \boldsymbol{\tau} : \nabla \mathbf{v}, \qquad \text{(B.53)}$$

where $e_K = \frac{1}{2} v^2 = \frac{1}{2} |\mathbf{v}|^2$ is the fluid kinetic energy per unit mass and $\mathbf{I}$ is the unity tensor.

### In terms of the temperature

For heat transfer problems, the following energy conservation equation is valid:

$$\rho c_p \left( \frac{\partial T}{\partial t} + \mathbf{v} \cdot \nabla T \right) = \nabla \cdot \lambda \nabla T - \nabla \cdot \mathbf{q}_r'' + q''' + \beta T \left( \frac{\partial p}{\partial t} + \mathbf{v} \cdot \nabla p \right) + \phi, \qquad \text{(B.54)}$$

where $\mathbf{q}_r''$ is the heat flux vector due to radiation, $q'''$ is the heat source, $\beta$ is the volumetric thermal expansion coefficient of the fluid, and $\phi \equiv \boldsymbol{\tau} : \nabla \mathbf{v} = \nabla \cdot (\boldsymbol{\tau} \cdot \mathbf{v}) - \mathbf{v} \cdot (\nabla \cdot \boldsymbol{\tau})$ is the energy dissipation term. For incompressible materials $c_p = c_v$ is valid.

*In the Cartesian coordinates* $(x,y,z)$:

$$\rho c_p \left( \frac{\partial T}{\partial t} + u\frac{\partial T}{\partial x} + v\frac{\partial T}{\partial y} + w\frac{\partial T}{\partial z} \right) =$$

$$\frac{\partial}{\partial x}\left( \lambda \frac{\partial T}{\partial x} \right) + \frac{\partial}{\partial y}\left( \lambda \frac{\partial T}{\partial y} \right) + \frac{\partial}{\partial z}\left( \lambda \frac{\partial T}{\partial z} \right)$$

$$-\left( \frac{\partial q_{rx}''}{\partial x} + \frac{\partial q_{ry}''}{\partial y} + \frac{\partial q_{rz}''}{\partial z} \right) + q'''$$

$$+\beta T \left( \frac{\partial p}{\partial t} + u\frac{\partial p}{\partial x} + v\frac{\partial p}{\partial y} + w\frac{\partial p}{\partial z} \right) + \phi$$

(B.55)

where $q_{rx}''$, $q_{ry}''$, and $q_{rz}''$ are components of the radiation heat flux vector $\mathbf{q}_r''$.

*In the cylindrical coordinates* $(r,\theta,z)$:

$$\rho c_p \left( \frac{\partial T}{\partial t} + u\frac{\partial T}{\partial r} + \frac{v}{r}\frac{\partial T}{\partial \theta} + w\frac{\partial T}{\partial z} \right) =$$

$$\frac{1}{r}\frac{\partial}{\partial r}\left( \lambda r \frac{\partial T}{\partial r} \right) + \frac{1}{r^2}\frac{\partial}{\partial \theta}\left( \lambda \frac{\partial T}{\partial \theta} \right) + \frac{\partial}{\partial z}\left( \lambda \frac{\partial T}{\partial z} \right)$$

$$-\left( \frac{1}{r}\frac{\partial (r q_{rr}'')}{\partial r} + \frac{1}{r}\frac{\partial q_{r\theta}''}{\partial \theta} + \frac{\partial q_{rz}''}{\partial z} \right) + q'''$$

$$+\beta T \left( \frac{\partial p}{\partial t} + u\frac{\partial p}{\partial r} + \frac{v}{r}\frac{\partial p}{\partial \theta} + w\frac{\partial p}{\partial z} \right) + \phi$$

(B.56)

where $q_{rr}''$, $q_{r\theta}''$, and $q_{rz}''$ are components of the radiation heat flux vector $\mathbf{q}_r''$ in the cylindrical coordinate system.

# C Correlations

In this Appendix we provide selected correlations for friction factors and heat transfer coefficients.

## C.1 FRICTION FACTORS

### DEFINITIONS

The Darcy friction factor:

$$\Lambda = \frac{4\tau_w}{\frac{1}{2}\rho U^2} = \frac{8\tau_w}{\rho U^2},$$

(C.1)

where $\tau_w$ is the wall shear stress, $\rho$ is the fluid density, and $U$ is the mean velocity of the fluid flowing in the channel.

The Fanning friction factor:

$$C_f = \frac{\tau_w}{\frac{1}{2}\rho U^2} = \frac{2\tau_w}{\rho U^2},$$

(C.2)

where all parameters have the same definitions as above.

Reynolds number:

$$\text{Re} = \frac{UD_h}{\nu} = \frac{\rho U D_h}{\mu},$$

(C.3)

where $D_h$ is the hydraulic diameter.

Hydraulic diameter:

$$D_h = \frac{4A}{P_w},$$

(C.4)

where $A$ is the cross-section area of a channel and $P_w$ is its wetted perimeter.

### LAMINAR FLOW FORMULAS

Circular tube:

$$\Lambda = \frac{64}{\text{Re}},$$

(C.5)

where Re is the Reynolds number. The formula is valid for $\text{Re} < 2000$.

## TURBULENT FLOW FORMULAS

*Blasius correlation:*

$$\Lambda = \frac{0.3164}{\text{Re}^{0.25}} , \tag{C.6}$$

where Re is the Reynolds number. The correlation is valid for $4 \cdot 10^3 < \text{Re} < 10^5$ [19].

*Colebrook correlation:*

$$\frac{1}{\sqrt{\Lambda}} = -2 \log_{10} \left( \frac{\varepsilon/D}{3.7} + \frac{2.51}{\text{Re}\sqrt{\Lambda}} \right) , \tag{C.7}$$

where Re is the Reynolds number, $D$ is the diameter in m, and $\varepsilon$ is the wall roughness in m. The correlation is valid for rough tubes and $\text{Re} > 4 \cdot 10^3$ [44].

*Moody correlation:*

$$\Lambda = 1.375 \times 10^{-3} \left[ 1 + \left( 2 \times 10^4 \frac{\varepsilon}{D} + \frac{10^6}{\text{Re}} \right)^{1/3} \right] , \tag{C.8}$$

where Re is the Reynolds number, $D$ is the diameter in m, and $\varepsilon$ is the wall roughness in m. The correlation is valid for $4000 < \text{Re} < 10^8$ and $0 < \varepsilon/D < 1$ [166].

*Haaland correlation:*

$$\Lambda = \left\{ -1.8 \log_{10} \left[ \left( \frac{\varepsilon/D}{3.7} \right)^{1.11} + \frac{6.9}{\text{Re}} \right] \right\}^{-2} , \tag{C.9}$$

where Re is the Reynolds number, $D$ is the diameter in m, and $\varepsilon$ is the wall roughness in m. The correlation is valid for $4 \cdot 10^3 < \text{Re} < 10^8$ and $10^{-6} < \varepsilon/D < 0.05$ [82].

## C.2 HEAT TRANSFER

### DEFINITIONS

*Film Temperature:*

$$T_f = \frac{1}{2}(T_b + T_w) , \tag{C.10}$$

where $T_b$ is the far-field or bulk temperature and $T_w$ is the wall temperature.

*Nusselt number:*

$$\text{Nu} = \frac{hD}{\lambda} , \tag{C.11}$$

where $h$ is the heat transfer coefficient, $D$ is the diameter, and $\lambda$ is the thermal conductivity.

*Rayleigh number:*

$$Ra_H = \frac{g\beta\Delta T H^3}{av},$$  (C.12)

where $\beta$ is the thermal expansion coefficient, $\Delta T$ is the scale of the temperature difference, $H$ is the height, $a$ is the thermal diffusivity, and $v$ is the kinematic viscosity.

## INTERNAL CONVECTION

### Natural Laminar Convection

The recommended heat transfer correlation for laminar natural convection in a rectangular enclosure heated and cooled from the side is as follows,

$$\overline{Nu_H} = 0.22 \left(\frac{Pr}{0.2+Pr}Ra_H\right)^{0.28}\left(\frac{L}{H}\right)^{0.09},$$  (C.13)

for

$$2 < \frac{H}{L} < 10, \quad Pr < 10^5, \quad Ra_H < 10^{13},$$

and

$$\overline{Nu_H} = 0.18 \left(\frac{Pr}{0.2+Pr}Ra_H\right)^{0.29}\left(\frac{L}{H}\right)^{-0.13},$$  (C.14)

for

$$1 < \frac{H}{L} < 2, \quad 10^{-3} < Pr < 10^5, \quad 10^3 < \frac{Pr}{0.2+Pr}Ra_H\left(\frac{L}{H}\right)^3.$$

Here $\overline{Nu_H} = \overline{q''}H/(\lambda\Delta T)$, $H$ is the height of the enclosure, $L$ is the length of the enclosure, $Pr$ is the Prandtl number, and $Ra_H$ is the Rayleigh number [29].

### Laminar Forced Convection

*Constant wall temperature:*

$$Nu = 3.658.$$  (C.15)

*Constant wall heat flux:*

$$Nu = 4.364.$$  (C.16)

*Hausen correlation:*

$$Nu = 3.66 + \frac{0.0668\frac{D}{L}RePr}{1+0.04(\frac{D}{L}RePr)^{2/3}},$$  (C.17)

where $Re$ is the Reynolds number, $Pr$ is the Prandtl number, $D$ is the tube diameter in m, and $L$ is the tube length in m [85].

*Sieder-Tate correlation:*

$$Nu = 1.86 \left(\frac{D}{L}RePr\right)^{1/3} \left(\frac{\mu_b}{\mu_w}\right)^{0.14}, \qquad (C.18)$$

where Re is the Reynolds number, Pr is the Prandtl number, $\mu_b$ is the fluid dynamic viscosity at the mean bulk temperature, and $\mu_w$ is the fluid dynamic viscosity at the wall temperature [203].

**Turbulent Forced Convection**

*Dittus-Boelter correlation:*

$$Nu = 0.023Re^{4/5}Pr^n, \qquad (C.19)$$

where Re is the Reynolds number and Pr is the Prandtl number. The Prandtl number exponent is $n = 0.3$ when the fluid is being cooled and $n = 0.4$ when the fluid is being heated. The correlation is valid for $0.7 < Pr < 120$, $2500 < Re < 1.24 \cdot 10^5$, and $L/D > 60$ [56].

*Sieder-Tate correlation:*

$$Nu = 0.027Re^{4/5}Pr^{1/3} \left(\frac{\mu_b}{\mu_w}\right)^{0.14}, \qquad (C.20)$$

where Re is the Reynolds number, Pr is the Prandtl number, $\mu_b$ is the fluid dynamic viscosity at the mean bulk temperature, and $\mu_w$ is the fluid dynamic viscosity at the wall temperature. The correlation is valid for $0.7 < Pr < 1.67 \cdot 10^4$ and $10^4 < Re$ [203].

*Colburn correlation:*

$$Nu = 0.023Re^{0.8}Pr^{1/3}, \qquad (C.21)$$

where Re is the Reynolds number and Pr is the Prandtl number. The correlation is valid for $0.5 < Pr < 3$ and $10^4 < Re < 10^5$ [40].

*Friend-Metzner correlation:*

$$Nu = \frac{(\Lambda/8)RePr}{1.2 + 11.8(\Lambda/8)^{0.5}(Pr-1)Pr^{-1/3}}, \qquad (C.22)$$

where $\Lambda$ is the Darcy friction factor, Re is the Reynolds number, and Pr is the Prandtl number. The correlation is valid for $50 < Pr < 600$ and $5 \cdot 10^4 < Re < 5 \cdot 10^6$ [71].

*Sandall et al. correlation:*

$$Nu = \frac{(\Lambda/8)RePr}{12.48Pr^{2/3} - 7.853Pr^{1/3} + 3.613\ln Pr + 5.8 + C}$$

$$C = 2.78\ln\left[\left(\frac{\Lambda}{8}\right)^{0.5}\frac{Re}{45}\right], \qquad (C.23)$$

where $\Lambda$–the Darcy friction factor, Re is the Reynolds number, and Pr is the Prandtl number. The correlation is valid for $0.5 < \text{Pr} < 2 \cdot 10^3$ and $10^4 < \text{Re} < 5 \cdot 10^6$ [191].

*Gnielinski correlation:*

$$\text{Nu} = \frac{(\Lambda/8)(\text{Re} - 1000)\text{Pr}}{1 + 12.7(\Lambda/8)^{1/2}(\text{Pr}^{2/3} - 1)}, \tag{C.24}$$

where $\Lambda$–the Darcy friction factor, Re is the Reynolds number, and Pr is the Prandtl number. The correlation is valid for $0.5 < \text{Pr} < 2 \cdot 10^3$ and $2300 < \text{Re} < 5 \cdot 10^6$ [77].

## EXTERNAL CONVECTION

### Natural Convection

*Crossflow around a horizontal cylinder:*

$$\overline{\text{Nu}}_D = \left\{ 0.6 + \frac{0.387\text{Ra}_D^{1/6}}{[1 + (0.559/\text{Pr})^{9/16}]^{8/27}} \right\}^2, \tag{C.25}$$

where $\text{Ra}_D$ is the Rayleigh number and Pr is the Prandtl number. This equation is valid for $10^{-5} < \text{Ra}_D < 10^{12}$ and the entire Prandtl number range [35].

*Flow around a sphere:*

$$\overline{\text{Nu}}_D = 2 + \frac{0.589\,\text{Ra}_D^{1/4}}{\left[ 1 + (0.469/\text{Pr})^{9/16} \right]^{4/9}}, \tag{C.26}$$

where $\text{Ra}_D$ is the Rayleigh number and Pr is the Prandtl number. This correlation is valid for $\text{Pr} \gtrsim 0.7$ and $\text{Ra}_D < 10^{11}$ [36].

### Forced Convection

*Crossflow around a single cylinder:*

$$\text{Nu} = 0.3 + \frac{0.62\text{Re}^{0.5}\text{Pr}^{1/3}}{\left[ 1 + (0.4/\text{Pr})^{2/3} \right]^{0.25}} \left[ 1 + \left( \frac{\text{Re}}{282000} \right)^{5/8} \right]^{0.8}, \tag{C.27}$$

where Re is the Reynolds number and Pr is the Prandtl number, both evaluated at the film temperature [37].

# D Data Tables

## D.1 STEAM-WATER PROPERTIES

The properties shown in this section are based on the IAPWS-IF97 standard. Density ($\rho$), specific enthalpy ($i$), dynamic viscosity ($\mu$), specific heat ($c_p$), thermal conductivity ($\lambda$), and specific entropy ($s$) are provided for sub-cooled water and superheated steam. For saturation conditions, the following properties are provided: pressure ($p$), temperature ($T_{sat}$), specific enthalpy of water ($i_f$), specific enthalpy of steam ($i_g$), latent heat ($i_{fg} = i_g - i_f$), specific entropy of water ($s_f$), specific entropy of steam ($s_g$), entropy of vaporization ($s_{fg} = s_g - s_f$), and surface tension ($\sigma$).

### Sub-cooled and Superheated Conditions

| $p$ (bar) | $T$ (°C) | $\rho$ (kg/m³) | $i$ (kJ/kg) | $\mu$ (μPa·s) | $c_p$ (kJ/kg K) | $\lambda$ (W/kg K) | $s$ (kJ/kg K) |
|---|---|---|---|---|---|---|---|
| 1 | 50 | 988.047 | 209.412 | 546.852 | 4.17956 | 0.64051 | 0.70375 |
| 1 | 70 | 977.779 | 293.074 | 403.900 | 4.18810 | 0.65961 | 0.95495 |
| 1 | 90 | 965.318 | 376.992 | 314.413 | 4.20502 | 0.67302 | 1.19263 |
| 1 | 95 | 961.894 | 398.030 | 297.286 | 4.21057 | 0.67555 | 1.25017 |
| 1 | 99.6059$^f$ | 958.637 | 417.436 | 282.947 | 4.21615 | 0.67759 | 1.30256 |
| 1 | 99.6059$^g$ | 0.59031 | 2674.95 | 12.2561 | 2.07594 | 0.02475 | 7.35881 |
| 1 | 105 | 0.58124 | 2686.09 | 12.4568 | 2.05460 | 0.02514 | 7.38847 |
| 1 | 110 | 0.57313 | 2696.32 | 12.6444 | 2.03992 | 0.02551 | 7.41536 |
| 1 | 130 | 0.54309 | 2736.72 | 13.4059 | 2.00391 | 0.02710 | 7.51814 |
| 1 | 150 | 0.51634 | 2776.59 | 14.1830 | 1.98566 | 0.02880 | 7.61467 |
| 10 | 130 | 935.211 | 546.882 | 213.084 | 4.26285 | 0.68525 | 1.63392 |
| 10 | 150 | 917.304 | 632.575 | 182.593 | 4.30857 | 0.68421 | 1.84137 |
| 10 | 170 | 897.586 | 719.320 | 159.605 | 4.36867 | 0.67886 | 2.04166 |
| 10 | 175 | 892.358 | 741.207 | 154.727 | 4.38647 | 0.67686 | 2.09077 |
| 10 | 179.886$^f$ | 887.127 | 762.683 | 150.248 | 4.40511 | 0.67465 | 2.13843 |
| 10 | 179.886$^g$ | 5.14539 | 2777.12 | 15.0220 | 2.71498 | 0.03540 | 6.58498 |
| 10 | 185 | 5.06580 | 2790.70 | 15.2439 | 2.60350 | 0.03549 | 6.61479 |
| 10 | 190 | 4.99205 | 2803.52 | 15.4608 | 2.52852 | 0.03564 | 6.64262 |
| 10 | 210 | 4.72720 | 2852.20 | 16.3254 | 2.36144 | 0.03662 | 6.74555 |
| 10 | 230 | 4.49848 | 2898.45 | 17.1867 | 2.27017 | 0.03803 | 6.83935 |
| 20 | 165 | 903.304 | 698.085 | 165.047 | 4.34830 | 0.68132 | 1.99095 |
| 20 | 185 | 882.210 | 785.746 | 146.059 | 4.42128 | 0.67287 | 2.18657 |
| 20 | 205 | 858.983 | 875.096 | 130.947 | 4.51844 | 0.66021 | 2.37744 |
| 20 | 210 | 852.803 | 897.760 | 127.628 | 4.54761 | 0.65637 | 2.42460 |
| 20 | 212.385$^f$ | 849.798 | 908.622 | 126.107 | 4.56234 | 0.65444 | 2.44702 |
| 20 | 212.385$^g$ | 10.0421 | 2798.38 | 16.1449 | 3.19036 | 0.04165 | 6.33916 |
| 20 | 215 | 9.95172 | 2806.59 | 16.2653 | 3.09115 | 0.04156 | 6.35603 |
| 20 | 220 | 9.78788 | 2821.67 | 16.4955 | 2.94874 | 0.04146 | 6.38676 |
| 20 | 240 | 9.21758 | 2877.21 | 17.4060 | 2.64811 | 0.04178 | 6.49719 |

DOI: 10.1201/9781003255000-D

| $p$ | $T$ | $\rho$ | $i$ | $\mu$ | $c_p$ | $\lambda$ | $s$ |
|-----|-----|--------|-----|-------|-------|-----------|-----|
| (bar) | (°C) | (kg/m³) | (kJ/kg) | ($\mu$Pa·s) | (kJ/kg K) | (W/kg K) | (kJ/kg K) |
| 20 | 260 | 8.74124 | 2928.47 | 18.3042 | 2.49094 | 0.04280 | 6.59522 |
| 30 | 185 | 882.895 | 786.231 | 146.307 | 4.41656 | 0.67368 | 2.18516 |
| 30 | 205 | 859.771 | 875.472 | 131.199 | 4.51220 | 0.66112 | 2.37580 |
| 30 | 225 | 834.169 | 966.942 | 118.797 | 4.64161 | 0.64424 | 2.56319 |
| 30 | 230 | 827.319 | 990.247 | 116.016 | 4.68102 | 0.63931 | 2.60974 |
| 30 | 233.858$^f$ | 821.895 | 1008.37 | 113.947 | 4.71380 | 0.63530 | 2.64562 |
| 30 | 233.858$^g$ | 15.0006 | 2803.26 | 16.9033 | 3.61228 | 0.04670 | 6.18579 |
| 30 | 235 | 14.9339 | 2807.35 | 16.9578 | 3.55258 | 0.04660 | 6.19384 |
| 30 | 240 | 14.6569 | 2824.56 | 17.1974 | 3.34354 | 0.04628 | 6.22755 |
| 30 | 260 | 13.7196 | 2886.42 | 18.1401 | 2.90697 | 0.04594 | 6.34585 |
| 30 | 280 | 12.9607 | 2942.16 | 19.0637 | 2.68542 | 0.04660 | 6.44851 |
| 50 | 215 | 848.994 | 921.516 | 125.226 | 4.55759 | 0.65517 | 2.46625 |
| 50 | 235 | 822.272 | 1014.05 | 113.872 | 4.70373 | 0.63636 | 2.65202 |
| 50 | 255 | 792.192 | 1110.06 | 104.056 | 4.91017 | 0.61276 | 2.83731 |
| 50 | 260 | 784.020 | 1134.77 | 101.771 | 4.97552 | 0.60602 | 2.88388 |
| 50 | 263.943$^f$ | 777.360 | 1154.50 | 100.008 | 5.03218 | 0.60046 | 2.92075 |
| 50 | 263.943$^g$ | 25.3509 | 2794.23 | 18.0327 | 4.43784 | 0.05564 | 5.97370 |
| 50 | 265 | 25.2219 | 2798.88 | 18.0857 | 4.35703 | 0.05545 | 5.98235 |
| 50 | 270 | 24.6503 | 2819.84 | 18.3371 | 4.04602 | 0.05469 | 6.02113 |
| 50 | 290 | 22.8018 | 2893.00 | 19.3191 | 3.36622 | 0.05318 | 6.15348 |
| 50 | 310 | 21.3827 | 2956.58 | 20.2736 | 3.02176 | 0.05315 | 6.26445 |
| 70 | 235 | 824.240 | 1014.38 | 114.398 | 4.68438 | 0.63860 | 2.64788 |
| 70 | 255 | 794.608 | 1109.91 | 104.626 | 4.88106 | 0.61545 | 2.83226 |
| 70 | 275 | 760.633 | 1210.24 | 95.8194 | 5.17298 | 0.58690 | 3.01868 |
| 70 | 280 | 751.244 | 1236.34 | 93.7016 | 5.26978 | 0.57878 | 3.06608 |
| 70 | 285.83$^f$ | 739.724 | 1267.44 | 91.2529 | 5.40039 | 0.56873 | 3.12199 |
| 70 | 285.83$^g$ | 36.5236 | 2772.57 | 18.9606 | 5.35404 | 0.06437 | 5.81463 |
| 70 | 290 | 35.6584 | 2793.98 | 19.1739 | 4.93602 | 0.06321 | 5.85279 |
| 70 | 295 | 34.7341 | 2817.70 | 19.4277 | 4.57120 | 0.06211 | 5.89473 |
| 70 | 315 | 31.8273 | 2899.57 | 20.4198 | 3.72521 | 0.05968 | 6.03644 |
| 70 | 335 | 29.6751 | 2969.28 | 21.3836 | 3.28458 | 0.05912 | 6.15302 |
| 90 | 255 | 796.965 | 1109.82 | 105.187 | 4.85349 | 0.61808 | 2.82731 |
| 90 | 275 | 763.650 | 1209.44 | 96.4513 | 5.12787 | 0.59017 | 3.01243 |
| 90 | 295 | 724.282 | 1316.03 | 88.1505 | 5.57120 | 0.55578 | 3.20336 |
| 90 | 300 | 713.071 | 1344.27 | 86.0630 | 5.73047 | 0.54590 | 3.25286 |
| 90 | 303.347$^f$ | 705.158 | 1363.65 | 84.6519 | 5.85416 | 0.53893 | 3.28657 |
| 90 | 303.347$^g$ | 48.7973 | 2742.88 | 19.8302 | 6.47619 | 0.07378 | 5.67901 |
| 90 | 305 | 48.2101 | 2753.35 | 19.9126 | 6.19242 | 0.07302 | 5.69714 |
| 90 | 310 | 46.6216 | 2782.61 | 20.1625 | 5.55785 | 0.07106 | 5.74754 |
| 90 | 330 | 41.9577 | 2878.87 | 21.1488 | 4.25988 | 0.06661 | 5.90996 |
| 90 | 350 | 38.7321 | 2957.22 | 22.1128 | 3.63702 | 0.06497 | 6.03781 |
| 110 | 270 | 775.183 | 1183.49 | 99.1686 | 5.01117 | 0.60069 | 2.96011 |
| 110 | 290 | 738.514 | 1287.02 | 90.9210 | 5.37032 | 0.56894 | 3.14725 |
| 110 | 310 | 693.719 | 1400.08 | 82.6809 | 6.00704 | 0.52945 | 3.34447 |
| 110 | 315 | 680.510 | 1430.72 | 80.5130 | 6.25681 | 0.51793 | 3.39679 |
| 110 | 318.081$^f$ | 671.796 | 1450.28 | 79.1380 | 6.44269 | 0.51042 | 3.42995 |
| 110 | 318.081$^g$ | 62.5239 | 2706.39 | 20.7156 | 7.91681 | 0.08464 | 5.55453 |

| $p$ | $T$ | $\rho$ | i | $\mu$ | $c_p$ | $\lambda$ | $s$ |
|---|---|---|---|---|---|---|---|
| (bar) | (°C) | (kg/m³) | (kJ/kg) | ($\mu$Pa·s) | (kJ/kg K) | (W/kg K) | (kJ/kg K) |
| 110 | 320 | 61.4344 | 2721.07 | 20.8020 | 7.40624 | 0.08328 | 5.57932 |
| 110 | 325 | 58.9813 | 2755.61 | 21.0326 | 6.48019 | 0.08036 | 5.63730 |
| 110 | 345 | 52.2055 | 2864.80 | 21.9762 | 4.72521 | 0.07375 | 5.81702 |
| 110 | 365 | 47.7809 | 2950.60 | 22.9199 | 3.93839 | 0.07069 | 5.95368 |
| 130 | 280 | 760.638 | 1233.53 | 95.6254 | 5.12483 | 0.58889 | 3.04664 |
| 130 | 300 | 721.600 | 1339.92 | 87.6056 | 5.55148 | 0.55462 | 3.23555 |
| 130 | 320 | 672.705 | 1458.02 | 79.2983 | 6.36059 | 0.51168 | 3.43800 |
| 130 | 325 | 657.886 | 1490.64 | 77.0374 | 6.70089 | 0.49904 | 3.49276 |
| 130 | 330.857$^f$ | 638.371 | 1531.40 | 74.2013 | 7.25793 | 0.48288 | 3.56058 |
| 130 | 330.857$^g$ | 78.2159 | 2662.89 | 21.6783 | 9.90715 | 0.09803 | 5.43388 |
| 130 | 335 | 74.6713 | 2700.35 | 21.8236 | 8.32392 | 0.09370 | 5.49568 |
| 130 | 340 | 71.2780 | 2738.92 | 22.0193 | 7.19537 | 0.08981 | 5.55886 |
| 130 | 360 | 62.2874 | 2858.09 | 22.8859 | 5.08166 | 0.08091 | 5.75029 |
| 130 | 380 | 56.6452 | 2949.64 | 23.7914 | 4.17422 | 0.07548 | 5.89272 |
| 150 | 295 | 735.696 | 1310.98 | 90.2953 | 5.35874 | 0.56769 | 3.18002 |
| 150 | 315 | 691.612 | 1423.51 | 82.3501 | 5.95672 | 0.52868 | 3.37462 |
| 150 | 335 | 632.628 | 1553.95 | 73.4367 | 7.32307 | 0.47855 | 3.59258 |
| 150 | 340 | 613.094 | 1592.27 | 70.7684 | 8.06472 | 0.46328 | 3.65534 |
| 150 | 342.158$^f$ | 603.514 | 1610.15 | 69.5045 | 8.52522 | 0.45616 | 3.68445 |
| 150 | 342.158$^g$ | 96.7109 | 2610.86 | 22.7932 | 12.9821 | 0.11579 | 5.31080 |
| 150 | 345 | 92.5891 | 2644.47 | 22.8229 | 10.8504 | 0.11062 | 5.36530 |
| 150 | 350 | 87.1027 | 2693.00 | 22.9353 | 8.78851 | 0.10406 | 5.44350 |
| 150 | 370 | 74.1123 | 2831.40 | 23.6517 | 5.68675 | 0.08878 | 5.66236 |
| 150 | 390 | 66.6294 | 2932.11 | 24.4949 | 4.52636 | 0.08154 | 5.81664 |
| 175 | 305 | 719.996 | 1363.23 | 87.2869 | 5.51155 | 0.55453 | 3.26519 |
| 175 | 325 | 672.722 | 1479.84 | 79.3640 | 6.23114 | 0.51278 | 3.46340 |
| 175 | 345 | 606.735 | 1619.24 | 70.0094 | 8.09451 | 0.45862 | 3.69248 |
| 175 | 350 | 583.284 | 1662.45 | 66.9931 | 9.31471 | 0.44189 | 3.76210 |
| 175 | 354.671$^f$ | 554.671 | 1710.76 | 24.5980 | 11.7161 | 0.42460 | 3.83933 |
| 175 | 354.671$^g$ | 126.154 | 2529.11 | 24.5959 | 20.3861 | 0.15041 | 5.14280 |
| 175 | 360 | 112.735 | 2612.11 | 24.3426 | 12.5541 | 0.13220 | 5.27448 |
| 175 | 365 | 104.987 | 2667.41 | 24.3260 | 9.88367 | 0.12202 | 5.36149 |
| 175 | 385 | 87.7186 | 2819.31 | 24.8116 | 6.12497 | 0.09637 | 5.59616 |
| 175 | 405 | 78.2387 | 2926.89 | 25.5561 | 4.80232 | 0.08825 | 5.75727 |
| 200 | 315 | 703.574 | 1416.48 | 84.3906 | 5.68601 | 0.54063 | 3.35047 |
| 200 | 335 | 652.646 | 1537.85 | 76.4101 | 6.56010 | 0.49605 | 3.55332 |
| 200 | 355 | 577.560 | 1689.10 | 66.3874 | 9.24850 | 0.43768 | 3.79782 |
| 200 | 360 | 548.088 | 1739.97 | 62.8131 | 11.4527 | 0.41979 | 3.87860 |
| 200 | 365.746$^f$ | 490.524 | 1827.10 | 27.5020 | 23.1986 | 0.40374 | 4.01538 |
| 200 | 365.746$^g$ | 170.699 | 2411.39 | 27.4892 | 45.6779 | 0.22650 | 4.92990 |
| 200 | 370 | 144.458 | 2526.33 | 26.3168 | 18.6702 | 0.17938 | 5.10937 |
| 200 | 375 | 130.248 | 2602.59 | 25.9146 | 12.7497 | 0.14516 | 5.22745 |
| 200 | 395 | 104.325 | 2783.66 | 25.9134 | 6.93650 | 0.10712 | 5.50299 |
| 200 | 415 | 91.6515 | 2902.98 | 26.4952 | 5.23792 | 0.09618 | 5.67907 |
| 220.64 | 325 | 685.279 | 1471.51 | 81.4013 | 5.91075 | 0.52493 | 3.43824 |
| 220.64 | 345 | 629.437 | 1599.29 | 73.2031 | 7.02522 | 0.47699 | 3.64828 |

| $p$ | $T$ | $\rho$ | $i$ | $\mu$ | $c_p$ | $\lambda$ | $s$ |
|-----|-----|--------|-----|-------|-------|-----------|-----|
| (bar) | (°C) | (kg/m³) | (kJ/kg) | (μPa·s) | (kJ/kg K) | (W/kg K) | (kJ/kg K) |
| 220.64 | 365 | 538.987 | 1770.57 | 61.8605 | 11.5669 | 0.41312 | 3.92076 |
| 220.64 | 370 | 494.900 | 1840.38 | 56.8630 | 17.8178 | 0.39475 | 4.02962 |
| 220.64 | 373.946$^c$ | 333.590 | 2068.59 | 40.4857 | 15508.6 | 0.80297 | 4.38272 |
| 220.64 | 375 | 210.605 | 2335.69 | 30.5518 | 66.7832 | 0.33572 | 4.79535 |
| 220.64 | 380 | 165.350 | 2497.57 | 27.9948 | 20.0991 | 0.18704 | 5.04437 |
| 220.64 | 400 | 121.890 | 2732.92 | 26.9019 | 8.10092 | 0.11983 | 5.40004 |
| 220.64 | 420 | 104.816 | 2867.92 | 27.2430 | 5.77696 | 0.10433 | 5.59782 |
| 250 | 335 | 668.367 | 1526.35 | 78.8446 | 6.11310 | 0.51075 | 3.52197 |
| 250 | 355 | 608.436 | 1659.94 | 70.5147 | 7.44975 | 0.46032 | 3.73786 |
| 250 | 375 | 505.649 | 1849.18 | 58.2685 | 13.6039 | 0.38691 | 4.03418 |
| 250 | 380 | 451.047 | 1935.30 | 52.4156 | 23.1002 | 0.38914 | 4.16646 |
| 250 | 384.863$^p$ | 317.491 | 2150.71 | 39.5717 | 71.1190 | 0.36068 | 4.49474 |
| 250 | 390 | 215.205 | 2395.46 | 31.7048 | 28.4665 | 0.22626 | 4.86555 |
| 250 | 395 | 184.181 | 2503.96 | 29.9468 | 17.2634 | 0.18248 | 5.02857 |
| 250 | 415 | 137.975 | 2730.55 | 28.4612 | 8.19086 | 0.12882 | 5.36331 |
| 250 | 435 | 118.460 | 2868.26 | 28.5967 | 5.92912 | 0.11239 | 5.56073 |

$^f$ Saturated liquid phase.

$^g$ Saturated vapor phase.

$^c$ Critical point temperature.

$^p$ Pseudo-critical point temperature.

## Saturated Conditions

| $p$ | $T_{sat}$ | $i_f$ | $i_g$ | $i_{fg}$ | $s_f$ | $s_g$ | $s_{fg}$ | $\sigma$ |
|-----|-----------|-------|-------|----------|-------|-------|----------|----------|
| (bar) | (°C) | (kJ/kg) | (kJ/kg) | (kJ/kg) | (J/kg K) | (J/kg K) | (J/kg K) | (mN/m) |
| 1 | 99.61 | 417.436 | 2674.95 | 2257.51 | 1302.56 | 7358.81 | 6056.25 | 58.99 |
| 2 | 120.2 | 504.684 | 2706.24 | 2201.56 | 1530.10 | 7126.86 | 5596.76 | 54.93 |
| 3 | 133.5 | 561.455 | 2724.89 | 2163.44 | 1671.76 | 6991.57 | 5319.80 | 52.20 |
| 4 | 143.6 | 604.723 | 2738.06 | 2133.33 | 1776.60 | 6895.42 | 5118.82 | 50.10 |
| 5 | 151.8 | 640.185 | 2748.11 | 2107.92 | 1860.60 | 6820.58 | 4959.98 | 48.35 |
| 6 | 158.8 | 670.501 | 2756.14 | 2085.64 | 1931.10 | 6759.17 | 4828.07 | 46.84 |
| 7 | 165.0 | 697.143 | 2762.75 | 2065.61 | 1992.08 | 6706.98 | 4714.90 | 45.51 |
| 8 | 170.4 | 721.018 | 2768.30 | 2047.28 | 2045.99 | 6661.54 | 4615.55 | 44.32 |
| 9 | 175.4 | 742.725 | 2773.04 | 2030.31 | 2094.40 | 6621.24 | 4526.83 | 43.22 |
| 10 | 179.9 | 762.683 | 2777.12 | 2014.44 | 2138.43 | 6584.98 | 4446.55 | 42.22 |
| 11 | 184.1 | 781.198 | 2780.67 | 1999.47 | 2178.86 | 6551.99 | 4373.12 | 41.28 |
| 12 | 188.0 | 798.499 | 2783.77 | 1985.27 | 2216.30 | 6521.69 | 4305.39 | 40.40 |
| 13 | 191.6 | 814.764 | 2786.49 | 1971.73 | 2251.18 | 6493.65 | 4242.46 | 39.58 |
| 14 | 195.0 | 830.132 | 2788.89 | 1958.76 | 2283.88 | 6467.52 | 4183.64 | 38.80 |
| 15 | 198.3 | 844.717 | 2791.01 | 1946.29 | 2314.68 | 6443.05 | 4128.37 | 38.06 |
| 16 | 201.4 | 858.610 | 2792.88 | 1934.27 | 2343.81 | 6420.02 | 4076.21 | 37.36 |
| 17 | 204.3 | 871.888 | 2794.53 | 1922.64 | 2371.46 | 6398.25 | 4026.79 | 36.69 |
| 18 | 207.1 | 884.614 | 2795.99 | 1911.37 | 2397.79 | 6377.60 | 3979.80 | 36.04 |
| 19 | 209.8 | 896.844 | 2797.26 | 1900.42 | 2422.94 | 6357.94 | 3934.99 | 35.42 |
| 20 | 212.4 | 908.622 | 2798.38 | 1889.76 | 2447.02 | 6339.16 | 3892.14 | 34.83 |
| 21 | 214.9 | 919.989 | 2799.36 | 1879.37 | 2470.13 | 6321.20 | 3851.06 | 34.26 |
| 22 | 217.3 | 930.981 | 2800.20 | 1869.22 | 2492.36 | 6303.95 | 3811.59 | 33.70 |

| $p$ | $T_{sat}$ | $i_f$ | $i_g$ | $i_{fg}$ | $s_f$ | $s_g$ | $s_{fg}$ | $\sigma$ |
|---|---|---|---|---|---|---|---|---|
| (bar) | (°C) | (kJ/kg) | (kJ/kg) | (kJ/kg) | (J/kg K) | (J/kg K) | (J/kg K) | (mN/m) |
| 23 | 219.6 | 941.626 | 2800.92 | 1859.30 | 2513.77 | 6287.37 | 3773.60 | 33.17 |
| 24 | 221.8 | 951.952 | 2801.54 | 1849.58 | 2534.44 | 6271.40 | 3736.95 | 32.65 |
| 25 | 224.0 | 961.983 | 2802.04 | 1840.06 | 2554.43 | 6255.97 | 3701.55 | 32.15 |
| 26 | 226.1 | 971.740 | 2802.45 | 1830.71 | 2573.77 | 6241.06 | 3667.29 | 31.66 |
| 27 | 228.1 | 981.241 | 2802.78 | 1821.54 | 2592.52 | 6226.62 | 3634.10 | 31.18 |
| 28 | 230.1 | 990.503 | 2803.02 | 1812.51 | 2610.73 | 6212.61 | 3601.89 | 30.72 |
| 29 | 232.0 | 999.542 | 2803.18 | 1803.63 | 2628.41 | 6199.01 | 3570.60 | 30.27 |
| 30 | 233.9 | 1008.37 | 2803.26 | 1794.89 | 2645.62 | 6185.79 | 3540.17 | 29.83 |
| 31 | 235.7 | 1017.00 | 2803.28 | 1786.28 | 2662.38 | 6172.92 | 3510.54 | 29.41 |
| 32 | 237.5 | 1025.45 | 2803.24 | 1777.79 | 2678.71 | 6160.37 | 3481.66 | 28.99 |
| 33 | 239.2 | 1033.72 | 2803.13 | 1769.41 | 2694.64 | 6148.14 | 3453.50 | 28.58 |
| 34 | 240.9 | 1041.83 | 2802.96 | 1761.14 | 2710.19 | 6136.19 | 3426.00 | 28.18 |
| 35 | 242.6 | 1049.78 | 2802.74 | 1752.97 | 2725.39 | 6124.51 | 3399.12 | 27.79 |
| 36 | 244.2 | 1057.57 | 2802.47 | 1744.90 | 2740.25 | 6113.09 | 3372.84 | 27.41 |
| 37 | 245.8 | 1065.23 | 2802.15 | 1736.91 | 2754.79 | 6101.92 | 3347.13 | 27.04 |
| 38 | 247.3 | 1072.76 | 2801.78 | 1729.02 | 2769.03 | 6090.97 | 3321.94 | 26.67 |
| 39 | 248.9 | 1080.15 | 2801.36 | 1721.21 | 2782.98 | 6080.24 | 3297.26 | 26.31 |
| 40 | 250.4 | 1087.43 | 2800.90 | 1713.47 | 2796.65 | 6069.71 | 3273.06 | 25.96 |
| 41 | 251.8 | 1094.58 | 2800.39 | 1705.81 | 2810.07 | 6059.38 | 3249.31 | 25.61 |
| 42 | 253.3 | 1101.63 | 2799.85 | 1698.22 | 2823.23 | 6049.23 | 3226.00 | 25.27 |
| 43 | 254.7 | 1108.57 | 2799.27 | 1690.70 | 2836.15 | 6039.25 | 3203.10 | 24.94 |
| 44 | 256.1 | 1115.40 | 2798.65 | 1683.25 | 2848.85 | 6029.45 | 3180.60 | 24.61 |
| 45 | 257.4 | 1122.14 | 2798.00 | 1675.85 | 2861.33 | 6019.80 | 3158.47 | 24.29 |
| 46 | 258.8 | 1128.79 | 2797.31 | 1668.52 | 2873.60 | 6010.30 | 3136.71 | 23.98 |
| 47 | 260.1 | 1135.34 | 2796.59 | 1661.24 | 2885.66 | 6000.95 | 3115.29 | 23.66 |
| 48 | 261.4 | 1141.81 | 2795.83 | 1654.02 | 2897.54 | 5991.74 | 3094.20 | 23.36 |
| 49 | 262.7 | 1148.20 | 2795.04 | 1646.85 | 2909.23 | 5982.66 | 3073.43 | 23.06 |
| 50 | 263.9 | 1154.50 | 2794.23 | 1639.73 | 2920.75 | 5973.70 | 3052.96 | 22.76 |
| 51 | 265.2 | 1160.73 | 2793.38 | 1632.65 | 2932.09 | 5964.87 | 3032.78 | 22.47 |
| 52 | 266.4 | 1166.88 | 2792.51 | 1625.62 | 2943.27 | 5956.15 | 3012.89 | 22.18 |
| 53 | 267.6 | 1172.96 | 2791.60 | 1618.64 | 2954.29 | 5947.55 | 2993.26 | 21.90 |
| 54 | 268.8 | 1178.98 | 2790.67 | 1611.69 | 2965.15 | 5939.05 | 2973.89 | 21.62 |
| 55 | 270.0 | 1184.92 | 2789.72 | 1604.79 | 2975.88 | 5930.65 | 2954.78 | 21.34 |
| 56 | 271.1 | 1190.81 | 2788.74 | 1597.93 | 2986.45 | 5922.35 | 2935.90 | 21.07 |
| 57 | 272.3 | 1196.63 | 2787.73 | 1591.10 | 2996.89 | 5914.15 | 2917.26 | 20.81 |
| 58 | 273.4 | 1202.39 | 2786.70 | 1584.31 | 3007.20 | 5906.04 | 2898.83 | 20.54 |
| 59 | 274.5 | 1208.09 | 2785.64 | 1577.55 | 3017.38 | 5898.01 | 2880.63 | 20.28 |
| 60 | 275.6 | 1213.73 | 2784.56 | 1570.83 | 3027.44 | 5890.07 | 2862.63 | 20.02 |
| 61 | 276.7 | 1219.32 | 2783.46 | 1564.14 | 3037.38 | 5882.21 | 2844.83 | 19.77 |
| 62 | 277.7 | 1224.86 | 2782.33 | 1557.48 | 3047.20 | 5874.42 | 2827.23 | 19.52 |
| 63 | 278.8 | 1230.34 | 2781.19 | 1550.84 | 3056.91 | 5866.71 | 2809.81 | 19.28 |
| 64 | 279.8 | 1235.78 | 2780.02 | 1544.24 | 3066.50 | 5859.08 | 2792.57 | 19.03 |
| 65 | 280.9 | 1241.17 | 2778.83 | 1537.66 | 3076.00 | 5851.51 | 2775.51 | 18.79 |
| 66 | 281.9 | 1246.51 | 2777.62 | 1531.11 | 3085.39 | 5844.01 | 2758.62 | 18.55 |
| 67 | 282.9 | 1251.81 | 2776.39 | 1524.58 | 3094.68 | 5836.58 | 2741.89 | 18.32 |
| 68 | 283.9 | 1257.06 | 2775.13 | 1518.07 | 3103.88 | 5829.20 | 2725.32 | 18.09 |
| 69 | 284.9 | 1262.27 | 2773.86 | 1511.59 | 3112.98 | 5821.89 | 2708.91 | 17.86 |
| 70 | 285.8 | 1267.44 | 2772.57 | 1505.13 | 3121.99 | 5814.63 | 2692.64 | 17.63 |
| 71 | 286.8 | 1272.57 | 2771.26 | 1498.69 | 3130.92 | 5807.43 | 2676.52 | 17.41 |
| 72 | 287.7 | 1277.65 | 2769.93 | 1492.27 | 3139.76 | 5800.29 | 2660.53 | 17.19 |
| 73 | 288.7 | 1282.70 | 2768.58 | 1485.87 | 3148.51 | 5793.19 | 2644.68 | 16.97 |

| $p$ | $T_{sat}$ | $i_f$ | $i_g$ | $i_{fg}$ | $s_f$ | $s_g$ | $s_{fg}$ | $\sigma$ |
|---|---|---|---|---|---|---|---|---|
| (bar) | (°C) | (kJ/kg) | (kJ/kg) | (kJ/kg) | (J/kg K) | (J/kg K) | (J/kg K) | (mN/m) |
| 74 | 289.6 | 1287.72 | 2767.21 | 1479.49 | 3157.19 | 5786.15 | 2628.96 | 16.75 |
| 75 | 290.5 | 1292.70 | 2765.82 | 1473.12 | 3165.78 | 5779.16 | 2613.37 | 16.54 |
| 76 | 291.4 | 1297.64 | 2764.41 | 1466.78 | 3174.30 | 5772.21 | 2597.91 | 16.33 |
| 77 | 292.4 | 1302.55 | 2762.99 | 1460.44 | 3182.74 | 5765.30 | 2582.56 | 16.12 |
| 78 | 293.2 | 1307.42 | 2761.55 | 1454.12 | 3191.12 | 5758.44 | 2567.33 | 15.91 |
| 79 | 294.1 | 1312.27 | 2760.09 | 1447.82 | 3199.42 | 5751.63 | 2552.21 | 15.71 |
| 80 | 295.0 | 1317.08 | 2758.61 | 1441.53 | 3207.65 | 5744.85 | 2537.20 | 15.51 |
| 81 | 295.9 | 1321.86 | 2757.12 | 1435.25 | 3215.82 | 5738.11 | 2522.29 | 15.31 |
| 82 | 296.7 | 1326.61 | 2755.60 | 1428.99 | 3223.92 | 5731.41 | 2507.49 | 15.11 |
| 83 | 297.6 | 1331.34 | 2754.07 | 1422.74 | 3231.96 | 5724.75 | 2492.79 | 14.91 |
| 84 | 298.4 | 1336.03 | 2752.52 | 1416.49 | 3239.93 | 5718.12 | 2478.19 | 14.72 |
| 85 | 299.3 | 1340.70 | 2750.96 | 1410.26 | 3247.85 | 5711.52 | 2463.67 | 14.53 |
| 86 | 300.1 | 1345.34 | 2749.38 | 1404.04 | 3255.70 | 5704.96 | 2449.26 | 14.34 |
| 87 | 300.9 | 1349.96 | 2747.78 | 1397.82 | 3263.50 | 5698.43 | 2434.92 | 14.15 |
| 88 | 301.7 | 1354.54 | 2746.16 | 1391.62 | 3271.25 | 5691.93 | 2420.68 | 13.96 |
| 89 | 302.5 | 1359.11 | 2744.53 | 1385.42 | 3278.94 | 5685.45 | 2406.52 | 13.78 |
| 90 | 303.3 | 1363.65 | 2742.88 | 1379.23 | 3286.57 | 5679.01 | 2392.44 | 13.60 |
| 91 | 304.1 | 1368.17 | 2741.22 | 1373.05 | 3294.16 | 5672.59 | 2378.44 | 13.41 |
| 92 | 304.9 | 1372.66 | 2739.53 | 1366.87 | 3301.69 | 5666.20 | 2364.51 | 13.24 |
| 93 | 305.7 | 1377.14 | 2737.83 | 1360.70 | 3309.18 | 5659.84 | 2350.66 | 13.06 |
| 94 | 306.5 | 1381.59 | 2736.12 | 1354.53 | 3316.61 | 5653.49 | 2336.88 | 12.88 |
| 95 | 307.3 | 1386.02 | 2734.38 | 1348.37 | 3324.00 | 5647.17 | 2323.17 | 12.71 |
| 96 | 308.0 | 1390.43 | 2732.64 | 1342.21 | 3331.35 | 5640.88 | 2309.53 | 12.54 |
| 97 | 308.8 | 1394.81 | 2730.87 | 1336.06 | 3338.65 | 5634.60 | 2295.95 | 12.37 |
| 98 | 309.5 | 1399.18 | 2729.09 | 1329.90 | 3345.90 | 5628.35 | 2282.44 | 12.20 |
| 99 | 310.3 | 1403.54 | 2727.29 | 1323.75 | 3353.12 | 5622.11 | 2268.99 | 12.03 |
| 100 | 311.0 | 1407.87 | 2725.47 | 1317.61 | 3360.29 | 5615.89 | 2255.60 | 11.86 |
| 101 | 311.7 | 1412.18 | 2723.64 | 1311.46 | 3367.42 | 5609.69 | 2242.27 | 11.70 |
| 102 | 312.5 | 1416.48 | 2721.79 | 1305.31 | 3374.52 | 5603.50 | 2228.99 | 11.54 |
| 103 | 313.2 | 1420.76 | 2719.93 | 1299.17 | 3381.57 | 5597.34 | 2215.77 | 11.38 |
| 104 | 313.9 | 1425.02 | 2718.04 | 1293.02 | 3388.59 | 5591.18 | 2202.59 | 11.22 |
| 105 | 314.6 | 1429.27 | 2716.14 | 1286.88 | 3395.57 | 5585.04 | 2189.48 | 11.06 |
| 106 | 315.3 | 1433.50 | 2714.23 | 1280.73 | 3402.51 | 5578.92 | 2176.40 | 10.90 |
| 107 | 316.0 | 1437.72 | 2712.30 | 1274.58 | 3409.42 | 5572.80 | 2163.38 | 10.75 |
| 108 | 316.7 | 1441.92 | 2710.35 | 1268.43 | 3416.30 | 5566.70 | 2150.40 | 10.59 |
| 109 | 317.4 | 1446.11 | 2708.38 | 1262.27 | 3423.14 | 5560.61 | 2137.47 | 10.44 |
| 110 | 318.1 | 1450.28 | 2706.39 | 1256.12 | 3429.95 | 5554.53 | 2124.58 | 10.29 |
| 111 | 318.8 | 1454.44 | 2704.39 | 1249.96 | 3436.73 | 5548.46 | 2111.73 | 10.14 |
| 112 | 319.4 | 1458.58 | 2702.37 | 1243.79 | 3443.48 | 5542.40 | 2098.92 | 9.988 |
| 113 | 320.1 | 1462.72 | 2700.34 | 1237.62 | 3450.20 | 5536.34 | 2086.14 | 9.841 |
| 114 | 320.8 | 1466.84 | 2698.28 | 1231.45 | 3456.89 | 5530.29 | 2073.41 | 9.695 |
| 115 | 321.4 | 1470.95 | 2696.21 | 1225.26 | 3463.55 | 5524.25 | 2060.70 | 9.550 |
| 116 | 322.1 | 1475.05 | 2694.12 | 1219.08 | 3470.18 | 5518.22 | 2048.03 | 9.406 |
| 117 | 322.7 | 1479.13 | 2692.02 | 1212.88 | 3476.79 | 5512.19 | 2035.40 | 9.264 |
| 118 | 323.4 | 1483.21 | 2689.89 | 1206.68 | 3483.37 | 5506.16 | 2022.79 | 9.123 |
| 119 | 324.0 | 1487.27 | 2687.75 | 1200.47 | 3489.93 | 5500.14 | 2010.21 | 8.983 |
| 120 | 324.7 | 1491.33 | 2685.58 | 1194.26 | 3496.46 | 5494.12 | 1997.66 | 8.844 |
| 121 | 325.3 | 1495.37 | 2683.40 | 1188.03 | 3502.97 | 5488.10 | 1985.13 | 8.707 |
| 122 | 325.9 | 1499.41 | 2681.20 | 1181.79 | 3509.45 | 5482.08 | 1972.63 | 8.570 |
| 123 | 326.6 | 1503.43 | 2678.98 | 1175.55 | 3515.91 | 5476.06 | 1960.15 | 8.435 |
| 124 | 327.2 | 1507.45 | 2676.74 | 1169.29 | 3522.35 | 5470.04 | 1947.69 | 8.301 |

| $p$ | $T_{sat}$ | $i_f$ | $i_g$ | $i_{fg}$ | $s_f$ | $s_g$ | $s_{fg}$ | $\sigma$ |
|---|---|---|---|---|---|---|---|---|
| (bar) | (°C) | (kJ/kg) | (kJ/kg) | (kJ/kg) | (J/kg K) | (J/kg K) | (J/kg K) | (mN/m) |
| 125 | 327.8 | 1511.46 | 2674.49 | 1163.02 | 3528.77 | 5464.02 | 1935.25 | 8.168 |
| 126 | 328.4 | 1515.47 | 2672.21 | 1156.74 | 3535.17 | 5458.00 | 1922.83 | 8.037 |
| 127 | 329.0 | 1519.46 | 2669.91 | 1150.45 | 3541.55 | 5451.98 | 1910.43 | 7.906 |
| 128 | 329.7 | 1523.45 | 2667.59 | 1144.14 | 3547.91 | 5445.95 | 1898.04 | 7.777 |
| 129 | 330.3 | 1527.43 | 2665.25 | 1137.82 | 3554.25 | 5439.92 | 1885.67 | 7.648 |
| 130 | 330.9 | 1531.40 | 2662.89 | 1131.49 | 3560.58 | 5433.88 | 1873.30 | 7.521 |
| 131 | 331.5 | 1535.37 | 2660.51 | 1125.14 | 3566.88 | 5427.84 | 1860.95 | 7.395 |
| 132 | 332.0 | 1539.33 | 2658.11 | 1118.78 | 3573.17 | 5421.79 | 1848.61 | 7.270 |
| 133 | 332.6 | 1543.29 | 2655.69 | 1112.40 | 3579.45 | 5415.73 | 1836.28 | 7.146 |
| 134 | 333.2 | 1547.24 | 2653.24 | 1106.00 | 3585.71 | 5409.66 | 1823.95 | 7.023 |
| 135 | 333.8 | 1551.19 | 2650.77 | 1099.58 | 3591.96 | 5403.59 | 1811.63 | 6.901 |
| 136 | 334.4 | 1555.14 | 2648.28 | 1093.15 | 3598.19 | 5397.50 | 1799.31 | 6.780 |
| 137 | 335.0 | 1559.08 | 2645.77 | 1086.70 | 3604.41 | 5391.41 | 1787.00 | 6.660 |
| 138 | 335.5 | 1563.01 | 2643.24 | 1080.22 | 3610.62 | 5385.30 | 1774.68 | 6.541 |
| 139 | 336.1 | 1566.95 | 2640.68 | 1073.73 | 3616.81 | 5379.18 | 1762.37 | 6.423 |
| 140 | 336.7 | 1570.88 | 2638.09 | 1067.21 | 3623.00 | 5373.05 | 1750.05 | 6.306 |
| 141 | 337.2 | 1574.81 | 2635.49 | 1060.68 | 3629.18 | 5366.90 | 1737.72 | 6.190 |
| 142 | 337.8 | 1578.74 | 2632.85 | 1054.12 | 3635.34 | 5360.74 | 1725.39 | 6.075 |
| 143 | 338.3 | 1582.66 | 2630.20 | 1047.53 | 3641.50 | 5354.56 | 1713.06 | 5.962 |
| 144 | 338.9 | 1586.59 | 2627.51 | 1040.93 | 3647.66 | 5348.37 | 1700.71 | 5.849 |
| 145 | 339.5 | 1590.51 | 2624.81 | 1034.29 | 3653.80 | 5342.15 | 1688.35 | 5.737 |
| 146 | 340.0 | 1594.44 | 2622.07 | 1027.63 | 3659.94 | 5335.92 | 1675.98 | 5.626 |
| 147 | 340.5 | 1598.37 | 2619.31 | 1020.95 | 3666.07 | 5329.67 | 1663.60 | 5.516 |
| 148 | 341.1 | 1602.29 | 2616.52 | 1014.23 | 3672.20 | 5323.40 | 1651.20 | 5.407 |
| 149 | 341.6 | 1606.22 | 2613.71 | 1007.49 | 3678.32 | 5317.11 | 1638.79 | 5.298 |
| 150 | 342.2 | 1610.15 | 2610.86 | 1000.71 | 3684.45 | 5310.80 | 1626.35 | 5.191 |
| 151 | 342.7 | 1614.08 | 2607.99 | 993.909 | 3690.57 | 5304.46 | 1613.90 | 5.085 |
| 152 | 343.2 | 1618.02 | 2605.09 | 987.073 | 3696.68 | 5298.11 | 1601.42 | 4.980 |
| 153 | 343.7 | 1621.96 | 2602.16 | 980.205 | 3702.80 | 5291.72 | 1588.92 | 4.875 |
| 154 | 344.3 | 1625.90 | 2599.21 | 973.303 | 3708.92 | 5285.31 | 1576.39 | 4.772 |
| 155 | 344.8 | 1629.85 | 2596.22 | 966.366 | 3715.04 | 5278.88 | 1563.84 | 4.669 |
| 156 | 345.3 | 1633.80 | 2593.20 | 959.395 | 3721.16 | 5272.41 | 1551.26 | 4.567 |
| 157 | 345.8 | 1637.76 | 2590.15 | 952.386 | 3727.28 | 5265.92 | 1538.64 | 4.467 |
| 158 | 346.3 | 1641.72 | 2587.06 | 945.341 | 3733.41 | 5259.40 | 1525.99 | 4.367 |
| 159 | 346.8 | 1645.69 | 2583.95 | 938.256 | 3739.54 | 5252.85 | 1513.31 | 4.268 |
| 160 | 347.4 | 1649.67 | 2580.80 | 931.132 | 3745.68 | 5246.27 | 1500.59 | 4.170 |
| 161 | 347.9 | 1653.66 | 2577.62 | 923.968 | 3751.82 | 5239.66 | 1487.84 | 4.072 |
| 162 | 348.4 | 1657.65 | 2574.41 | 916.762 | 3757.97 | 5233.01 | 1475.04 | 3.976 |
| 163 | 348.9 | 1661.65 | 2571.16 | 909.513 | 3764.13 | 5226.33 | 1462.20 | 3.881 |
| 164 | 349.4 | 1665.66 | 2567.88 | 902.220 | 3770.30 | 5219.61 | 1449.32 | 3.786 |
| 165 | 349.9 | 1669.68 | 2564.57 | 894.882 | 3776.48 | 5212.86 | 1436.39 | 3.693 |
| 166 | 350.3 | 1673.75 | 2561.25 | 887.498 | 3782.72 | 5206.13 | 1423.42 | 3.600 |
| 167 | 350.8 | 1677.80 | 2557.85 | 880.052 | 3788.93 | 5199.29 | 1410.36 | 3.508 |
| 168 | 351.3 | 1681.86 | 2554.41 | 872.551 | 3795.16 | 5192.40 | 1397.25 | 3.417 |
| 169 | 351.8 | 1685.94 | 2550.93 | 864.993 | 3801.40 | 5185.47 | 1384.07 | 3.327 |
| 170 | 352.3 | 1690.04 | 2547.41 | 857.377 | 3807.67 | 5178.49 | 1370.82 | 3.237 |
| 171 | 352.8 | 1694.15 | 2543.85 | 849.701 | 3813.95 | 5171.46 | 1357.51 | 3.149 |
| 172 | 353.3 | 1698.27 | 2540.23 | 841.963 | 3820.26 | 5164.38 | 1344.12 | 3.061 |
| 173 | 353.7 | 1702.42 | 2536.58 | 834.160 | 3826.59 | 5157.24 | 1330.65 | 2.975 |
| 174 | 354.2 | 1706.58 | 2532.87 | 826.291 | 3832.95 | 5150.05 | 1317.11 | 2.889 |
| 175 | 354.7 | 1710.76 | 2529.11 | 818.351 | 3839.33 | 5142.80 | 1303.47 | 2.804 |

| $p$ | $T_{sat}$ | $i_f$ | $i_g$ | $i_{fg}$ | $s_f$ | $s_g$ | $s_{fg}$ | $\sigma$ |
|---|---|---|---|---|---|---|---|---|
| (bar) | (°C) | (kJ/kg) | (kJ/kg) | (kJ/kg) | (J/kg K) | (J/kg K) | (J/kg K) | (mN/m) |
| 176 | 355.1 | 1714.97 | 2525.31 | 810.339 | 3845.73 | 5135.48 | 1289.75 | 2.720 |
| 177 | 355.6 | 1719.19 | 2521.45 | 802.252 | 3852.17 | 5128.10 | 1275.93 | 2.637 |
| 178 | 356.1 | 1723.45 | 2517.53 | 794.087 | 3858.64 | 5120.65 | 1262.01 | 2.554 |
| 179 | 356.5 | 1727.72 | 2513.56 | 785.839 | 3865.14 | 5113.13 | 1247.99 | 2.473 |
| 180 | 357.0 | 1732.02 | 2509.53 | 777.507 | 3871.68 | 5105.54 | 1233.86 | 2.392 |
| 181 | 357.4 | 1736.35 | 2505.44 | 769.085 | 3878.26 | 5097.87 | 1219.61 | 2.312 |
| 182 | 357.9 | 1740.71 | 2501.28 | 760.570 | 3884.88 | 5090.11 | 1205.23 | 2.233 |
| 183 | 358.4 | 1745.10 | 2497.06 | 751.958 | 3891.54 | 5082.27 | 1190.73 | 2.155 |
| 184 | 358.8 | 1749.53 | 2492.77 | 743.245 | 3898.24 | 5074.34 | 1176.10 | 2.078 |
| 185 | 359.3 | 1753.99 | 2488.41 | 734.424 | 3905.00 | 5066.31 | 1161.31 | 2.002 |
| 186 | 359.7 | 1758.48 | 2483.98 | 725.491 | 3911.81 | 5058.19 | 1146.38 | 1.927 |
| 187 | 360.1 | 1763.02 | 2479.46 | 716.441 | 3918.67 | 5049.96 | 1131.28 | 1.852 |
| 188 | 360.6 | 1767.60 | 2474.86 | 707.266 | 3925.60 | 5041.61 | 1116.02 | 1.779 |
| 189 | 361.0 | 1772.22 | 2470.18 | 697.961 | 3932.59 | 5033.16 | 1100.57 | 1.706 |
| 190 | 361.5 | 1776.89 | 2465.41 | 688.518 | 3939.64 | 5024.57 | 1084.93 | 1.634 |
| 191 | 361.9 | 1781.61 | 2460.54 | 678.928 | 3946.78 | 5015.86 | 1069.08 | 1.563 |
| 192 | 362.3 | 1786.39 | 2455.57 | 669.185 | 3953.99 | 5007.01 | 1053.02 | 1.494 |
| 193 | 362.8 | 1791.22 | 2450.50 | 659.277 | 3961.28 | 4998.01 | 1036.72 | 1.425 |
| 194 | 363.2 | 1796.12 | 2445.31 | 649.194 | 3968.67 | 4988.85 | 1020.18 | 1.357 |
| 195 | 363.6 | 1801.08 | 2440.01 | 638.925 | 3976.16 | 4979.52 | 1003.36 | 1.290 |
| 196 | 364.1 | 1806.12 | 2434.57 | 628.457 | 3983.75 | 4970.01 | 986.265 | 1.223 |
| 197 | 364.5 | 1811.23 | 2429.00 | 617.776 | 3991.46 | 4960.31 | 968.857 | 1.158 |
| 198 | 364.9 | 1816.43 | 2423.29 | 606.866 | 3999.29 | 4950.41 | 951.116 | 1.094 |
| 199 | 365.3 | 1821.71 | 2417.42 | 595.710 | 4007.26 | 4940.27 | 933.015 | 1.031 |
| 200 | 365.7 | 1827.10 | 2411.39 | 584.287 | 4015.38 | 4929.90 | 914.523 | 0.969 |
| 201 | 366.2 | 1832.60 | 2405.17 | 572.575 | 4023.66 | 4919.26 | 895.606 | 0.908 |
| 202 | 366.6 | 1838.21 | 2398.76 | 560.548 | 4032.12 | 4908.34 | 876.223 | 0.848 |
| 203 | 367.0 | 1843.96 | 2392.13 | 548.178 | 4040.77 | 4897.10 | 856.332 | 0.789 |
| 204 | 367.4 | 1849.85 | 2385.28 | 535.429 | 4049.64 | 4885.52 | 835.878 | 0.731 |
| 205 | 367.8 | 1855.90 | 2378.16 | 522.261 | 4058.76 | 4873.56 | 814.800 | 0.674 |
| 206 | 368.2 | 1862.13 | 2370.76 | 508.630 | 4068.15 | 4861.18 | 793.027 | 0.619 |
| 207 | 368.6 | 1868.56 | 2363.04 | 494.478 | 4077.85 | 4848.32 | 770.473 | 0.565 |
| 208 | 369.0 | 1875.23 | 2354.97 | 479.737 | 4087.89 | 4834.93 | 747.035 | 0.512 |
| 209 | 369.4 | 1882.16 | 2346.49 | 464.327 | 4098.35 | 4820.93 | 722.586 | 0.460 |
| 210 | 369.8 | 1889.40 | 2337.54 | 448.147 | 4109.26 | 4806.23 | 696.973 | 0.410 |
| 211 | 370.2 | 1896.99 | 2328.06 | 431.066 | 4120.73 | 4790.72 | 669.994 | 0.361 |
| 212 | 370.6 | 1905.02 | 2317.94 | 412.917 | 4132.85 | 4774.24 | 641.393 | 0.313 |
| 213 | 371.0 | 1913.57 | 2307.05 | 393.479 | 4145.77 | 4756.60 | 610.829 | 0.268 |
| 214 | 371.4 | 1922.77 | 2295.22 | 372.446 | 4159.69 | 4737.52 | 577.829 | 0.224 |
| 215 | 371.8 | 1932.81 | 2282.19 | 349.378 | 4174.89 | 4716.61 | 541.718 | 0.181 |
| 216 | 372.2 | 1943.96 | 2267.57 | 323.610 | 4191.81 | 4693.28 | 501.468 | 0.141 |
| 217 | 372.6 | 1956.70 | 2250.75 | 294.048 | 4211.16 | 4666.55 | 455.392 | 0.104 |
| 218 | 372.9 | 1971.88 | 2230.56 | 258.683 | 4234.27 | 4634.66 | 400.388 | 0.069 |
| 219 | 373.3 | 1991.44 | 2204.48 | 213.041 | 4264.15 | 4593.70 | 329.550 | 0.038 |
| 220 | 373.7 | 2021.91 | 2164.20 | 142.293 | 4310.86 | 4530.84 | 219.977 | 0.012 |

## D.2 PROBABILITY TABLES

Table D.1 contains values of a function $1 - \Phi(z)$, where $\Phi(z)$ is given by Eq. (11.11).

## TABLE D.1
## Complementary Cumulative Standard Normal Distribution[1]

| z | 0 | 0.01 | 0.02 | 0.03 | 0.04 | 0.05 | 0.06 | 0.07 | 0.08 | 0.09 |
|---|---|---|---|---|---|---|---|---|---|---|
| 0.0 | $.5000_{+0}$ | $.4960_{+0}$ | $.4920_{+0}$ | $.4880_{+0}$ | $.4840_{+0}$ | $.4800_{+0}$ | $.4760_{+0}$ | $.4720_{+0}$ | $.4681_{+0}$ | $.4641_{+0}$ |
| 0.1 | $.4601_{+0}$ | $.4562_{+0}$ | $.4522_{+0}$ | $.4482_{+0}$ | $.4443_{+0}$ | $.4403_{+0}$ | $.4364_{+0}$ | $.4325_{+0}$ | $.4285_{+0}$ | $.4246_{+0}$ |
| 0.2 | $.4207_{+0}$ | $.4168_{+0}$ | $.4129_{+0}$ | $.4090_{+0}$ | $.4051_{+0}$ | $.4012_{+0}$ | $.3974_{+0}$ | $.3935_{+0}$ | $.3897_{+0}$ | $.3859_{+0}$ |
| 0.3 | $.3820_{+0}$ | $.3782_{+0}$ | $.3744_{+0}$ | $.3707_{+0}$ | $.3669_{+0}$ | $.3631_{+0}$ | $.3594_{+0}$ | $.3556_{+0}$ | $.3519_{+0}$ | $.3482_{+0}$ |
| 0.4 | $.3445_{+0}$ | $.3409_{+0}$ | $.3372_{+0}$ | $.3335_{+0}$ | $.3299_{+0}$ | $.3263_{+0}$ | $.3227_{+0}$ | $.3191_{+0}$ | $.3156_{+0}$ | $.3120_{+0}$ |
| 0.5 | $.3085_{+0}$ | $.3050_{+0}$ | $.3015_{+0}$ | $.2980_{+0}$ | $.2945_{+0}$ | $.2911_{+0}$ | $.2877_{+0}$ | $.2843_{+0}$ | $.2809_{+0}$ | $.2775_{+0}$ |
| 0.6 | $.2742_{+0}$ | $.2709_{+0}$ | $.2676_{+0}$ | $.2643_{+0}$ | $.2610_{+0}$ | $.2578_{+0}$ | $.2546_{+0}$ | $.2514_{+0}$ | $.2482_{+0}$ | $.2450_{+0}$ |
| 0.7 | $.2419_{+0}$ | $.2388_{+0}$ | $.2357_{+0}$ | $.2326_{+0}$ | $.2296_{+0}$ | $.2266_{+0}$ | $.2236_{+0}$ | $.2206_{+0}$ | $.2176_{+0}$ | $.2147_{+0}$ |
| 0.8 | $.2118_{+0}$ | $.2089_{+0}$ | $.2061_{+0}$ | $.2032_{+0}$ | $.2004_{+0}$ | $.1976_{+0}$ | $.1948_{+0}$ | $.1921_{+0}$ | $.1894_{+0}$ | $.1867_{+0}$ |
| 0.9 | $.1840_{+0}$ | $.1814_{+0}$ | $.1787_{+0}$ | $.1761_{+0}$ | $.1736_{+0}$ | $.1710_{+0}$ | $.1685_{+0}$ | $.1660_{+0}$ | $.1635_{+0}$ | $.1610_{+0}$ |
| 1.0 | $.1586_{+0}$ | $.1562_{+0}$ | $.1538_{+0}$ | $.1515_{+0}$ | $.1491_{+0}$ | $.1468_{+0}$ | $.1445_{+0}$ | $.1423_{+0}$ | $.1400_{+0}$ | $.1378_{+0}$ |
| 1.1 | $.1356_{+0}$ | $.1335_{+0}$ | $.1313_{+0}$ | $.1292_{+0}$ | $.1271_{+0}$ | $.1250_{+0}$ | $.1230_{+0}$ | $.1210_{+0}$ | $.1190_{+0}$ | $.1170_{+0}$ |
| 1.2 | $.1150_{+0}$ | $.1131_{+0}$ | $.1112_{+0}$ | $.1093_{+0}$ | $.1074_{+0}$ | $.1056_{+0}$ | $.1038_{+0}$ | $.1020_{+0}$ | $.1002_{+0}$ | $.9852_{-1}$ |
| 1.3 | $.9680_{-1}$ | $.9509_{-1}$ | $.9341_{-1}$ | $.9175_{-1}$ | $.9012_{-1}$ | $.8850_{-1}$ | $.8691_{-1}$ | $.8534_{-1}$ | $.8379_{-1}$ | $.8226_{-1}$ |
| 1.4 | $.8075_{-1}$ | $.7926_{-1}$ | $.7780_{-1}$ | $.7635_{-1}$ | $.7493_{-1}$ | $.7352_{-1}$ | $.7214_{-1}$ | $.7078_{-1}$ | $.6943_{-1}$ | $.6811_{-1}$ |
| 1.5 | $.6680_{-1}$ | $.6552_{-1}$ | $.6425_{-1}$ | $.6300_{-1}$ | $.6178_{-1}$ | $.6057_{-1}$ | $.5937_{-1}$ | $.5820_{-1}$ | $.5705_{-1}$ | $.5591_{-1}$ |
| 1.6 | $.5479_{-1}$ | $.5369_{-1}$ | $.5261_{-1}$ | $.5155_{-1}$ | $.5050_{-1}$ | $.4947_{-1}$ | $.4845_{-1}$ | $.4745_{-1}$ | $.4647_{-1}$ | $.4551_{-1}$ |
| 1.7 | $.4456_{-1}$ | $.4363_{-1}$ | $.4271_{-1}$ | $.4181_{-1}$ | $.4092_{-1}$ | $.4005_{-1}$ | $.3920_{-1}$ | $.3836_{-1}$ | $.3753_{-1}$ | $.3672_{-1}$ |
| 1.8 | $.3593_{-1}$ | $.3514_{-1}$ | $.3437_{-1}$ | $.3362_{-1}$ | $.3288_{-1}$ | $.3215_{-1}$ | $.3144_{-1}$ | $.3074_{-1}$ | $.3005_{-1}$ | $.2937_{-1}$ |
| 1.9 | $.2871_{-1}$ | $.2806_{-1}$ | $.2742_{-1}$ | $.2680_{-1}$ | $.2618_{-1}$ | $.2558_{-1}$ | $.2499_{-1}$ | $.2441_{-1}$ | $.2385_{-1}$ | $.2329_{-1}$ |
| 2.0 | $.2275_{-1}$ | $.2221_{-1}$ | $.2169_{-1}$ | $.2117_{-1}$ | $.2067_{-1}$ | $.2018_{-1}$ | $.1969_{-1}$ | $.1922_{-1}$ | $.1876_{-1}$ | $.1830_{-1}$ |
| 2.1 | $.1786_{-1}$ | $.1742_{-1}$ | $.1700_{-1}$ | $.1658_{-1}$ | $.1617_{-1}$ | $.1577_{-1}$ | $.1538_{-1}$ | $.1500_{-1}$ | $.1462_{-1}$ | $.1426_{-1}$ |
| 2.2 | $.1390_{-1}$ | $.1355_{-1}$ | $.1320_{-1}$ | $.1287_{-1}$ | $.1254_{-1}$ | $.1222_{-1}$ | $.1191_{-1}$ | $.1160_{-1}$ | $.1130_{-1}$ | $.1101_{-1}$ |
| 2.3 | $.1072_{-1}$ | $.1044_{-1}$ | $.1017_{-1}$ | $.9903_{-2}$ | $.9641_{-2}$ | $.9386_{-2}$ | $.9137_{-2}$ | $.8894_{-2}$ | $.8656_{-2}$ | $.8424_{-2}$ |
| 2.4 | $.8197_{-2}$ | $.7976_{-2}$ | $.7760_{-2}$ | $.7549_{-2}$ | $.7343_{-2}$ | $.7142_{-2}$ | $.6946_{-2}$ | $.6755_{-2}$ | $.6569_{-2}$ | $.6387_{-2}$ |
| 2.5 | $.6209_{-2}$ | $.6036_{-2}$ | $.5867_{-2}$ | $.5703_{-2}$ | $.5542_{-2}$ | $.5386_{-2}$ | $.5233_{-2}$ | $.5084_{-2}$ | $.4940_{-2}$ | $.4798_{-2}$ |
| 2.6 | $.4661_{-2}$ | $.4527_{-2}$ | $.4396_{-2}$ | $.4269_{-2}$ | $.4145_{-2}$ | $.4024_{-2}$ | $.3907_{-2}$ | $.3792_{-2}$ | $.3681_{-2}$ | $.3572_{-2}$ |
| 2.7 | $.3466_{-2}$ | $.3364_{-2}$ | $.3264_{-2}$ | $.3166_{-2}$ | $.3071_{-2}$ | $.2979_{-2}$ | $.2890_{-2}$ | $.2802_{-2}$ | $.2717_{-2}$ | $.2635_{-2}$ |
| 2.8 | $.2555_{-2}$ | $.2477_{-2}$ | $.2401_{-2}$ | $.2327_{-2}$ | $.2255_{-2}$ | $.2185_{-2}$ | $.2118_{-2}$ | $.2052_{-2}$ | $.1988_{-2}$ | $.1926_{-2}$ |
| 2.9 | $.1865_{-2}$ | $.1807_{-2}$ | $.1750_{-2}$ | $.1694_{-2}$ | $.1641_{-2}$ | $.1588_{-2}$ | $.1538_{-2}$ | $.1489_{-2}$ | $.1441_{-2}$ | $.1394_{-2}$ |
| 3.0 | $.1349_{-2}$ | $.1306_{-2}$ | $.1263_{-2}$ | $.1222_{-2}$ | $.1182_{-2}$ | $.1144_{-2}$ | $.1106_{-2}$ | $.1070_{-2}$ | $.1035_{-2}$ | $.1000_{-2}$ |
| 3.1 | $.9676_{-3}$ | $.9354_{-3}$ | $.9042_{-3}$ | $.8740_{-3}$ | $.8447_{-3}$ | $.8163_{-3}$ | $.7888_{-3}$ | $.7621_{-3}$ | $.7363_{-3}$ | $.7113_{-3}$ |
| 3.2 | $.6871_{-3}$ | $.6636_{-3}$ | $.6409_{-3}$ | $.6189_{-3}$ | $.5976_{-3}$ | $.5770_{-3}$ | $.5570_{-3}$ | $.5377_{-3}$ | $.5190_{-3}$ | $.5009_{-3}$ |
| 3.3 | $.4834_{-3}$ | $.4664_{-3}$ | $.4500_{-3}$ | $.4342_{-3}$ | $.4188_{-3}$ | $.4040_{-3}$ | $.3897_{-3}$ | $.3758_{-3}$ | $.3624_{-3}$ | $.3494_{-3}$ |
| 3.4 | $.3369_{-3}$ | $.3248_{-3}$ | $.3131_{-3}$ | $.3017_{-3}$ | $.2908_{-3}$ | $.2802_{-3}$ | $.2700_{-3}$ | $.2602_{-3}$ | $.2507_{-3}$ | $.2415_{-3}$ |
| 3.5 | $.2326_{-3}$ | $.2240_{-3}$ | $.2157_{-3}$ | $.2077_{-3}$ | $.2000_{-3}$ | $.1926_{-3}$ | $.1854_{-3}$ | $.1784_{-3}$ | $.1717_{-3}$ | $.1653_{-3}$ |
| 3.6 | $.1591_{-3}$ | $.1530_{-3}$ | $.1473_{-3}$ | $.1417_{-3}$ | $.1363_{-3}$ | $.1311_{-3}$ | $.1261_{-3}$ | $.1212_{-3}$ | $.1166_{-3}$ | $.1121_{-3}$ |
| 3.7 | $.1078_{-3}$ | $.1036_{-3}$ | $.9961_{-4}$ | $.9573_{-4}$ | $.9201_{-4}$ | $.8841_{-4}$ | $.8495_{-4}$ | $.8162_{-4}$ | $.7841_{-4}$ | $.7532_{-4}$ |
| 3.8 | $.7234_{-4}$ | $.6948_{-4}$ | $.6672_{-4}$ | $.6407_{-4}$ | $.6151_{-4}$ | $.5905_{-4}$ | $.5669_{-4}$ | $.5441_{-4}$ | $.5222_{-4}$ | $.5012_{-4}$ |
| 3.9 | $.4809_{-4}$ | $.4614_{-4}$ | $.4427_{-4}$ | $.4247_{-4}$ | $.4074_{-4}$ | $.3907_{-4}$ | $.3747_{-4}$ | $.3593_{-4}$ | $.3445_{-4}$ | $.3303_{-4}$ |
| 4.0 | $.3167_{-4}$ | $.3035_{-4}$ | $.2909_{-4}$ | $.2788_{-4}$ | $.2672_{-4}$ | $.2560_{-4}$ | $.2453_{-4}$ | $.2350_{-4}$ | $.2251_{-4}$ | $.2156_{-4}$ |
| 4.1 | $.2065_{-4}$ | $.1978_{-4}$ | $.1894_{-4}$ | $.1813_{-4}$ | $.1736_{-4}$ | $.1662_{-4}$ | $.1591_{-4}$ | $.1523_{-4}$ | $.1457_{-4}$ | $.1394_{-4}$ |
| 4.2 | $.1334_{-4}$ | $.1276_{-4}$ | $.1221_{-4}$ | $.1168_{-4}$ | $.1117_{-4}$ | $.1068_{-4}$ | $.1022_{-4}$ | $.9773_{-5}$ | $.9344_{-5}$ | $.8933_{-5}$ |
| 4.3 | $.8539_{-5}$ | $.8162_{-5}$ | $.7801_{-5}$ | $.7455_{-5}$ | $.7124_{-5}$ | $.6806_{-5}$ | $.6503_{-5}$ | $.6212_{-5}$ | $.5933_{-5}$ | $.5667_{-5}$ |
| 4.4 | $.5412_{-5}$ | $.5168_{-5}$ | $.4935_{-5}$ | $.4711_{-5}$ | $.4497_{-5}$ | $.4293_{-5}$ | $.4097_{-5}$ | $.3910_{-5}$ | $.3732_{-5}$ | $.3561_{-5}$ |
| 4.5 | $.3397_{-5}$ | $.3241_{-5}$ | $.3091_{-5}$ | $.2949_{-5}$ | $.2812_{-5}$ | $.2682_{-5}$ | $.2557_{-5}$ | $.2438_{-5}$ | $.2324_{-5}$ | $.2216_{-5}$ |

[1] Notation $n_e$ used in the table is a shorthand for $n \cdot 10^e$. For example: $.1457_{-4} = 0.1457 \cdot 10^{-4}$.

## TABLE D.2
### Factors for Two-Sided Tolerance Limit

| n | 90% Confidence That Percentage of Population Between Limits is | | | 95% Confidence That Percentage of Population Between Limits is | | | 99% Confidence That Percentage of Population Between Limits is | | |
|---|---|---|---|---|---|---|---|---|---|
|   | 90% | 95% | 99% | 90% | 95% | 99% | 90% | 95% | 99% |
| 2 | 15.98 | 18.80 | 24.17 | 32.02 | 37.67 | 48.43 | 160.2 | 188.5 | 242.3 |
| 3 | 5.847 | 6.919 | 8.947 | 8.380 | 9.916 | 12.86 | 18.93 | 22.40 | 29.06 |
| 4 | 4.166 | 4.943 | 6.440 | 5.369 | 6.370 | 8.299 | 9.398 | 11.15 | 14.53 |
| 5 | 3.494 | 4.152 | 5.423 | 4.275 | 5.079 | 6.634 | 6.612 | 7.855 | 10.26 |
| 6 | 3.131 | 3.723 | 4.870 | 3.712 | 4.414 | 5.775 | 5.337 | 6.345 | 8.301 |
| 7 | 2.902 | 3.452 | 4.521 | 3.369 | 4.007 | 5.248 | 4.613 | 5.488 | 7.187 |
| 8 | 2.743 | 3.264 | 4.278 | 3.163 | 3.732 | 4.891 | 4.147 | 4.936 | 6.468 |
| 9 | 2.626 | 3.125 | 4.098 | 2.967 | 3.532 | 4.631 | 3.822 | 4.550 | 5.966 |
| 10 | 2.535 | 3.018 | 3.959 | 2.839 | 3.379 | 4.433 | 3.582 | 4.265 | 5.594 |
| 20 | 2.152 | 2.564 | 3.368 | 2.310 | 2.752 | 3.615 | 2.659 | 3.168 | 4.161 |
| 30 | 2.025 | 2.413 | 3.170 | 2.140 | 2.549 | 3.350 | 2.385 | 2.841 | 3.733 |
| ∞ | 1.645 | 1.960 | 2.576 | 1.645 | 1.960 | 2.576 | 1.645 | 1.960 | 2.576 |

# References

1. A.W. Adamson. Potential distortion model for contact angle and spreading. II. Temperature dependent effects. *Journal of Colloid and Interface Science*, 44(2):273–281, 1973.

2. W.S. Alioshin, N.M. Kuznetsov, and A.A. Sarkisov. *Nuclear Marine Propulsion*. Sudostroienie, Leningrad, USSR, 1968.

3. M.A. Amidu, S. Jung, and H. Kim. Direct experimental measurement for partitioning of wall heat flux during subcooled flow boiling: Effect of bubble areas of influence factor. *International Journal of Heat Mass Transfer*, 127:515–533, 2018.

4. D.L.J. Anderson, R.L. Judd, and H. Merte, Jr. Site activation phenomena in saturated pool nucleate boiling. In *Fluids Engineering, Heat Transfer, and Lubrication Conf., ASME Paper 70-HT-14*, Detroit, MI, 1970.

5. H. Anglart. Initial entrained fraction at onset of annular flow. In *Procedings of the 18th International Topical Meeting on Nuclear Reactor Thermal Hydraulics (NURETH-18)*, Portland, OR, August 18–23, 2019.

6. H. Anglart and D. Caraghiaur. CFD modeling of boiling annular-mist flow for dryout investigations. *Multiphase Science and Technology*, 23(2–4):223–251, 2011.

7. H. Anglart and O. Nylund. CFD application to prediction of void distribution in two-phase bubbly flows in rod bundles. *Nclear Engineering and Design*, 163:81–98, 1996.

8. R. Aris. *Vectors, Tensors, and the Basic Equations of Fluid Mechanics*. Dover Publications, New York, 1962.

9. D. Baron, L. Hallstadius, K. Kulacsy, R. Largenton, and J. Noirot. Fuel Performance of Light Water Reactors (Uranium Oxide and MOX. In R.J.M. Konings and R.E. Stoller, editors, *Comprehensive Nuclear Materials*, chapter 2.02, pages 35–70. Elsevier, Oxford, second edition, 2020.

10. N. Basu, G.R Warrier, and V.K. Dhir. Onset of nucleate boiling and active nucleation site density during subcooled flow boiling. *Journal of Heat Transfer*, 124:717–728, 2002.

11. J.A. Beattie and O.C. Bridgeman. A new equation of state for fluids. II. Application to helium, neon, argon, hydrogen, nitrogen, oxygen, air and methane. *Journal of the American Chemical Society*, 50:3133–3138, 1928.

12. K.M. Becker. An analytical and experimental study of burnout conditions in vertical round ducts. Technical Report AE-178, AB Atomenergi, Studsvik, Sweden, 1965.

13. A. Bejan. *Convection Heat Transfer*. Wiley, New York, second edition edition, 1995.

14. O. Beneš and R.J.M. Konings. Molten Salt Reactor Fuel and Coolant. In R.J.M. Konings and R.E. Stoller, editors, *Comprehensive Nuclear Materials*, chapter 5.18, pages 609–644. Elsevier, Oxford, second edition, 2020.

15. R.J. Benjamin and A.R. Balakrishnan. Nucleation site density in pool boiling of saturated pure liquids: Effect of surface microroughness and surface and liquid physical properties. *Experimental Thermal and Fluid Science*, 15(1):32–42, 1997.

16. P.J. Berenson. Transition boiling heat transfer. In *Proceedings of the 4th National Heat Transfer Conference, AIChE Preprint 18*, Buffalo, NY, 1960.

17. A.E. Bergles and W.M. Rohsenow. The determination of forced convection surface-boiling heat transfer. *Journal of Heat Transfer, Transactions of the ASME*, 86:365–372, 1964.

18. A.A. Bishop, R.O. Sandberg, and L.S. Tong. Forced convection heat transfer to water at near-critical temperatures and super-critical pressures. In *AIChE–I. Chem. E. Joint Meeting, Paper 2–7*, London, UK, June 1965.

19. H. Blasius. *Mitteilungen über Forschungsarbeiten auf dem Gebiete des Ingenieurwesens*, chapter Das Aehnlichkeitsgesetz bei Reibungsvorgängen in Flüssigkeiten, pages 1–41. Springer Berlin Heidelberg, 1913.

20. V.M. Borishanskii, M.A. Gotovskii, and E.V. Firsova. Heat transfer to liquid metal flowing longitudinally in wetted bundles of rods. *Soviet Atomic Energy*, 27:1347–1350, 1969.

21. R.W. Bowring. A simple but accurate round tube, uniform heat flux, dryout correlation over the pressure range 0.7–17 MN/m$^2$ (100–2500 PSIA). Technical Report AEEW–R 789, United Kingdom Atomic Energy Authority, Winfrith, Dorchester, Dorset, 1972.

22. R.W. Bowring. WSC2 - a subchannel dryout correlation for water-cooled clusters over the pressure range 3.4–15.9 MPa (500–2300 psia). Technical Report AEEW–R 983, United Kingdom Atomic Energy Authority, Winfrith, Dorchester, Dorset, 1979.

23. B.E. Boyack *et al.* Quantifying reactor safety margins part 1: An overview of the core scaling, applicability, and uncertainty evaluation methodology. *Nuclear Engineering and Design*, 119:1–15, 1990.

24. L.A. Bromley. Heat transfer in stable film boiling. *Chemical Engineering Progress*, 46(5):221–238, 1950.

25. L.A. Bromley, N.R. LeRoy, and J.A. Robbers. Heat transfer in forced convection film boiling. *Industrial & Engineering Chemistry*, 45(12):2639–2646, 1953.

26. N. R. Brown, A. T. Nelson, and K. A. Terrani. Accident-Tolerant Fuel. In R.J.M. Konings and R.E. Stoller, editors, *Comprehensive Nuclear Materials*, chapter 5.20, pages 684–706. Elsevier, Oxford, second edition, 2020.

27. K.A. Brownlee. *Statistical Theory and Methodology in Science and Engineering*. Wiley, New York, 1965.

28. V.P. Carey. *Liquid-Vapor Phase-Change Phenomena*. Hemisphere Publishing Corporation, Bristol, PA, 1992.

29. I. Catton. Natural convection in enclosures. In *6th Int. Heat Transfer Conf.*, volume 6, pages 13–43, Toronto, 1979.

30. G.P Celata. Letter to the Editor. On the application method of critical heat flux correlations. *Nuclear Engineering and Design*, 163:241–242, 1996.

31. Y.A. Cengel and M.A. Boles. *Thermodynamics: An Engineering Approach*. McGraw-Hill Book Company, New York, 2010.

32. J.C. Chen. Correlation for boiling heat transfer to saturated fluids in convective flow. *Industrial and Engineering Chemistry Process Design and Development*, 5(3):322–339, 1966.

33. E. Chibowski. Surface free energy of sulfur–Revisited I. Yellow and orange samples solidified against glass surface. *Journal of Colloid and Interface Science*, 319(2):505–513, 2008.

34. Y.C. Chiew and E.D. Glandt. The effect of structure on the conductivity of a dispersion. *Journal of Colloid and Interface Science*, 94(1):90–104, 1983.

35. S.W. Churchill. Empirical expressions for the shear stress in turbulent flow in commercial pipe. *AIChE Journal*, 19(2):375–376, 1973.

36. S.W. Churchill. Free convection around immersed bodies. In E.U. Schlunder, editor, *Heat Exchanger Design Handbook*, chapter 2. Hemisphere, New York, 1983.

37. S.W. Churchill and M. Bernstein. A correlating equation for forced convection from gases and liquids to a circular cylinder in crossflow. *Journal of Heat Transfer*, 99(2):300–306, 1977.

38. S.W. Churchill and H.H.S. Chu. Correlating equations for laminar and turbulent free convection from a vertical plate. *International Journal of Heat and Mass Transfer*, 18:1323–1329, 1975.

39. S.W. Churchill and H. Ozoe. Correlations for forced convection with uniform heating in flow over a plate and in developing and fully developed flow in a tube. *Journal of Heat Transfer*, 95:78–84, 1973.

40. A.P. Colburn. A method of correlating forced convection heat-transfer data and a comparison with fluid friction. *International Journal of Heat and Mass Transfer*, 7(12):1359–1384, 1964.

41. R. Cole. Bubble frequencies and departure volumes at subatmospheric pressures. *AIChE Journal*, 13(4):779–783, 1967.

42. R. Cole. Boiling nucleation. *Advances in Heat Transfer*, 10:86–164, 1974.

43. R. Cole and W.R. Rohsenow. Correlation of bubble departure diameters for boiling of saturated liquids. *Chemical Engineering Progress Symposium Series*, 65(92):211–213, 1968.

44. C.F. Colebrook. Turbulent flow in pipes, with particular reference to the transition region between the smooth and rough pipe laws. *Journal of the ICE*, 11(4):133–156, 1939.

45. J.G. Collier. The problem of burnout in liquid cooled nuclear reactors. Technical Report AERE-R 3698, Atomic Energy Research Establishment, Harwell, Berkshire, UK, 1961.

46. M.G. Cooper. Heat flow rates in saturated nucleate pool boiling - a wide-ranging examination using reduced properties. *Advances in Heat Transfer*, 16:157–239, 1984.

47. M.G. Cooper and A.J.P. Lloyd. The microlayer in nucleate pool boiling. *International Journal of Heat and Mass Transfer*, 12:895–913, 1969.

48. W.E. Cummins. AP1000 Design Control Document. Technical Report APP-GW-GL-700, rev. 14, Westinghouse Electric Company LLC, Pittsburgh, PA, 2004.

49. J.H. Cushman. Multiphase transport equations I: General equation for macroscopic local space-time homogeneity. *Transport Theory and Statistical Physics*, 12:35–71, 1983.

50. E.J. Davis and G.H. Anderson. The incipience of nucleate boiling in forced convection flow. *AIChE Journal*, 12:774–780, 1966.

51. V.H.M. Del Valle and D.R.B. Kenning. Subcooled flow boiling at high heat flux. *International Journal of Heat Mass Transfer*, 28:1907–1920, 1985.

52. J.M. Delhaye. Jump conditions and entropy sources in two-phase systems. Local instant formulation. *International Journal of Multiphase Flow*, 1:395–409, 1974.

53. J.M. Delhaye, M. Giot, and M.L. Riethmuller. *Thermohydraulics of Two-Phase Systems for Industrial Design and Nuclear Engineering*. Hemisphere Publishing Corp., New York, 1999.

54. P.A. Demkowicz, B. Liu, and J.D. Hunn. Coated particle fuel: Historical perspectives and current progress. *Journal of Nuclear Materials*, 515:434–450, 2019.

55. J.J. DiNuno, F.D. Anderson, R.E. Baker, and R.L. Waterfield. Calculation of distance factors for power and test reactor sites. Technical Report TID-14844, U.S. Atomic Energy Commission, Washington, D.C., 1962.

56. F.W. Dittus and L.M.K. Boelter. Heat transfer in automobile radiators of the tubular type. *University of California Publications in Engineering*, 2(13):443–461, 1930.

57. V.E. Doroshchuk and F.P. Lantsman. Selecting magnitudes of critical heat fluxes with water boiling in vertical uniformly heated tubes. *Thermal Engineering (USSR)*, 17(12):18–21, 1970.

58. R.S. Dougall and W.M. Rohsenow. Film boiling on the inside of vertical tubes with upward flow of the fluid at low qualities. Technical Report 9079-26, Massachusetts Institute of Technology, Cambridge, MA, 1963.

59. D.A. Drew and S.L. Passman. *Theory of Multicomponent Fluids*. Springer, New York, 1999.

60. J.J. Duderstadt and L.J. Hamilton. *Nuclear Reactor Analysis*. Wiley, Boca Raton, FL, 1976.

61. S.A. Ebrahim, F.-B. Cheung, S.M. Bajorek, K. Tien, and C.L. Hoxie. Heat transfer correlation for film boiling during quenching of micro-structured surfaces. *Nuclear Engineering and Design*, 398, 2022.

62. M.S. El-Genk, S.D. Bedrose, and D.V. Rao. Forced and combined convection of water in a vertical seven-rod bundle with P/D=1.38. *International Journal of Heat and Mass Transfer*, 33(6):1289–1297, 1990.

63. M.S. El-Genk, B. Su, and Z. Guo. Experimental studies of forced, combined and natural convection of water in vertical nine-rod bundles with square lattice. *International Journal of Heat and Mass Transfer*, 36(9):2359–2374, 1993.

64. W. Fan, H. Li, and H. Anglart. A study of rewetting and conjugate heat transfer influence on dryout and post-dryout phenomena with multi-domain coupled CFD. *International Journal of Heat and Mass Transfer*, 163:120503, 2020.

65. G.K. Filonenko. Hydraulic resistance in pipelines. *Teploenergetika*, 4:40–44, 1954.

66. J. Fink. Thermophysical properties of uranium dioxide. *Journal of Nuclear Materials*, 279:1–18, 2000.

67. S. Fischer, T. Gambaryan-Roisman, and P. Stephan. On the development of a thin evaporating liquid film at a receding liquid/vapour-interface. *International Journal of Heat and Mass Transfer*, 88:346–356, 2015.

68. H.K. Forster and N. Zuber. Growth of a vapour bubble in superheated liquid. *Journal of Applied Physics*, 25:474–478, 1954.

69. H.K. Forster and N. Zuber. Dynamic of vapor bubbles and boiling heat transfer. *AIChE Journal*, 1(4):531–535, 1955.

70. Framatome. Epr Design Description. Technical report, Framatome ANP, Inc., Lynchburg, VA, August 2005.

71. W.L. Friend and A.B. Metzner. Turbulent heat transfer inside tubes and the analogy among heat, mass, and momentum transfer. *AIChE Journal*, 4(4):393–402, 1958.

72. W. Fritz. Berechnung des maximalvolume von dampfblasen. *Physikalische Zeitschrift*, 36:379–388, 1935.

73. R.F. Gaertner. Distribution of active sites in the nucleate boiling of liquids. *AIChE Chemical Engineering Progress Symposium Series*, 59(41):52, 1963.

74. R.F. Gaertner. Photographic study of nucleate pool boiling on a horizontal surface. *Transactions of ASME, Journal of Heat Transfer*, 87:17–29, 1965.

75. J. Garnier and S. Begej. Ex-reactor determination of thermal gap conductance between uranium dioxide and Zircaloy-4. Technical Report NUREG/CR-0330, Pacific Northwest Laboratory, Richland WA, 1980.

76. L.F. Glushchenko, S.I. Kalachev, and O.F. Gandzyuk. Determining the conditions of existence of deteriorated heat transfer at supercritical pressure of the medium. *Thermal Engineering*, 2:69–72, 1972.

77. V. Gnielinski. New equation for heat and mass transfer in turbulent pipe and channel flow. *International Chemical Engineering*, 16:359–368, 1976.

78. E.E. Gonzo. Estimating correlations for the effective thermal conductivity of granular materials. *Chemical Engineering Journal*, 90:299–302, 2002.

79. D.C. Groeneveld. Post-dryout heat transfer at reactor operating conditions. Technical Report AECL-4513, Chalk River Nuclear Laboratories, Chalk River, Ontario, 1973.

80. D.C. Groeneveld *et al*. The 1995 look-up table for critical heat flux in tubes. *Nuclear Engineering and Design*, 163:1–23, 1996.

81. D.C. Groeneveld *et al*. The 2006 CHF look-up table. *Nuclear Engineering and Design*, 237:1909–1922, 2007.

82. S.E. Haaland. Simple and explicit formulas for the friction factor in turbulent pipe flow. *Journal of Fluids Engineering*, 105:89–90, 1983.

83. D.L. Hagrman, G.A. Reymann, and R.E. Mason (eds.). MATPRO-Version 11 (Revision 2) A handbook of materials properties for use in the analysis of light water reactor fuel rod behavior. Technical Report NUREG/CR-0497, TREE-1280, Rev 2, Idaho National Engineering Laboratory, August 1981.

84. C.Y. Han and P. Griffith. The mechanism of heat transfer in nuclear pool boiling - Part I. Bubble initiation, growth and departure. *International Journal of Heat and Mass Transfer*, 8:887–904, 1965.

85. H. Hausen. Darstellung des warmeuberganges in rohren durch verallgeminerte potenzbeziehungen. *Z. Ver deutsch. Ing Beih. Verfahrenstech.*, 4:91–98, 1943.

86. J.E. Hench and J.C. Gillis. Correlation of critical heat flux data for application to boiling water reactor conditions. Technical Report EPRI NP-1898, Electric Power Research Institute, Palo Alto, CA, 1981.

87. R.E. Henry. A correlation for minimum film boiling temperature. *Chemical Engineering Progress Symposium Series*, 70(138):81–90, 1974.

88. G.F. Hewitt. Some calculations on holdup, heat transfer, and nucleation for steam-water flow in a 0.5 cm bore tube. Technical Report AERE-R-3984, Harwell AERE, Richland, Washington, 1962.

89. G.F. Hewitt and A. Govan. Phenomenological modelling of non-equilibrium flows with phase change. *Internatioal Journal of Heat and Mass Transfer*, 33(3):229–242, 1990.

90. C.W. Hirt and B.D. Nichols. Volume of fluid (VOF) method for the dynamics of free boundaries. *Journal of Computational Physics*, 39:201–225, 1981.

91. G.L. Hofman, L.C. Walters, and T.H. Bauer. Metallic fast reactor fuels. *Progress in Nuclear Energy*, 31(1):83–110, 1997.

92. J. Hou, M. Avramova, and K. Ivanov. Best-estimate plus uncertainty framework for multiscale, multiphisics light water reactor core analysis. *Science and Technology of Nuclear Installations*, 2020:18, 2020.

93. Y.Y. Hsu. On the size range of active nucleation cavities on a heating surface. *Journal of Heat Transfer*, 84C(3):207–216, 1962.

94. Y.Y. Hsu and J.W. Westwater. Approx. theory for film boiling on vertical surfaces. In *AIChE Chemical Engineering Progress Symposium Series*, number 30 in 56, page 15, New York, 1960. AIChE.

95. IAEA. Thermophysical properties of materials for water cooled reactors. Technical Report IAEA-TECDOC-949, International Atomic Energy Agency, Vienna, 1997.

96. IAEA. Safety of nuclear power plants: Design. Technical Report NS-R-1, International Atomic Energy Agency, Vienna, 2000.

97. IAEA. Safety of nuclear power plants: Operation. Technical Report NS-R-2, International Atomic Energy Agency, Vienna, 2000.

98. IAEA. Environmental consequences of the Chernobyl accidentand and their recommendation: Twenty years of experience. Technical Report STI/PUB/1239, International Atomic Energy Agency, Vienna, 2006.

99. IAEA. Fundamental safety principles. Safety fundamentals. Technical Report SF-1, International Atomic Energy Agency, Vienna, 2006.

100. IAEA. Thermophysical properties database of materials for light water reactors and havy water reactors. Final report of a coordinated research project 1999–2005. Technical Report IAEA-TECDOC-1496, International Atomic Energy Agency, Vienna, 2006.

101. IAEA. Accident analysis for nuclear power plants with modular high temperature gas cooled reactors. Technical Report SRS-54, International Atomic Energy Agency, Vienna, 2008.

102. IAEA. Best estimate safety analysis for nuclear power plants: Uncertainty evaluation. Technical Report SRS-52, International Atomic Energy Agency, Vienna, 2008.

103. IAEA. Thermophysical properties of materials for nuclear engineering: A tutorial and collection of data. Technical report, International Atomic Energy Agency, Vienna, 2008.

104. IAEA. Liquid metal coolants for fast reactors cooled by sodium, lead, and lead-bismuth eutectic. Technical Report NP-T-1.6, International Atomic Energy Agency, Vienna, 2012.

105. IAEA. The Fukushima Daiichi accident: Report by the Director General. Technical Report STI/PUB/1710, International Atomic Energy Agency, Vienna, 2015.

106. IAEA. Safety of nuclear power plants: Design. Technical Report SSR-2/1, International Atomic Energy Agency, Vienna, 2016.

107. IAEA. *IAEA Safety Glossary*. International Atomic Energy Agency, 2018.

108. IAEA. Design of the reactor core for nuclear power plants. Technical Report SSG-52, International Atomic Energy Agency, Vienna, 2019.

109. IAEA. Deterministic safety analysis for nuclear power plants. Technical Report SSG-2/1, International Atomic Energy Agency, Vienna, 2019.

110. IAEA. Operational limits and conditions and operating procedures for nuclear power plants: Safety guide. Technical Report NS-G-2.2, International Atomic Energy Agency, Vienna, 2020.

111. F. Inasaka and H. Nariai. Evaluation of subcooled critical heat flux correlations for tubes with and without internal twisted tapes. *Nuclear Engineering and Design*, 163:225–239, 1996.

112. M. Ishii. *Thermo-Fluid Dynamic Theory of Two-Phase Flow*. Collection de la Direction des Etudes et Researches d'Electricite de France, Paris, France, 1975.

113. M. Ishii and T. Hibiki. *Thermo-Fluid Dynamics of Two-Phase Flow*. Springer, New York, 2011.

114. J.D. Jackson. Fluid flow and convective heat transfer to fluids at supercritical pressure. *Nuclear Engineering and Design*, 264:24–40, 2013.

115. W. Jiang, J.D. Hales, B.W. Spencer, B.P. Collin, A.E. Slaughter, S.R. Novascone, A. Toptan, and K.A. Gamble. TRISO particle fuel performance and failure analysis with BISON. *Journal of Nuclear Materials*, 548:152795, 2021.

116. J.D. Jones and C.M. Leung. An improvement in the calculation of turbulent friction in smooth concentric annuli. *Journal of Fluids Engineering*, 103:615–623, 1981.

117. B. Kader. Temperature and concentration profiles in fully turbulent boundary layers. *International Journal of Heat and Mass Transfer*, 24:1541–1544, 1981.

118. B. Kader. Heat and mass transfer in pressure-gradient boundary layers. *International Journal of Heat and Mass Transfer*, 34:2837–2857, 1991.

119. M. Kato, S. Maeda, T. Abe, and K. Asakura. Uranium Oxide and MOX Production. In R.J.M. Konings and R.E. Stoller, editors, *Comprehensive Nuclear Materials*, chapter 2.01, pages 1–34. Elsevier, Oxford, second edition, 2020.

120. W. Kays, M. Crawford, and B. Weigand. *Convective Heat and Mass Transfer*. McGraw Hill, New York, fourth edition, 2003.

121. E.H. Kennard. *Kinetic theory of gases*. McGraw-Hill Book Company, Inc., New York, 1938.

122. D.E. Kim and J. Park. Experimental study of critical heat flux in pool boiling using visible-ray optics. *International Journal of Heat and Mass Transfer*, 169:120937, 2021.

123. H. Kim and H. Park. Bubble dynamics and induced flow in a subcooled nucleate pool boiling with varying subcooled temperature. *International Journal of Multiphase Flow*, 183:122054, 2022.

124. H.G. Kim and J.C. Lee. Development of a generalized critical heat flux correlation through the alternating conditional expectation algorithm. *Nuclear Science and Engineering*, 127:300–316, 1997.

125. J. Kim. Review of nucleate pool boiling bubble heat transfer mechanisms. *International Journal of Multiphase Flow*, 35(12):1067–1076, 2009.

126. Y.A. Kirichenko and P.S. Chernyakov. Determination of the first critical thermal flux on flat heaters. *Journal of Engineering Physics and Thermophysics*, 20:699–703, 1971.

127. J.F. Klausner, R. Mei, M. Bernhard, and L.Z. Zeng. Vapor bubble departure in forced convection boiling. *International Journal of Heat and Mass Transfer*, 36(3):651–662, 1993.

128. P.S. Klebanoff. Characteristics of turbulence in a boundary layer with zero pressure gradient. Technical Report 1247, National Advisory Committee for Aeronautics, Washington, 1955.

129. A.I. Klemin, L.N. Polianin, and M.M. Strigulin. *The Thermal-Hydraulic Analysis and Thermal Reliability of Nuclear Reactors (in Russian)*. Atomizdat, Moscow, USSR, 1980.

130. R.A. Knief. *Nuclear Engineering: Theory and Technology of Commercial Nuclear Power*. American Nuclear Society, La Grange Park, IL, second edition, 2008.

131. G. Kocamustafaogullari. Pressure dependence of bubble departure diameter for water. *International Communications in Heat and Mass Transfer*, 10(6):501–509, 1983.

132. D.L. Koch and J.L. Brady. A non-local description of advection-diffusion with application to dispersion in porous media. *Journal of Fluid Mechanics*, 180:387–403, 1987.

133. R.J.M. Konings and R. Stoller, editors. *Comprehensive Nuclear Materials*. Elsevier, Oxford, second edition, July 2020.

134. H.W. Kottowski. Activation energy of nucleation. *Progress in Heat and Mass Transfer*, 7:299–324, 1973.

135. E.A. Krasnoshchekov and V.S. Protopopov. Experimental study of heat exchange in carbon dioxide in the supercritical range at high temperature drops. *Teplofizika Vysokikh Temperatur*, 4(3):389–398, 1966.

136. N. Kurul and M.Z. Podowski. Multi-dimensional effects in forced convection sub-cooled boiling. In *9th Heat Transfer Conference*, volume 2, pages 21–26, Jerusalem, Israel, 1990. Hemisphere Publishing Corporation.

137. S. S. Kutateladze. On the transition to film boiling under natural convection. *Kotloturbostroenie*, 3:10–12, 1948.

138. S. S. Kutateladze. Boiling and bubbling heat transfer under free convection of liquid. *International Journal of Heat Mass Transfer*, 22:281–299, 1979.

139. R.T. Lahey, Jr. and F.J. Moody. *The thermal-hydraulics of a boiling water nuclear reactor*. American Nuclear Society, La Grange Park, Illinois, 1977.

140. D.D. Lanning, C.E. Beyer, and K.J. Geelhood. FRAPCON-3 updates, including mixed-oxide properties. Technical Report NUREG/CR-6534, Pacific Nortwest Laboratory, Richland, Washington, 2005.

141. R.A. Lee and K.H. Haller. Supercritical water heat transfer developments and applications. In *Proceedings of 5th International Heat Transfer Conference*, volume IV(B 7.7), pages 335–339, Japan, 1974.

142. E.J. LeFevre. Laminar free convection from a vertical plane surface. In *Proceedings of the Ninth International Congress of Applied Mechanics*, Brussels, 1956.

143. G.S. Lellouche *et al.* Quantifying reactor safety margins part 4: Uncertainty evaluation of LBLOCA analysis based on TRAC-PF1/MOD 1. *Nuclear Engineering and Design*, 119:67–95, 1990.

144. L.L. Levitan and F.P. Lantsman. Investigating burnout with flow of a steam-water mixture in a round tube. *Thermal Engineering (USSR)*, 22(1):102–105, 1970.

145. H. Li and H. Anglart. CFD model of diabatic annular two-phase flow using the Eulerian-Lagrangian approach. *Annals of Nuclear Energy*, 77:415–424, 2015.

146. H. Li and H. Anglart. Prediction of dryout and post-dryout heat transfer using a two-phase CFD model. *International Journal of Heat and Mass Transfer*, 99:839–850, 2015.

147. H. Li and H. Anglart. CFD prediction of droplet deposition in steam-water annular flow with flow obstacle effect. *Nuclear Engineering and Design*, 321:173–179, 2017.

148. Y.-Y. Li, Y.-J. Chen, and Z.-H. Liu. A uniform correlation for predicting pool boiling heat transfer on a plane surface with surface characteristics effects. *International Journal of Heat and Mass Transfer*, 77:809–817, 2014.

149. J.H. Lienhard. Corresponding states correlations for the spinodal and homogeneous nucleation temperatures. *Journal of Heat Transfer*, 104:379–381, 1982.

150. J.H. Lienhard. On the two regimes of the nucleate boiling. *Journal of Heat Transfer*, 107:262–264, 1985.

151. G.G. Loomis. Summary of the semiscale program (1965–1986). Technical Report NUREG/CR-4945, Idaho National Engineering Laboratory, Idaho Falls, ID, 1987.

152. H.G. Lyall. A comparison of helium and $CO_2$ as reactor coolants. *Journal of Nuclear Energy*, 26:49–60, 1973.

153. G. Markoczy. Convective heat transfer in rod clusters with turbulent axial coolant flow. Part I: mean values over the rod perimeter. *Warme- und Stoffubertraugung*, 5:204–212, 1972.

154. G.T. Mazuzan and J.S. Walker. Controlling the atom. Technical Report NUREG-1610, University of California Press, Oakland, CA, 1984.

155. R. Mei and J.F. Klausner. Unsteady force on a spherical bubble at finite reynolds number with small flactuations in the free stream velocity. *Physics of Fluids*, 4:63–70, 1992.

156. R. Mei and J.F. Klausner. Shear lift force on spherical bubbles. *International Journal of Heat and Fluid Flow*, 15(1):62–65, 1994.

157. B.B. Mikic and W.M. Rohsenow. A new correlation of pool boiling data including the effect of heating surface characteristics. *Journal of Heat Transfer*, 91:245–250, 1969.

158. B.B. Mikic, W.M. Rohsenow, and P. Griffith. On bubble growth rates. *International Journal of Heat and Mass Transfer*, 13:657–666, 1970.

159. G.H. Miller and E.G. Puckett. A high-order Godunov method for multiple condensed phases. *Journal of Computational Physics*, 128:134–164, 1996.

160. G.K Miller, D.A Petti, J.T. Maki, D.L. Knudson, and W.F. Skerjanc. Parfume theory and model basis report. Technical Report INL/EXT-08-14497, Idaho National Laboratory, Idaho Falls, Idaho, 2018.

161. S.M. Modro, S.N. Aksan, V.T. Berta, and A.B. Wahba. Review of LOFT Large Break Experiments. Technical Report NUREG/IA-0028, U.S. Nuclear Regulatory Commission, Washington, D.C., 1989.

162. S. Moghaddam and K. Kiger. Physical mechanisms of heat transfer during single bubble nucleate boiling of FC-72 under saturation conditions. II: Theoretical analysis. *International Journal of Heat and Mass Transfer*, 52(5-6):1295–1303, 2009.

163. R. Moissis and P.J. Berenson. On the hydrodynamic transition in nucleate boiling. *Journal of Heat Transfer*, 85:221–229, 1963.

164. S. Mokry, I. Pioro, A. Farah, K. King, S. Gupta, W. Peiman, and P. Kirillov. Development of supercritical water heat-transfer correlation for vertical bare tubes. *Nuclear Engineering and Design*, 241:1126–1136, 2011.

165. L.F. Moody. Friction factors for pipe flow. *Transactions of ASME*, 66:671–684, 1944.

166. L.F. Moody. An approximate formula for pipe friction factors. *Transactions of ASME*, 69:1005–1006, 1947.

167. A.T. Motta, A. Couet, and R.J. Comstock. Corrosion of zirconium alloys used for nuclear fuel cladding. *Annual Review of Material Research*, 45:311–343, 2015.

168. K. Nishikawa, Y. Fujita, S. Uchida, and H. Ohta. Effect of heating surface orientation on nucleate boiling heat transfer. In *Proceedings of ASME-JSME Thermal Engineering Joint Conference*, volume 1, pages 129–136, Honolulu, 1983. ASME, New York.

169. R.H. Notter and C.A. Sleicher. A solution to the turbulent Graetz problem, III. Fully developed and entry region heat transfer rates. *Chemical Engineering Science*, 27:2073–2093, 1972.

170. NRC. Review of operational errors and system misalignments identified during the Three Mile Island incident. NRC Bulletin No. 79-06, April 1979.

171. S. Nukiyama. The maximum and minimum values of heat Q transmitted from metal to boiling water under atmospheric pressure. *Journal of the Japan Society of Mechanical Engineers*, 37:367–374, 1934.

172. OECD-NEA. Nuclear fuel safety criteria technical review. Technical Report NEA-7072, Organization for Economic Co-operation and Development - Nuclear Energy Agency, Paris, France, 2012.

173. A.P. Ornatskii, L.F. Glushchenko, and S.I. Kalachev. Heat transfer with rising and falling flows of water in tubes of small diameter at supercritical pressure. *Thermal Engineering*, 18(5):137–141, 1971.

174. S.J. Osher and R.P. Fedkiw. Level set methods: an overview and some recent results. *Journal of Computational Physics*, 169:463–502, 2001.

175. S.J. Osher and J.A. Sethian. Front propagating with curvature dependent speed: Algorithms based on Hamilton-Jacobi formulations. *Journal of Computational Physics*, 79:12–49, 1988.

176. T. Pavlov, L. Vlahovic, D. Staicu, R.J.M. Konings, M.R. Wenman, P. Van Uffen, and R.W. Grimes. A new numerical method and modified apparatus for the simultaneous evaluation of thermo-physical properties above 1500 K: A case study on isostatically pressed graphite. *Thermochimica Acta*, 652:39–52, 2017.

177. I.L. Pioro. Experimental evaluation of constants for the rohsenow pool boiling correlation. *International Journal of Heat and Mass Transfer*, 42:2003–2013, 1999.

178. M.S. Plesset and M.S. Zwick. The growth of a vapour bubbles in superheated liquids. *Journal of Applied Physics*, 25:493–500, 1954.

179. S.B. Pope. *Turbulent Flows*. Cambridge University Press, Cambridge, UK, 2011.

180. M. Quintard and S. Whitaker. Two-phase flow in heterogeneous porous media i: The influence of large spatial and temporal gradients. *Transport in Porous Media*, 5:341–379, 1990.

181. M. Quintard and S. Whitaker. Two-phase flow in heterogeneous porous media ii: Numerical experiments for flow perpendicular to a stratified system. *Transport in Porous Media*, 5:429–472, 1990.

182. J.M. Ramilison and J.H. Lienhard. Transition boiling heat transfer and the film transition regime. *Journal of Heat Transfer*, 109:746–752, 1987.

183. J.M. Ramilison, P. Sadasivan, and J.H. Lienhard. Surface factors influencing burnout on flat heaters. *Journal of Heat Transfer*, 114:287–290, 1992.

184. D.G. Reddy and C.F. Fighetti. Parametric study of CHF data. Volume 2: A generalized subchannel CHF correlation for PWR and BWR fuel assemblies. Technical Report EPRI NP-2609, Electric Power Research Institute, Palo Alto, CA, 1983.

185. H. Reichardt. Vollständige Darstellung der turbulenten Geschwindigkeitsverteilung in glatten Leitungen. *ZAMM. Zeitschrift für angewandte Mathematik und Physik*, 31:208–219, 1951.

186. P.J. Roache. Quantification of uncertainty in computational fluid dynamics. *Annual Review of Fluid Mechanics*, 29:123–160, 1997.

187. H.A. Roberts and R.W. Bowring. Boiling nomenclature - a plea for consistency. *Nuclear Power*, 4(33):122, 1959.

188. W.M. Rohsenow. A method of correlating heat transfer data for surface boiling of liquids. *Transactions of ASME*, 74:969–976, 1952.

189. A.M. Ross and R.L. Stoute. Heat transfer coefficient between $UO_2$ and Zircaloy-2. Technical Report AECL-1552, Atomic Energy of Canada Limited, Chalk River, Ontario, 1962.

190. Z.R. Rosztoczy. Root causes of the Three Mile Island Accident. *Nuclear News*, 62(3):29–32, 2019.

191. O.C. Sandall, O. T. Hanna, and P. R. Mazet. A new theoretical formula for turbulent heat and mass transfer with gases or liquids in tube flow. *The Canadian Journal of Chemical Engineering*, 58(4):443–447, 1980.

192. T. Sato and H. Matsumura. On the conditions of incipient subcooled boiling with forced convection. *Bulletin of JSME*, 726:392–398, 1964.

193. H. Schlichting and K. Gersten. *Boundary Layer Theory*. Springer, Berlin Heidelberg, 8 edition, 2000.

194. B.R. Sehgal, editor. *Nuclear safety in Light Water Reactors: Severe accident phenomenology*. Elsevier/Academic Press, Amsterdam, The Netherlands, 2012.

195. J.A. Sethian. Evolution, implementation and application of level set and fast marching methods for advancing fronts. *Journal of Computational Physics*, 169:503–555, 2001.

196. J.A. Sethian and P. Smereka. Level set methods for fluid interfaces. *Annual Review of Fluid Mechanics*, 35:241–372, 2003.

197. R.K. Shah. A correlation for hydrodynamic rntry length solutions for circular and non-circular ducts. *Journal of Fluids Engineering*, 100:177–179, 1978.

198. R.K. Shah and M.S. Bhatti. Laminar convective heat transfer in ducts. In S. Kakac, R.K. Shah, and W. Aung, editors, *Handbook of Single-Phase Convective Heat Transfer*, Chapter 3. Wiley, New York, 1987.

199. R.K. Shah and A.L London. *Laminar Flow Forced Convection in Ducts*. Academic, New York, 1978.

200. M.E. Shitsman. Deteriorated heat transfer regimes at supercritical pressures. *Teplofizika Vysokikh Temperatur*, 1(2):267–275, 1963.

201. M.P. Short, D. Hussey, B.K. Kendrick, T.M. Besmann, C.R. Stanek, and S. Yip. Multiphysics modelling of porous CRUD deposits in nuclear reactors. *Journal of Nuclear Materials*, 443:579–587, 2013.

202. D.E. Shropshire. Lessons learned from Gen I carbon dioxide cooled reactors. In *Proceedings of ICONE 12: Twelfth International Conference on Nuclear Engineering*, Arlington, Virginia, April 2004.

203. E.N. Sieder and G.E. Tate. Heat transfer and pressure drop of liquids in tubes. *Industrial and Engineering Chemistry*, 28(12):1429–1435, 1936.

204. L.J. Siefken, Coryell E.W., E.A. Harvego, and J.K. Hohorst (eds.). SCDAP/RELAP5/MOD 3.3 code manual. MATPRO-A library of materials properties for light-water-reactor accident analysis. Technical Report NUREG/CR-6150, Vol. 4, Rev. 2, INEL-96/0422, Idaho National Engineering and Environmental Laboratory, January 2001.

205. J.C. Slattery. *Interfacial Transport Phenomena*. Springer-Verlag, New York, 1990.

206. L.L. Snead, T. Nozawa, Y. Katoh, T.-S. Byun, S. Kondo, and D.A. Petti. Handbook of SiC properties for fuel performance modeling. *Journal of Nuclear Materials*, 371:329–377, 2007.

207. V. Sobolev and P. Schuurmans. Thermophysical Properties of Liquid Metal Coolants: Na, Pb, Pb-Bi(e). In R.J.M. Konings and R.E. Stoller, editors, *Comprehensive Nuclear Materials*, chapter 7.15, pages 457–481. Elsevier, Oxford, second edition, 2020.

208. B. Söderquist, S. Hedberg, and G. Strand. Critical heat flux measurements in vertical 8 mm round tubes. Technical Report KTH/NEL/R–56–SE, KTH Royal Institute of Technology, Stockholm, Sweden, March 1994.

209. J.-W. Song, D.-L. Zheng, and L.-W. Fan. Temperature dependence of contact angles of water on a stainless steel surface at elevated temperatures and pressures: In situ characterization and thermodynamic analysis. *Journal of Colloid and Interface Science*, 561:870–880, 2020.

210. D.B. Spalding. A single formula for the law of the wall. *Journal of Applied Mechanics*, 28:455–457, 1961.

211. H. Steiner, A. Kobor, and L. Gebhard. A wall heat transfer model for subcooled boiling flow. *International Journal of Heat and Mass Transfer*, 48:4161–4173, 2005.

212. M.A. Styrikovich, T.Kh. Margulova, and Z.L. Miropolskii. Problems in the development of designs of supercritical boilers. *Teploenergetika*, 14(6):4–7, 1967.

213. H.S. Swenson, J.R. Carver, and C.R. Kakarala. Heat transfer to supercritical water in smooth bore tubes. *Journal of Heat Transfer, Trans. ASME*, 87:477–483, 1965.

214. R. Tadmor. Line energy and the relation between advancing, receding, and Young contact angles. *Langmuir*, 20(18):7659–7664, 2004.

215. T. Tanaka, K. Miyazaki, and T. Yabuki. Observation of heat transfer mechanisms in saturated pool boiling of water by high-speed infrared thermometry. *International Journal of Heat and Mass Transfer*, 170:121006, 2021.

216. T.G. Theofanous, T.N. Dinh, J.P. Tu, and A.T. Dinh. The boiling crisis phenomenon. Part II: dryout dynamics and burnout. *Experimental Thermal and Fluid Science*, 26:793–810, 2002.

217. T.G. Theofanous, J.P. Tu, A.T. Dinh, and T.N. Dinh. The boiling crisis phenomenon. Part I: nucleation and nucleate boiling heat transfer. *Experimental Thermal and Fluid Science*, 26:775–792, 2002.

218. N.E. Todreas and P. Hajzler. Letter to the Editor on the paper by D.C. Groneveld and L.K.H. Leung, The 1995 look-up table for critical heat flux in tubes. *Nuclear Engineering and Design*, 163:25–26, 1996.

219. N.E. Todreas and M.S. Kazimi. *Nuclear systems I - Thermal Hydraulic Fundamentals*. Hemisphere Publishing Corporation, New York, 1990.

220. N.E. Todreas, M.S. Kazimi, and M. Massoud. *Nuclear systems II - Elements of Thermal Hydraulic Design*. CRC Press, Boca Raton, FL, 2021.

221. L.S. Tong and Y.S. Tang. *Boiling Heat Transfer and Two-Phase Flow*. Taylor & Francis, Washington D.C., 1997.

222. A. Toptan, D.J. Kropaczek, and M.N. Avramova. Gap conductance modeling I: Theoretical considerations for single- and multi-component gases in curvlinear coordinates. *Nuclear Engineering and Design*, 353:110283, 2019.

223. G. Tryggvason, R. Scardovelli, and S. Zaleski. *Direct Numerical Simulations of Gas-Liquid Multiphase Flows*. Cambridge University Press, Cambridge, UK, 2011.

224. P.V. Tsvetkov, A. Waltar, and D. Todd. Introductory Design Considerations. In A. Waltar, D. Todd, and P.V. Tsvetkov, editors, *Fast Spectrum Reactors*, chapter 2, pages 23–38. Springer, New York, 2012.

225. H.C. Unal. Maximum bubble diameter, maximum bubble-growth time and bubble-growth rate during the subcooled nucleate flow boiling of water up to 17.7 MN/m$^2$. *International Journal of Heat and Mass Transfer*, 19:643–649, 1976.

226. A. Urbano, S. Tanguy, G. Huber, and C. Colin. Direct numerical simulation of nucleate boiling in micro-layer regime. *International Journal of Heat and Mass Transfer*, 123:1128–1137, 2018.

227. S.J.D. Van Stralen, M.S. Sohal, R. Cole, and W.M. Sluyter. Bubble growth rates in pure and binary systems: combimed effect of relaxation and evaporation microlayers. *International Journal of Heat and Mass Transfer*, 18:453–467, 1975.

228. G.B. Wallis. Flooding velocities for air and water in vertical tubes. Technical Report AEEW-R123, UK Atomic Energy Authority, Harwell, UK, 1961.

229. G.B. Wallis. *One-dimensional Two-phase Flow*. McGraw-Hill Book Company, New York, 1969.

230. Z.U.A. Warsi. *Fluid Dynamics. Theoretical and Computational Approaches*. Taylor & Francis, Boca Raton FL, 2006.

231. R.L. Webb. *Principles of Enhanced Heat Transfer*. Wiley, New York, 1994.

232. J. Weisman. Heat transfer to water flowing parallel to tube bundles. *Nuclear Science and Engineering*, 6(1):78–79, 1959.

233. F.M. White. *Viscous Fluid Flow*. McGraw-Hill Book Company, New York, second edition, 1991.

234. W. Wieselquist, M. Williams, D. Wiarda, M. Pigni, and U. Mertyurek. Overview of nuclear data uncertainty in Scale and application to light water reactor uncertainty analysis. Technical Report ORNL/TM-2017/706, Oak Ridge National Laboratory, Oak Ridge, TN, 2018.

235. D.C. Wilcox. *Turbulence Modeling for CFD*. DCW Industries, La Canada, CA, third edition, 2010.

236. G.E. Wilson *et al.* Quantifying reactor safety margins part 2: Characterization of important contributors to uncertainty. *Nuclear Engineering and Design*, 119:17–31, 1990.

237. G. Windecker and H. Anglart. Phase distribution in a BWR fuel assembly and evaluation of a multidimensional multifield model. *Nuclear Technology*, 134:49–61, 2001.

238. W. Wulff *et al.* Quantifying reactor safety margins part 3: Assessment and ranging of parameters. *Nuclear Engineering and Design*, 119:33–65, 1990.

239. K. Yamagata, K. Nishikawa, and S. Hasegawa. Forced convective heat transfer to supercritical water flowing in tubes. *International Journal of Heat and Mass Transfer*, 15:2575–2593, 1972.

240. F. Yin, T.K. Chen, and H.X. Li. An investigation on heat transfer to supercritical water in inclined upward smooth tubes. *Heat Transfer Engineering*, 27:44–52, 2006.

241. S.H. Yoon, J.H. Kim, Y.W. Hwang, M.S. Kim, K. Min, and Y. Kim. Heat transfer and pressure drop characteristics during the in-tube cooling process of carbon dioxide in the supercritical region. *International Journal of Refrigeration*, 26:857–864, 2003.

242. R. Zhang, Y. Huang, B. Shen, and R.Z. Wang. Flow and heat transfer characteristics of supercritical nitrogen in a vertical mini-tube. *International Journal of Thermal Science*, 50:287–295, 2011.

243. N. Zuber. On the stability of boiling heat transfer. *Trans. ASME*, 80:711–720, 1958.

244. N. Zuber. *Hydrodynamic aspects of boiling heat transfer*. PhD thesis, University of California, Los Angeles, CA, 1959.

245. N. Zuber *et al.* Quantifying reactor safety margins part 5: Evaluation of scale-up capabilities of best estimate codes. *Nuclear Engineering and Design*, 119:97–107, 1990.

246. N. Zuber *et al.* Quantifying reactor safety margins part 6: A physically based method of estimating PWR large break loss of coolant accident PCT. *Nuclear Engineering and Design*, 119:109–117, 1990.

247. A. Zukauskas. *Convective Transfer in Heat Exchangers (in Russian)*. Nauka, Moscow, 1982.

# Index

Printed in the United States
by Baker & Taylor Publisher Services